THE
FLAME
OF
REASON

CHRISTER STURMARK is a Swedish author and public intellectual. He is the founder and publisher of Fri Tanke ('Free Thought') and often appears on television and in public gatherings to defend science and to argue against dogmas, superstition and pseudoscientific thinking. He lives on the island of Lidingö, just outside Stockholm, with his twelve-year-old son Leo and his wife, the novelist Victoria Larm. He enjoys playing piano, guitar and chess.

DOUGLAS HOFSTADTER is a professor of cognitive science and comparative literature at Indiana University. His first book, *Gödel, Escher, Bach*, won the 1980 Pulitzer Prize for non-fiction. He is known for his writing on minds and machines, analogy and translation.

CHRISTER STURMARK

WITH DOUGLAS HOFSTADTER

THE FLAME OF REASON

Clear Thinking for the Twenty-First Century

An Apollo Book

First published in the UK in 2022 by Head of Zeus
This paperback edition first published in 2023 by Head of Zeus,
part of Bloomsbury Plc

9 7 5 3 1 2 4 6 8

A catalogue record for this book is available from
the British Library.

ISBN (PB): 9781803280998
ISBN (E): 9781803280967

Printed and bound in Great Britain by
CPI Group (UK) Ltd, Croydon CR0 4YY

Head of Zeus Ltd
5–8 Hardwick Street
London EC1R 4RG

WWW.HEADOFZEUS.COM

CONTENTS

PART II
THE PATHWAY TO A NEW ENLIGHTENMENT

FOREWORD

Douglas Hofstadter

It is a great pleasure and a great honor to write the foreword to this book. And it has also been a great pleasure and a great honor to translate this book into English—a task in which I have been engaged, with love, for several years.

I had never heard of Christer Sturmark until early 2016, when, via e-mail, he invited me to participate in a small symposium in Stockholm on the topic "limits to knowledge." He mentioned several people who would probably be there, some of whom I knew or had heard of, and I was intrigued. I happened to have lived in Stockholm long ago and was very fond of the city, so it was a temptation I couldn't resist. I might add that this unknown gentleman's invitation was written in a very lively and genuinely friendly manner, which also helped tip the balance.

I just told you that I had never heard of Christer, and that's what I sincerely believed at the time, but when I arrived in Stockholm, I found out that I was quite wrong. Christer showed me an enthusiastic fan letter (a genuine *postal* letter) that he had written to me back in the early 1980s, when he was a teenaged rock musician, and then he showed me my reply to him, and his reply to me, and my second reply to him . . . I had completely forgotten all that! Clearly, my long-ago reactions to his style must have been similar to my much more recent reactions. In 2016, Christer was still just as boyish and ebullient as he had been as a very young man.

I had an excellent time at his symposium in Stockholm in 2016, especially getting to know the Viennese mathematician and writer Karl Sigmund, who later became a great friend. But Karl wasn't the only good friend I made thanks to Christer. The other friend was Christer himself. Here's what happened.

The day the symposium ended, Christer invited a handful of participants to dinner at his house on the lovely island of Lidingö, just to the east of

Stockholm, across the Lilla Värtan Strait. I'll never forget how that evening his former wife and still great friend Gunilla Backman sang one of my own songs for us, accompanied by mathematician Anders Karlqvist at the piano. It practically moved me to tears.

And then Christer, who had just discovered with great surprise that I spoke some Swedish, spontaneously offered me a copy of his brand-new book *Upplysning i det tjugoförsta århundradet* ("Enlightenment in the Twenty-First Century"). It was a hefty tome, but Christer smiled and said, "Don't worry—it's in pretty simple Swedish. You'll easily be able to read it." What he meant was that it wasn't written in pompous academic jargon or in an obscure Swedish dialect or in ancient verse or anything of the sort. In actual fact, it was written in quite sophisticated Swedish and not at all a piece of cake for me to read. Luckily, though, I could make sense of most of what I saw without using a dictionary.

At the outset, I had no idea what the book was about, but on my flight home, I paged through it and got very intrigued. I could see that it was giving a personal vision of an idealistic way for human beings to live together and get along in a world filled with conflict; in fact, it was an eloquent paean for tolerance, clear thinking, and belief in science. In some ways, it reminded me of the classic book *Fads and Fallacies in the Name of Science* by Martin Gardner, a hero of mine. For me, reading that book in my teenage years had been a life-changing experience.

The flavor of Christer's book deeply appealed to my lifelong sense of idealism and belief in science, and once I was home I suddenly had the idea that maybe I could translate it into English. In so doing, not only would I be doing Christer a favor but also I would be doing myself a favor, since translating his book would force me to work hard on my Swedish, a language I've loved ever since 1966, when I lived in Sweden for half a year. Moreover, assuming that the book got published, I would hopefully be doing a favor for people in the English-speaking world, by making a valuable set of idealistic ideas available to them. Those were all excellent reasons for my suggestion, and to my delight, Christer was thrilled with my offer. And why shouldn't he have been? After all, an author whom he had once admired was now offering to be his translator, purely out of friendship! What could have surprised and gratified him more than that?

In fall 2016, I eagerly plunged into the task. During the course of several months, I translated roughly a page a day, altogether doing maybe one-third of the book. At that point, my sabbatical year from Indiana University was looming, and I had already been planning to spend the three winter months at the great university in Uppsala. It struck me that a very appropriate project to undertake during my three Uppsala months would be to complete the translation right there, only an hour or so away from Stockholm, so that Christer and I could meet from time to time and discuss all sorts of details.

Indeed, that's just what happened. During the cold, dark months from December 2017 through February 2018, my wife Baofen and I took the train several times from Uppsala to Stockholm, then made our way out to Christer's home on the far side of Lidingö (itself on the far side of Stockholm), and there we all had a wonderful time getting to know each other better. And in this way, a casual lecture invitation turned into a great friendship.

Now let me say some things about Christer's book itself, since that's what people expect from a foreword—and rightly so.

The Flame of Reason (as the author poetically renamed it in English) sprang out of Christer's youthful passion for logic, math, and science. As a boy, he was fascinated by the universe's paradoxicality, strangeness, and magic. But he soon came to see that although there was plenty of mystery, there was also a way to study and to penetrate much of the mystery—namely, through science and mathematics. He threw himself into the study of those disciplines, and also into related activities, like chess (which he still loves to play, especially with his young son Leo).

Out of this intense engagement with the world of ideas came a conviction that there is a kind of truth that transcends all dogmas, all superstitions, and all religions, and that if humanity as a whole were to embrace that sort of truth, it would open up a marvelous period of enlightenment and could even bring about world peace.

Throughout the years, Christer's involvement with science itself—computer science at the start, but then other sciences—gradually turned into a kind of *crusade* (if I dare use that inflammatory word)—a crusade for a science-based tolerance of people of all races, lifestyles, cultures, and belief systems.

Christer discovered that there was already a worldwide movement of kindred spirits who thought along these same lines—namely, *secular humanists*—people who believe in benevolence towards all humans, not for religious reasons but out of a belief in the power of tolerance and clear thinking, and also out of a vivid sense of our collective fragility on this tiny blue-green sphere spinning its way among billions of stars, themselves among billions of galaxies. In other words, a sense of profound humility inspired Christer (and other secular humanists) to try to get along with the other beings on this planet, rather than falling victim to blind, prejudice-driven hatreds and engaging in constant vicious battles with supposed "enemies."

Eventually, Christer, ever the idealist, decided to found his own publishing firm—Fri Tanke ("Free Thought")—and to publish high-quality books in Swedish, as well as a magazine called *Sans* ("Sense"), which would explain and advocate science, logic, and the philosophy of secular humanism, while arguing against pseudoscience, superstitions, and fundamentalist religions.

Christer's dream came true, thanks in part to help from his good friend Björn Ulvaeus (of ABBA fame), and some years ago he became an influential Swedish publisher. Among the many books issued by Fri Tanke are numerous translations of books written by some of the thinkers I most admire, such as Richard Dawkins, Daniel Dennett, Rebecca Goldstein, Mikhail Gorbachev, Andrew Hodges, Steven Pinker, and numerous others.

As the years passed, Christer's increasing visibility led him to become a well-known speaker and television personality in Sweden, representing the ideas of secular humanism most of all. He often could be seen arguing against astrologers who claimed everyone's fate was predestined in the stars, or debating with religious clerics who insisted that Darwinian evolution was a hoax, or defending his own credo that the philosophy of atheism should be every bit as respected by Swedish laws as organized religions are. Though unfailingly polite, Christer was always passionate about publicly defending all these stances in which he believed so deeply, no matter how hard it was to do so. My hat is off to him for his fervor and his courage!

While I was living in Uppsala and interacting with Christer quite frequently, I soon discovered that he was also practicing what he preached. He was a crusader for immigrants in Sweden, especially for people who had fled to Sweden as refugees from religious persecution. I saw firsthand how Christer gave enormous personal help to a Bangladeshi blogger who had fled his native land because of the threat of death from terrorists who hated free thinking and tolerance—the same kinds of extremists who had brutally attacked and nearly killed the teenaged girl Malala Yousafzai, who was an activist for education for girls in Pakistan.

In short, I came to admire Christer as a thinker, writer, publisher, and human being. We are now close friends, and part of what binds us is that we share a kind of youthful idealistic hope for humanity. His book—this book—expresses his sense of idealism clearly, concretely, and enjoyably.

Christer, like me, loves examples and stories to get his points across, and he uses them well in every chapter of this book.

Moreover—and here, perhaps, I detect a little trace of my own influence on him—he likes playing with form when he writes. Just as in *Gödel, Escher, Bach*, I alternated between more serious chapters and more playful dialogues, so Christer, in his book, alternates between more serious chapters and more playful "interludes," in which he allows himself a bit more liberty in expressing purely personal feelings.

As a translator of several previous books, I had quite a lot of experience in translation, and part of my style as a translator is that of taking lots of liberties—I call this "poetic lie-sense." Here and there, I allow myself to say things somewhat differently from the way they were expressed in the original, as I think it will work more effectively that way in English. I also sometimes take the liberty to suggest adding (or possibly dropping) a few ideas here and there. In this book, no less than in earlier books I'd translated, I used this brash style—but I had the advantage that I could always ask the original author if the small changes that I was suggesting were acceptable to him. To my gratification, Christer nearly always gave me a green light. This made for a very pleasant and easygoing relationship between author and translator, and during these past few years we've had great fun in our give-and-take as partners in the realization of this radical transplantation, into my language and my culture (namely, the English language and the American culture), of Christer's original book, which of course was deeply rooted in his own native soil of the Swedish language and culture.

Now that it's done, I will miss our delightful author/translator interactions, but it's high time for me to move on to new projects—and although I will no longer be translating Christer's book, our friendship will continue to grow and flourish. Most of all, I fervently hope that Christer's dream of a truly open society with universal tolerance and a universal reverence for science will come to exist, aided by his own contributions—especially by this highly stimulating and deeply personal book.

November 17, 2020
Bloomington, Indiana

FOREWORD

Christer Sturmark

Sometimes, in my darkest hours, I am worried that my fellow humans are slowly but surely losing the capacity of clear and independent thinking, of reason and rationality.

The global era of liberal ideas and values, which I see as having begun with the fall of the Berlin Wall and the end of the Cold War, seems tragically to have come to an end.

Many developments in the world at the start of this, the third decade of the twenty-first century, are making me increasingly worried. The world has been hit by the global pandemic of COVID-19, caused by a new coronavirus. Frantic work by scientists throughout the world has resulted in the development of vaccines with a speed never seen before—certainly an impressive proof of the power of science and collaboration. But at the same time, we are seeing an increase in anti-vaccination movements, often (but not always) founded on conspiracy theories about a "New World Order"— a nonexistent organization supposedly controlled by an elite (often believed to be Jewish, thus following classic antisemitic trends) having the goal of creating a world government that would control all of humanity through mind-controlling microchips and vaccines, and would reduce the world population to one-tenth of its current size.

The roots of the New World Order conspiracy theory in the United States can be traced to the militant antigovernment Right and the end-of-the-world brand of fundamentalist Christianity that fears the emergence of an Antichrist.

The QAnon conspiracy is another example of bizarre thinking that is on the rise in the United States: Its believers allege that there is an organization of pedophiles who worship Satan and who are running a global sex-trafficking business that involves politicians, police, and government institutes. No one can be trusted.

But conspiracy theories are just a small part of the problem: Throughout the world, ideas are spreading to the effect that certain words or pictures that make fun of various belief systems should be banned, and that those who break these principles should be punished by death.

Many of us remember the fatwa on Salman Rushdie pronounced by the Iranian leader Ayatollah Khomeini in 1989, by which he ordered Muslims throughout the world to kill Rushdie.

These kinds of ideas are spreading fast, and freedom of speech is under threat in an increasing number of countries throughout the world.

On October 16, 2020, the French middle-school teacher Samuel Paty was beheaded by a young Muslim. Paty had shown cartoons of the Islamic prophet Muhammad in class, during a discussion about the enlightenment idea of freedom of expression.

In Poland, a very conservative and Catholic regime has come to power. In Hungary, the trend is much the same. The freedom of speech and independent press are severely limited in these countries.

In 2018, a coalition government was elected in Italy, jointly ruled by a populist party (the Five Star Movement, or Movimento Cinque Stelle) and a group of right-wing extremists belonging to the Lega (formerly Lega Nord), a party founded on the idea of ejecting from Italy all parts of the country from Naples southward. In Austria and Russia, right-wing populism and conservative moral values are quickly gaining ground.

In the Philippines, President Rodrigo Duterte refuses to respect human and legal rights in his fight against drugs and Islamist jihadists. He has often proudly stated that he personally killed criminal suspects, and he has systematically supported death squads carrying out extrajudicial killings of drug users and other criminals. Many street children were among the victims. Duterte has also encouraged his soldiers to rape women.

Recent developments in China are also highly worrisome. Information technology is widely used to monitor and control the citizens, and as the economy is developing, the country is becoming more authoritarian, rather than less so. And in Turkey, Islamism and nationalistic ideas have taken over, while in India, Hindu nationalists have come to power.

And last but certainly not least, the United States of America, formerly a beacon of enlightenment and hope for much of the world, has been led for four years (2017–2020) by a president who seems unfit to run anything. That era is over now, but the country is deeply polarized in a way that will make it difficult to heal for a long time.

~

So, what is the antidote to all this?

What is needed, I believe, is a revival of what I call "enlightenment values." I believe we need to revive the art of clear thinking and bring about a renaissance of secular ethics.

This book is my attempt to contribute to such a development. I believe that one must begin with oneself, and work on a small scale. If each human being, whether young or old, were to decide to try to help build a new world in an open-minded way and were to try to be a bit more systematic and clear thinking, we would be well on the way toward my vision.

I also harbor a hope that such ideals could come to be included in the school systems throughout the world. Today, many schools could be said to be in a state of crisis. It's not so much due to problems of discipline and behavior but, rather, to a loss of perspective about the nature of knowledge and understanding. Students in schools and universities throughout the world need to be exposed to a more philosophical approach; they need to have more exercises in careful and clear thinking, greater awareness of life's complexities, and deeper probing into the nature of human existence. Only when we truly recognize ourselves as reflecting, conscious humans can we fully participate in life and improve it, not only for ourselves but also for others.

We also crucially need to realize that ethical and moral values do not have to stem from religion. Ethics is a long-standing branch of philosophy, and it has no indispensable link with religion. Indeed, moral values can be solidly grounded in a totally nonreligious, secular, and humanist fashion. We have to let our children know that a scientific outlook on the world is the most fascinating one there is, and that science, together with a humanist form of ethics, can ground a personal worldview.

This book grew out of my concern about rationality and enlightenment during the first twenty years of the twenty-first century, in which religious fundamentalism, pseudoscience, cultural relativism, post-truth relativism, conspiracy theories, and other antiscientific attitudes have been spreading like wildfire throughout the world.

This book project started a few years ago, when I wrote a book in Swedish about the necessity for a new enlightenment for the twenty-first century—an enlightenment that would bring back reason, clear thinking, and ethics and tolerance grounded in secular humanism.

The book you are now reading grew out of that book, in a wonderful collaboration with my intellectual hero from my teenage days, and now, as of a few years, my good friend, Professor Douglas Hofstadter. His book *Gödel, Escher, Bach: An Eternal Golden Braid* changed my life when I was in my early twenties, and it turned me away from the (passionate but unreasonable) ambition to become a pop star (or at least a pop musician, neither of which happened) and turned me toward mathematics, philosophy, computer science, and eventually writing and science publishing. You'll read more about that in chapter 1.

The story of how I came to collaborate with Douglas Hofstadter on this book is well told in Doug's foreword, so I will not repeat it here. His language skills and versatile mind made it possible for him to first translate my entire Swedish manuscript to English,[1] and then we worked together for three months while Doug visited my alma mater in Uppsala (where I studied computer science) to create this co-written version of my "enlightenment manifesto" for an English-speaking audience. I am thrilled that the book will also be published in Chinese, Russian, and Korean, so hopefully my enlightenment message will reach a much larger audience than just the tiny set of speakers of Swedish.

This book consists of two parts. Part I comes from taking a "microperspective." In it, my aim is to give the reader tools with which to think more clearly and more effectively in the everyday world. I want to provide keys and insights that will allow complex, sophisticated, multidimensional thoughts to bloom and to reach their full potential.

Part II takes an overarching "macro-perspective." Here my aim is to present a political and philosophical vision of a new *Age of Enlightenment*— a vision of a free and secular world in which people are not limited or oppressed by dogmas or superstitions, an open society without racism or sexism or other prejudices, a society where human rights occupy center stage.[2]

My hope is that this book can furnish people with tools with which to draw deeper conclusions and make wiser decisions, thereby helping them develop their capacities of reflection and analysis. In the final analysis, what it all comes down to is *the art of thinking clearly*. In part, this means that one's thoughts and reasoning should be lucid and sharp, not blurry and sloppy. But it also means that one should think things through carefully, not just leap to snap judgments based on little evidence. I hope that this book's usefulness will flow naturally out of my discussion of how we can learn to ground our thought processes in a solid understanding of how we acquire knowledge and process it.

I am also aware that some people will inevitably take my position as a critique of religious beliefs. However, I wish to emphasize that this is truly not my attitude. Throughout these past twenty years, I've made many friends, both religious and nonreligious ones—among them rabbis, imams, Jesuits, scientists, and philosophers. My thoughts have been greatly enriched by all these friends and acquaintances.

I have the highest respect for everyone's personal beliefs. I know that religion can play a very central role in a person's life, especially in periods of grief and despair. Although I personally do not believe in any kind of life after death or supernatural being or creator, I know that many religious people have such a comforting belief, and I fully respect that—but alas, I am unable to offer any replacement for it. Hopefully, though, even if I cannot provide any support for the idea that we continue to live on in any reasonable sense after we die, I will be able to offer a convincing reason to focus primarily on our life on this earth. (Even if there were a life after this one, that would not be a good reason to be less than fully engaged with one's life right now here on earth.)

Finally, I want to encourage my readers to read this book with a critical eye. After all, my whole idea of a "new enlightenment" is precisely that of examining claims and judgments in a critical manner. Please do not be put off by the fact that this book deals, from time to time, with philosophy. Many people think that philosophy is difficult or even opaque, but I think this is a misconception. Is it hard to play the piano? Well, it all depends on whether you are trying to play Bach or "Twinkle, Twinkle, Little Star." Much the same could be said concerning philosophy. I actually talk quite often about philosophical and existential questions with my eleven-year-old son Leo, and the questions he asks me are quite similar to the questions we grown-ups ask. Philosophizing can be done by people of any age. It just depends on how you approach the questions, and which aspects of philosophy you tackle.

In this book, between successive chapters you will always find an "Interlude" (like a dash of lemon sherbet served between two heavier courses of a long meal) on some topic or other. These interludes are a bit freer in form than the chapters and are sometimes more personal; they can also pose riddles without answering them.

Personally, I feel that a reawakening through a "new enlightenment" can contribute to a more open society, to a more democratic way of life, and to more favorable living conditions for all human beings. In my opinion, in today's globalized world, a secular vision of humanity and a secular body of ethical principles are the most promising avenues to bring about peaceful co-existence among peoples. We must take responsibility for life on earth, in the here and now. Too many people today suffer from religious or superstitious oppression of one form or another. The secular vision I am proposing is thus, aside from being a plan to bring enlightenment, a plan to bring liberation.

I have called the book *The Flame of Reason: Clear Thinking for the Twenty-First Century* because my dream is for it to be an optimistic manifesto for a new Age of Enlightenment.

January 1, 2021
Stockholm, Sweden

PRELUDE

Concerning Yesterday's World and Today's World

> I do not believe in revealed religion.
> I will have nothing to do with your immortality;
> we are miserable enough in this life,
> without the absurdity of speculating upon another.

—Lord Byron, letter to Thomas Moore, March 8, 1822

It was in a hotel room in Brazil that the Austrian writer Stefan Zweig (1881–1942) wrote his famous description of the collapse of the Old World. That book was his final desperate manifesto, and after he had finished writing it, he took his own life, and his wife took hers as well. The book first appeared in print in Sweden in 1942 under the title *The World of Yesterday* (actually, it was printed in German, and its title was *Die Welt von Gestern*). Zweig had for a long time been one of Europe's most popular writers, but Nazism had driven him into exile, and in Germany his books were burned.

The World of Yesterday grew out of Zweig's sense that the world around him had gone crazy. He had lived through a period in which the ethical compass and the belief in the future that had pervaded Europe up until World War I were suddenly uprooted and replaced by fanaticism and irrationality. As he puts it (as translated from the German by Anthea Bell):

> All the pale horses of the apocalypse have stormed through my life: revolution and famine, currency depreciation and terror, epidemics and emigration; I have seen great mass ideologies grow before my eyes and spread, Fascism in Italy, National Socialism in Germany, Bolshevism in Russia, and above all the ultimate pestilence that has poisoned the flower of our European culture, nationalism in general.
>
> I have been a defenseless, hapless witness of the unimaginable relapse of mankind into what was believed to be long-forgotten barbarism, with

its deliberate program of inhuman dogma. It was for our generation, after hundreds of years, to see again wars without actual declarations of war, concentration camps, torture, mass theft, and the bombing of defenseless cities, bestiality unknown for the last fifty generations, and it is to be hoped that future generations will not see them again.[1]

Then Zweig goes on to point out the paradoxical aspect of the times he had lived through:

Yet paradoxically, at the same time as our world was turning the moral clock back a thousand years, I have also seen mankind achieve unheard-of feats in the spheres of technology and the intellect, instantly outdoing everything previously achieved in millions of years: the conquest of the air with the airplane, words traveling all over the world at the moment when they are spoken, the conquest of space, the splitting of the atom, the defeat of even the most insidious diseases. Almost daily, things still impossible yesterday have become possible. Never until our time has mankind as a whole acted so diabolically, or made such almost divine progress.[2]

A Global Psychosis

Aside from the Eurocentric perspective in Zweig's description, it is hard not to see parallels with the world of today. The internet has linked most of the people of the world together through websites, e-mail, and social media like Facebook and Twitter.

Zweig's description concerns a Europe that had moved rapidly from a marvelous period of astonishing scientific progress, deep belief in the future, and a spirit of enlightenment to a nightmarish period of irrationality and fanaticism. And exactly that is happening once again today, but now on a global scale. Never has scientific progress been so impressive as nowadays. Never has it been so easy to dig up information and knowledge as today. Never has it been so hard for dictatorships and other totalitarian regimes to keep their populations in the murk of ignorance. Never has it been so simple to make oneself be seen, heard, or read by a global audience.

And yet, simultaneously, the world has been hit by a form of mass psychosis. On a daily basis, homosexual people are being killed or imprisoned, thanks to certain people's interpretations of God's will. Women are dying because they have been denied abortions. People are being stoned to death, or are having their hands chopped off, because of the way they happen to conceive of divine laws. Religious fundamentalists post videos of beheadings on the internet, urging viewers to join them in their holy war. People are hoodwinked into thinking that God can cure deadly diseases through miracles. Myths about witches or demons that must be exorcised result in cases of maltreatment or death, in which even children are too often the victims.

Since Ayatollah Khomeini's seizure of power in Iran in 1979, since the fall of the Berlin Wall and the end of the Cold War in 1989, and most of all, since the attack on the World Trade Center in September 2001, it seems that irrationality, superstition, and fanaticism have been making ever greater gains all around us. Examples are manifold; the list that follows is a tiny but representative sampler.

In November 2015, a series of terrorist attacks took place in Paris and the city's northern suburbs. There were several mass shootings and a suicide bombing at cafés and restaurants. The attackers killed 130 people, including 90 who were attending a concert at the Bataclan theater. The Islamic State (usually called "ISIS," or as they call themselves, "The Caliphate") claimed responsibility for the attacks.

On Bastille Day in France—July 14, 2016—a nineteen-ton truck was deliberately driven into a crowd of people celebrating on the Promenade des Anglais in Nice. Eighty-six people died and 459 were injured. The Islamic State claimed responsibility for the attacks.

A similar attack with a truck took place in Stockholm on April 7, 2017, killing five people and injuring many more. The driver had sworn allegiance to the Islamic State.

In September 2017, Indian journalist Gauri Lankesh was shot dead outside her home in Bangalore. Lankesh was known as a fierce critic of Hindu nationalist organizations in her state. In 2016, she had been convicted of libel because of an article she wrote, accusing members of the Bharatiya Janata party of theft.

In 2016, a band of militiamen in Kansas, calling themselves "Crusaders," were charged in a terrorist plot against Somali Muslim immigrants. They were planning a bomb attack on an apartment complex where many Muslim immigrants lived, and which was also home to a mosque.

In 2018, a meat seller in Uttar Pradesh, India, was beaten up by the police because of allegations that he had slaughtered cows (cows are believed to be holy in the Hindu religion), and he later died in a Delhi hospital. Since 2010, 28 Indians, of whom 24 were Muslims, have been killed, and 124 have been injured, in acts of violence sparked by disputes over the allegedly sacred cows.

In 2018, a Buddhist mob attacked mosques and Muslim-owned businesses in Sri Lanka, resulting in two deaths. Sri Lanka is a Buddhist-majority country. Since 2012, tensions and violence there have been fueled by hard-line Buddhists and their organization BBS, or Bodu Bala Sena (meaning "Buddhist Power Force").

In 2012, Savita Halappanavar had a miscarriage in the seventeenth week of her pregnancy. She was residing in Ireland, and the doctors in the hospital refused to remove the fetus. The explanation they gave her was: "This is a Catholic country." She died of blood poisoning one week later.

In February 2014, Uganda's president Yoweri Museveni signed a law that permitted the death penalty for homosexuals. At the same time it also became mandatory to report anyone whom one believed to be homosexual. The reasoning behind this Bible-inspired law was eagerly supported by the Anglican Church. On the day after the law came into force, a Ugandan newspaper published the names and photographs of 200 alleged homosexuals, every one of whom, from that moment on, lived under the constant threat of death. (In August 2014, the law was annulled, but for purely formal reasons, when Uganda's constitutional court ruled that too few members of parliament had been present when the law was passed. Some Christian churches in Uganda are continuing, however, to push for the law to be passed again. Homosexuality is still a crime in Uganda.)

In March 2013, the Catholic Church in Burundi asked the state to prevent a splinter group of the church from making its monthly pilgrimage. The pilgrims made their way to a spot where they thought that the Virgin Mary was making an apparition, but the church claimed that the apparition was false and wanted to prevent the group from going there. This resulted in monthly skirmishes between the police and the pilgrims. On one of these occasions, the police shot into the group, killing ten people and injuring thirty-five.

In 2014, in Sudan, twenty-seven-year-old Mariam Yahia Ibrahim was sentenced to death by hanging for having declared herself Christian rather than Muslim. (In 1983, Sudan had instituted Sharia law, which is described in some detail in chapter 10.) Though she was eight months pregnant, the young woman was nonetheless sentenced to be whipped one hundred times before her execution was to take place. After a huge international outcry, she was liberated and flown to the United States.

In 2014, Brunei took the first step toward instituting Sharia law. This initial step involved the criminalization of extramarital pregnancy, of missing Friday prayers, and of proselytizing for any religion other than Islam, with the punishment ranging from fines to prison terms. In the second step, theft and the consumption of alcohol by Muslims were made punishable by whippings or possibly by cutting off of limbs or other body parts. These laws came into force in 2015. In 2016 came the third step, in which adultery, sexual intercourse between homosexuals, and blasphemy against the prophet Muhammad would all be punished by stoning to death.

In June 2018, the secular Bangladeshi blogger Shahzahan Bachchu was killed by unknown attackers. This, unfortunately, is just one example in a series of killings of secular activists in Bangladesh. Earlier, in 2015, the secular publisher Faisal Arefin Dipan was hacked to death. Prior to that, the secular blogger Avijit Roy was also murdered. Also in 2015, the blogger Niloy Neel, who defended atheist views, was hacked to death in his home in Dhaka, the capital city. Some Bangladeshi secular bloggers managed to flee to Sweden, and in 2017 and 2018, they were granted temporary asylum there.

In 2017 and 2018, many members of the Muslim Rohingya people in Myanmar were attacked and killed by Buddhist extremists. Luckily, many Rohingas managed to escape to Bangladesh, where today they are living in refugee camps.

In 2015, an ultra-Orthodox Jewish anti-gay extremist attacked a Gay Pride festival in Jerusalem, stabbing one person to death and injuring several others. The attacker had been released from prison just three weeks before the attack, after having served ten years for a similar attack he made in 2005.

In January 2015, militant Islamists invaded a meeting of the editors of the French satirical magazine *Charlie Hebdo*, killing twelve people and injuring many more. The victims' "crime" involved the publication of satirical drawings that criticized various religions, including Islam. After the attack, the perpetrators fled and were eventually killed, after an intense police search that lasted for three days. Linked to this attack was another attack on a Jewish shop in Paris, in which several people were taken hostage. Some of the hostages died when the French police attempted to liberate them.

In that same month, several people were killed by two suicide bombers in a market in northeast Nigeria. The perpetrators were two teenage girls, each wearing a bomb strapped to her waist, who blew themselves up, along with their victims.

What is it that drives people to carry out all these fanatical acts, and what gives rise to the bizarre beliefs underlying such acts?

One can't help but wonder what makes so many people say that they love and revere a god who seems to accept the fact that people act so barbarically.

The god whom they claim to love also lets vast numbers of innocent people suffer inconceivably as a result of all sorts of natural catastrophes. What kind of god is it that lets children on vacation in Thailand lose their parents in a gigantic tidal wave? And what kind of morality does a person have who thanks God for having saved them in a bus crash, when people sitting all around them died? At the same time as their god allows (or causes) disasters to take place every single day throughout the world, many believers presume that this god is merely reacting with justifiable wrath at such supposedly evil acts as stem-cell research and acts of love committed by homosexuals.

This god also seems unable to tolerate not being believed in by some people. In many countries, the "crime" of denying God's existence merits the legal system's most severe punishments. God would certainly seem to have a peculiar set of priorities. And so I wonder: Why on earth should billions of people respect or worship such a god?

Fatal Superstition

Irrational thinking throughout the world crops up not only in expressions of religious extremism. Magical thinking and other superstitious notions are also widespread, and moreover, they can be life threatening.

In October 2014, a fifty-five-year-old woman in the Indian state of Chhattisgarh was tortured and murdered by her own family members because they suspected that she, through witchcraft, had brought illness upon a family member. Chili powder was smeared all over her eyes and into her ears, and then she was caned to death.

A twelve-year-old girl in Brazil who was suffering from cancer died because her father believed more in a "miracle doctor" supposedly having healing powers than in the modern treatment of cancer using chemotherapy and radiation.

A British family who were members of the Christian sect of Jehovah's Witnesses refused to let their child have a blood transfusion because that would go against their religious convictions. Jehovah's Witnesses believe that both the Old Testament (Genesis 9:4; Leviticus 17:10; Deuteronomy 12:23) and the New Testament (Acts 15:28–29) command them to abstain from using the blood of others. In March 2014, however, an English court ruled that the parents did not have the authority to stop the blood transfusion from taking place.

This was good, but in many other countries, parents' superstitious beliefs are accorded higher respect than their children's lives. In thirty-seven U.S. states, there are laws that clearly state that if parents deny a child medical treatment because of religious convictions, they cannot be held legally responsible for any harm that might come to the child.

The theory of evolution lies at the base of modern medical research and practice. Yet nearly half of all U.S. citizens believe that Darwin's theory is false and that humans, exactly as they are today, were created by God. To them it is inconceivable that there could exist any kind of link between humans and apes, or that humans might have changed throughout time.

A full one-third of the American populace believes in ghosts and telepathy. One American in four believes in astrology, and also in the idea that Jesus will come back to earth within the next fifty years. Some American politicians believe that we don't need to worry our heads over global warming because Jesus will solve all our problems when he returns to earth.

In May 2014, the Hindu nationalist Narendra Modi, of the Bharatiya Janata Party (BJP), became prime minister of India. At that time, the BJP parliament member Ramesh Pokhriyal Nishank stated in a political debate that as trology was far ahead of science, and that science was in fact "a pygmy" compared to astrology.

Swedish Extremism and Superstition

Of course, Sweden is not immune to religious extremism and superstition. Since summer 2012, hundreds of Swedes traveled to Syria in order to fight hand in hand alongside religious jihadists.

Two parents in the town of Borås concluded that their twelve-year-old daughter was possessed by evil spirits. They sought help from the pastor in a local free church (a non-state-sponsored Christian church) in order to drive out the demons. The girl was severely mishandled, and both the parents and the pastor were sentenced to jail terms. The year was not 1613 but 2013. Not too long after that, yet another case of torture in a suspected case of exorcism created a stir in the same city.

At the same time as children in Sweden are mistreated by exorcists, Swedish television channels happily show entertainment programs based on just such belief in spirits, featuring "haunted houses" and seances with mediums who claim to be able to communicate with the souls of dead people. The responsible officials on Channel TV4 consider it all to be harmless fun.

A teenage girl in northern Sweden was living with a man who regularly beat her. She sought advice from a fortuneteller, who divined her future using Tarot cards. He stated that she was simply going through a troublesome patch in her relationship, but that the Tarot cards showed that her partner would become much warmer to her if they had children together. The fortuneteller's advice made her stop seeking help. Her trust in the Tarot cards meant that she would continue to be victimized for many more years.

A Swedish parliament member belonging to the Green Party requested that the parliament's investigation bureau look into suspicious contrail lines in the sky, which were referred to as "chemtrails." (The "chemtrail" interpretation of contrail lines stems from a conspiracy theory that claims that white lines in the sky, left by jet planes as they fly, are part of a secret attempt to control the weather, human behavior, and countless other phenomena. In actual fact, contrail lines arise when the hot gases expelled by the jet engines come in contact with cold air at high altitudes. The lines are just streams of ice crystals.) The Green Party's previous leader claimed that the CIA and Russia lay behind these mysterious displays in the sky, and a parliament member belonging to the Center Party claimed that airplanes were spraying out chemicals in an attempt to control the weather.

Sweden's small but vocal Christian Values Party (www.kristnavarde-partiet.se) seeks to totally ban abortion. It also aims to overturn the ban on caning children and does its best to obstruct *in vitro* fertilization.[3] The party also seeks to block same-sex marriage and to deny same-sex couples the right to adopt children. Its members also believe that the teaching of evolution in schools is harmful.

In southern Sweden, there is a school supported by the Christian doomsday sect known as the Plymouth Brothers. (The Plymouth Brothers took their name from the city of Plymouth in England, where the movement's first gathering took place in 1832. In Sweden, this cult was established in the 1870s, and today it boasts roughly four hundred members.) Sect members do not allow their children to go to public schools and even refuse to eat at the same table as nonmembers of the sect. Their children should have as little contact as possible with the surrounding society, while they are awaiting the second coming of Jesus. Women are not allowed to have short hair, and they must not work outside the home after marriage. Sect members are not allowed to watch television or listen to the radio. They must not study at a university or vote in any election. And yet, the Swedish parliament decided that the sect would receive federal support for its school, in the name of religious freedom. That these children are thereby exposed to extensive brainwashing seems not to matter in any way.

Religion, the "New Age" movement, quack medicine, and superstition are not harmless. What we humans believe is a matter of major importance. And yet, politicians, journalists, and society in general are often inconsistent in their attitudes toward different sorts of beliefs.

Suppose that you have a good friend who at a certain point becomes more and more skeptical about taking his diabetes medicine. You grow very worried about his behavior and confront him sternly. He explains that he won't take his medicine because he thinks it has been poisoned by the CIA. They are out to get him, he says—they want to kill him.

You call up a doctor and relate what just happened. The doctor investigates your friend and reports the diagnosis: paranoid delusions, psychic disturbances, mental illness. Your friend is offered treatment.

Another good friend just broke off her long-term relationship with her live-in partner. You think this is a mistake and ask her why she did this. She replies that after attending a meeting of the Christian group called "The Word of Life," she realizes that she no longer can live with her partner because they aren't married. She fearfully insists that she will be punished forever in hell for having had sex before marriage. Moreover, she won't ever let a swear word cross her lips, because then she will suffer the same fate.

You call up the doctor once again and ask him to check out your friend. It turns out he gives the same diagnosis to her: paranoid delusions, psychic disturbances, mental illness. But no—society says instead that she is just "reborn." She has simply become religious, that's all.

Society judges these two friends of yours in totally different ways, although it is even less likely that hell is awaiting friend #2 when she dies than that the CIA has poisoned friend #1's diabetes medicine. After all, the CIA at least *exists*, and it is even known to have poisoned certain people that it considered to be a threat.

It's obvious that society's general measuring rods are not based on what is sensible and what is not sensible. Other factors play a role in determining what is considered to be reasonable behavior, and what is not.

Why do people seek certainty in religions, in the New Age movement, or in superstitions? Why do people believe in such strange ideas and dogmas? The question is complicated, and there are many answers—not just sociological answers but also political ones, psychological ones, and ones coming from evolutionary biology.

Our ever-more globalized world can give rise to feelings of insecurity, rootlessness, and identity crisis. For some people, the solution is to seek security in the warm fuzzy feelings of the New Age movement. The New Age has become Generation X's religion: quack medical claims such as astrology, homeopathy, *feng shui*, Tarot cards, *Reiki* ("universal life energy" in Japanese) healing, iris-based diagnosis techniques, Chinese alternative medicine, magnet therapy, psychosynthesis, *chakra* balancing, Kirlian photography, and other pseudoscientific ideas are spreading like crazy, particularly with the help of the media, in a ceaseless flood.

Courses in "leadership and personal development" are launched by cynical exploiters, as are pyramid-style get-rich-quick schemes on the web, inspired by the world of the New Age movement. All sorts of quack treatments for serious illnesses are hawked with the promise of bringing fantastic, rapid effects. Pseudoscientific and New Age ideas are luring more and more people, and many are not only tricked but also exploited—sometimes with tragic consequences.

When believers are confronted with the fact that there is nothing to support these theories, nothing that would indicate that they are true, they often point to a criticism inspired by New Age thinking—namely, criticism of the very concept of truth. Claims are not true or untrue, so this idea runs—they are merely true *for certain people* or true *for certain cultures*. "Such-and-so may not be true for *you*, but it's true for *me*!"

The notion that all truth is relative (i.e., that what is true varies from person to person), and that no set of cultural values is better than any other, is encouraged in certain intellectual circles, as well as in Swedish cultural debates. Unfortunately, such wishy-washy ideas do nothing to help in the fight against global warming or to raise our general moral consciousness. In fact, quite the contrary, they fool exactly those people who most need to be involved in the project of a new enlightenment. Such a relativist stance is really the result of intellectual laziness on the part of people who prefer to glibly slip from one idea to another rather than reaching conclusions through careful deliberation. And what the relativists don't seem to understand is that if there is no objective truth, then you can't even be *wrong* about anything! Relativism is thus a conveniently self-reinforcing belief system.

And yet, among relativists, prejudices flourish against people who have chosen to seek clarity and a scientific basis for their beliefs. Scientifically inclined people often run into relativist or New Age objections such as this:

"You only believe in what science can explain! How can you be so narrow-minded?" Or else, "Well, what about *love*? Don't you believe in love? Obviously you can't *explain* love, so that means you can't believe in it!"

To defend their blurry ways of thinking, some relativists and New Agers try to make themselves immune to criticism, and they shy away from reconsidering their views in light of new facts that crop up. Behind such weird insults as "You're nothing but a science fundamentalist!," there is often a great well of ignorance of—and a contempt and an intolerance for—science and its practitioners.

Sweden is a quite secular country. Religion no longer plays a large role in our society, especially in comparison with many other countries. Most Swedish citizens consider themselves nonreligious. The Swedish Lutheran Church (once officially the state religion) today has such an unclear and blurry message to offer that few people care at all about its teachings anymore.

But Sweden is also part of Europe, and in Europe as a whole the situation is considerably more serious. Many contemporary European politicians are insisting that Europe should have a Christian set of values, and in countries like Poland, fundamentalist Christians actually are running the government.

In the United States, despite the long-standing official separation of church and state, the conservative Christian movement exerts major influences on the country's politics. The same holds for many Catholic countries and, of course, for the rapidly evolving political form of Islam, which everywhere seeks to install Sharia law as the law of the land and make the Koran become the ultimate guiding principle for all human beings.

What We Need Is a New Age of Enlightenment

Religious beliefs have very serious consequences as far as society and politics are concerned. Opinions about such matters as women's rights, abortion laws, stem-cell research, contraceptives, children's rights, animal rights, euthanasia, homosexuality, marriage, science, and so forth are all deeply affected by religious views throughout the world.

Today, international politics is pervaded by a conflict of ideas between those who seek a new secular enlightenment and those who cling to a conservative view of the world. This conflict not only affects world politics in the highest degree but also touches the lives of all ordinary people. Both in Europe and in the rest of the world, a war is being waged against terrorism and religious fundamentalism. The once-promising Arab Spring ran out of gas and turned into a cold winter, and the rise and fall of the Islamic State showed the world the most gruesome violence it had seen in a long time.

The worst forms of religious fundamentalism, whose consequences we see in such lands as Uganda, Sudan, Syria, and Iraq, don't have a great deal of support in our part of the world. But the basic attitudes underlying fundamentalism, such as the claim that the principles of morality were dictated directly by God, are very widespread, even in a progressive (and mostly nonreligious) country like Sweden.

We truly need to put this all behind us, for these are ideas that belong to a bygone era. In their place we need to formulate a new enlightenment-oriented secular humanism, in which what is deeply *human* is found at the very core. We need to reawaken the basic values and ideals that defined the original age of enlightenment. We need to accept the idea that the world we inhabit is part of nature, and that it has no trace of supernatural or magical forces. Only when such a worldview predominates will respect for people and their relations occupy center stage, as opposed to people's relations to one or more gods. Ethical questions should be detached from religion. This doesn't mean that the questions become any easier—just that ideas are tested and judged without being profoundly tainted and constrained by religious dogmas.

Such a form of secular humanism builds on the power of free thought—the power to investigate and understand the natural world. Although not *everything* can be investigated or understood, the sincere quest for knowledge and understanding establishes a flexible, nondogmatic attitude toward the world. Curiosity and openness lie at the core of such an attitude. The scientific method of careful and open-minded testing, as well as science's creative and reflective ways of thinking, provides key tools. What clear, science-inspired thinking helps us understand, among many other things, is that a person can be good and can be motivated to carry out morally good actions without ever bowing to, or being limited by, supposedly divine forces.

This attitude is also characterized by a desire to move toward a secular structure for society and a secular form of politics, in which all people in a given situation are treated with equal respect and consideration, independently of what beliefs or cultural background they might have. In such a society, the laws, the norms, and the public places do not reflect religious presumptions of any sort but, rather, a human set of ethical principles not linked to any religion. Religion and politics are kept apart. Every person can believe what they wish to believe, as long as it does not infringe on others, and as long as citizens are not compelled to submit to a state religion.

Such a secular vision of humanity represents a strong belief in human beings, in their abilities, and in their potential to grow and change. It represents the idea that goodness in human life comes from *within*, rather than being imposed from on high.

Today, the global society forged by the internet is taking shape quickly. Not just companies but countries are becoming increasingly globalized, and, of course, individuals are doing so as well. People throughout the world can raise their voices and be heard, and compete in terms of knowledge and skills as never before. Thanks to social media, isolated individuals can play a larger role than ever in political developments. Today, more people have greater access to information and knowledge than ever was dreamed of before, and more people are concerned about the world situation. More people have the chance, through their own actions, to make a difference.

Each one of us, as an individual, matters. It is thus vitally important that each of us should choose, in a conscious and reflective manner, our own views of reality, of the world, and of humanity. And this means that it is crucial for us all to train ourselves in the art of thinking clearly.

Part I

THE ART OF THINKING CLEARLY

CHAPTER ONE

TO MEET THE WORLD
WITH AN OPEN MIND

Concerning the Tools and Compass
Needed in the Quest for Knowledge

Judge a man by his questions rather than his answers.—Pierre-Marc-Gaston, duc de Lévis[1]

Would we humans have been just as curious about the world and the universe if the sky above us had always been cloudy? I wonder. The great existential questions were most likely first posed when people looked up at the sky on a starry night. What is there out there? Where do *I* come from? Why do I exist? Why is there something at all, instead of just nothing? What does it mean to be human? What should I believe in? How am I going to decide on my moral values? And lastly, what is the meaning of life?

Each person making the bumpy trip from childhood to adulthood runs into these same eternal questions in a fresh way, and continues, for the rest of their life's trip, to reflect about them.

My Own Bumpy Trip

As a child I loved to read tales of fantasy. One of my favorites was *Alice in Wonderland* by Lewis Carroll (1832–1898).[2] Wonderland was a fantastic place where anything was possible, but Alice was a skeptic. I always think of the time when, in the story, she meets the White Queen. Alice innocently says, "One can't believe impossible things," to which the White Queen haughtily retorts, "I daresay you haven't had much practice. When I was younger, I always did it for half an hour a day. Why, sometimes I've believed as many as six impossible things before breakfast."

When I was a child, I thought that Alice had a boring attitude. *Obviously* one should believe in what is impossible! Curiosity was already then a strong driving force in my life.

When I was little, I thought I could move things with the force of thought alone, thanks to the Israeli magician Uri Geller. In 1972, Geller appeared on television in the United States, and millions of viewers saw keys bending and watches suddenly stopping without his ever touching them. He succeeded in convincing many people that he had true parapsychological powers, allowing him to make things move by the pure power of concentration, but skeptical magicians, especially Ray Hyman and James Randi, revealed how he depended on trickery. Their debunking, however, was not known to the general public, and certainly not to me as a child.

I remember Uri Geller so clearly from my childhood, and, of course, I believed in his magical powers. Many was the time I sat there staring at a matchbox, concentrating all my powers on it, trying to make it move—just a little bit, just a tiny bit, even just a millimeter! I so yearned for it to happen! Maybe I just wanted to feel that I was different and special. Why not believe it, when I *wanted* to believe it? But it never worked. At some point, I started to suspect that things wouldn't become true merely because I *wanted* them to be true. Today I know that Uri Geller was a faker. He certainly faked me out.

As I grew older, I started to devour books about physics and other sciences. Among my favorite readings were fantasy and science fiction novels. I thought that *The Foundation Trilogy* by Isaac Asimov and *Lord of the Rings* by J. R. R. Tolkien were masterworks. Gradually, my growing knowledge about the world started competing with my fascination with "magical" phenomena. The child in me dearly wanted to believe in magic, parapsychological phenomena, gods, and other supernatural beings, but I was inexorably growing older.

I grew up in the tiny town—really just a village—of Mariefred, Sweden (about thirty miles west of Stockholm), which had an extremely limited supply of things for kids to do, aside from the usual sports. But sports were not for me. I have never played soccer or ice hockey in my life. When I turned twelve, I became obsessed with mathematics. While other kids were playing soccer, I would sit all alone in my room and plot curves on graph paper. I soon started to program a Texas Instruments TI59 pocket calculator, and then a computer (an ABC 80). I really was quite the nerd.

When I reached the age of fifteen, I came across the writings of the British philosopher Bertrand Russell (1872–1970). I first read his memoirs, which made a strong impression on me. Russell was not just a philosopher; he was also a political activist, and he wanted to have an impact on others. He certainly had a decisive influence on my choices in life as a teenager. At some point, I read his book *Why I Am Not a Christian* (1958), which paved the way for my interest in philosophy and sparked my interest in society and in political engagement. In 1950, Russell was awarded the Nobel Prize in Literature "in recognition of his varied and significant writings in which he champions humanitarian ideals and freedom of thought."[3]

At around this age, I started, reluctantly, to realize that it was intellectually dishonest to believe in something just because I wanted it to be true. I also began to understand that it could even be immoral to believe in something without a good reason to do so. If I can believe in anything I want, then my treatment of other people may be totally arbitrary.

I started to see that one has to be able to *justify* one's beliefs. One has to have a plausible reason for taking a statement as true. And thus I eventually gave up my belief in magic and in parapsychological phenomena; however, I continued to love performing magic tricks, and I still do some magic today.

Taking the place of magical thinking in my mind was my discovery that the real world is filled with wonderful mysteries. Unsolved riddles in the world of science fascinated me in a way that no magical wishful thinking ever had. Mysteries inside mathematics, physics, chemistry, and biology were much more interesting to me, since they were *genuine* mysteries. They were just as magical, though in a more abstract sense of the word.

I'll never forget my first experience with logical paradoxes. It was April Fool's Day. That morning, an older friend of mine who liked math said to me, "Today is April Fool's Day, and I'm going to trick you in a way that no one has ever tricked you before!" This was a little scary-sounding—and challenging. Therefore, all day long I paid extremely close attention to everything he said and did, but nothing seemed like a trap at all. Eventually evening fell. I tried to look back on the day and figure out when or how he had tried to trick me, but I couldn't recall anything suspicious. I thought and thought, and when bedtime came, I had a hard time falling asleep. The next day I ran into my friend again and told him with annoyance that I hadn't slept the night before, because he had broken his promise of tricking me.

Then he said triumphantly, "So you expected I would trick you at some point yesterday?"

"Yes."

"But I *didn't* trick you, did I?"

"No."

"But you *believed* I was going to?"

"Yes."

"Well, then—I fooled you, didn't I? You gullibly fell for what I told you in the morning!"

On hearing this, I realized that my friend had indeed fooled me as never before—by *not* fooling me. Or rather, he *didn't* fool me at all, hence he fooled me. Many years later, I found out that my friend had borrowed this paradoxical trick from a book he'd read by the American philosopher and logician Raymond Smullyan.[4]

My teenage rebellion made me want to stop being a nerd and to get at least a few points for being cool, especially with girls. The solution I hit on was to learn to play the guitar and to try starting a rock band. This was my main activity while I was in high school in the larger but still very small town of Strängnäs, about ten miles northwest of Mariefred. I played hooky pretty often, taking the commuter train into Stockholm, and there I got very involved with music and going out to nightclubs.

At age twenty, I was playing guitar in a rock band and was pretty sure that I was going to be a musician. Our group, called "Heroes," had put out a record, and at this point I started making frequent trips to London with the goal of establishing contacts in the music world and hopefully getting a recording contract there. (If in our band's name you pick up on the influence of David Bowie, you aren't wrong.) In the end, however, I found that London's nightclub life and music scene weren't my cup of tea, after all, and were even causing me to lose my bearings. This was no good, and after a while, I decided to go back home to Sweden.

On my trip home, I happened to buy a copy of Douglas Hofstadter's 1979 book *Gödel, Escher, Bach: An Eternal Golden Braid*. I gobbled the entire book down during the trip home and throughout the next couple of weeks. It deals with art, music, mathematics, and philosophy in a rich contrapuntal fashion, and it left an indelible impression on me. It opened my eyes onto a whole new world. *This* was what I wanted my life to be about! My earlier teenage passion for mathematics was reawakened, and all at once I decided to study math, philosophy, and computer science at Uppsala University. And I realized that I wanted to weave all these interests together, with music being just a hobby instead. And actually, being a nerd wasn't so uncool after Hofstadter's book became well known. There was hope even for us nerds!

Today I have to admit that I may occasionally have gone a bit overboard with my nerdity, but I am unbelievably grateful for the thinking tools that I acquired during my university days.[5]

To Have an Open Mind

The most enjoyable trip one can take is the trip to knowledge and insights, driven by curiosity and a sense of wonder. What's amazing about such a trip is that the farther you travel, the more you find there is to discover. The more you understand, the more you realize how little of all that is in principle understandable you actually do understand.

To take such a trip, it's best to have the proper gear in one's backpack. Certain types of equipment are needed in order to avoid falling into traps and going astray, but above all one needs a compass to find the right direction. This book is about such tools, and it tries to supply its readers with such a compass. But it also deals with the traps and blind alleys that one can so easily wind up in.

An important tool is openness. People should be open minded. But what does "to be open" really mean? I have often broached this issue with various people, and I've observed that the concept "open" is frequently misinterpreted. So let me give an example.

You and a friend are looking at a TV show about a supposedly haunted house. Taking part in the show is a "medium" who holds seances and claims to speak with the deceased—people who have passed over to "the other side." Afterward, the two of you discuss the program. Your friend claims that ghosts and spirits of dead people really exist, and that there are people who can communicate with them.

You reply that you believe there are no such ghosts or spirits, and that so-called "mediums" either are charlatans or are fooling themselves. Your friend then says, "Come on! You should be more open minded! Don't be so narrow! Just be more open-minded about whether ghosts exist or not."

How often have I wound up in just that position! Sometimes it's been in a debate on television, other times just in ordinary chit-chat. But there is something weird going on in the earlier conversation. Which of the two is more open minded? Can you tell if you look solely at what they believe? One's degree of openness shouldn't have to do with *what* one believes, should it? Shouldn't it instead be measured by how willing one is to *change* one's opinion, on the basis of new facts and pieces of evidence?

The skeptic who thinks that ghosts don't exist can certainly be the less open minded of the two. That person might rigidly insist, "I will never, ever believe in ghosts, even if I encounter one in the flesh in broad daylight!" Clearly, this is not a very open state of mind. But if one day it were proven that human souls continue to exist in another realm after death, and that it was possible to set up communication channels with these "spirits," then the skeptic might change their mind and might start believing in talking to the dead.

But it could also be the ghost-believer whose mind is less open. Just imagine this person saying, "I once witnessed a highly strange event that I couldn't explain, and I won't ever believe that there is a scientific explanation of what I saw. I'll always believe in ghosts!" In such a case, it's sensible to say that this person is very dogmatic—the opposite of open minded.

Suppose that my friend and I are discussing life on other planets. I say that I think there is life on other planets, but my friend says this is silly. Which of us is the more open-minded one? Clearly it's impossible to answer the question if one looks solely at what we believe on this topic.

Let's take another example. Suppose I have a small, closed black box sitting on my lap, and I say to my two friends Adam and Eve, "Do you think there's an apple in this box? Just make a guess, using all your intuition and common sense. . . . So now—do you think there's an apple in here?"

Adam swears that he believes there is an apple in the box, while Eve believes the box is empty. Simply knowing that one of them believes there *is* an apple, while the other believes there is *not* one, we can't determine which of the two is more open minded. In sum: one's degree of openness is not a function of *what* one believes, but of *how willing* one is to change one's mind, when new facts and discoveries come to light.

Today, however, one frequently runs into people who claim that if Person A believes in ghosts, flying saucers, and alien abductions while Person B is a skeptic of such things, then clearly A is more open minded than B. What sense is there behind such a view? Does one get credit for open-mindedness simply because one believes in implausible, crazy-sounding ideas? Surely that doesn't make sense.

Let's try another thought experiment that clearly shows that looking at beliefs alone cannot be the right way to determine someone's degree of open-mindedness. In contemporary Europe there are a lot of people on the extreme right, including some who go so far as to deny that the Holocaust took place during World War II. This is a completely implausible stance. But should we say that these people are *more open minded than we are*, precisely because they hold this belief, while we simpletons are convinced that the Holocaust *did* take place? No; that would be not just silly but deeply foolish.

How open minded are those folks who staunchly assert that no human has ever stood on the moon, and that the supposed moon landing in 1969 was all fakery, just filmed in a Hollywood studio? Such an attitude is hardly more open minded than believing that a moon landing actually did happen.

The trait of "openness" should instead be thought of as being a close cousin to curiosity. To be open-minded means to be constantly sensitive and always on the lookout for ways to reevaluate one's values and one's ideas, whenever one runs into new facts or viewpoints.

More than just open-mindedness is needed to allow an idea to be thought through carefully, however. One also needs a criterion for what is sensible to think of as possibly true. Swedish philosopher Ingemar Hedenius stated a very simple but helpful principle that he called "the principle of intellectual honesty": Believe in an idea if and only if you have good reasons to think it is true.

This implies that you should be ready to explore all conceivable alternatives in order to choose which among them is the most plausible or reasonable. Such an attitude would constitute the best and truest kind of openness, in contrast to someone who will gladly swallow any random idea that's thrown at them. Such naïve gullibility hardly merits the label "open"; it is simply a symptom of an immature thinking style. In the United States, there is a slogan that says, "Don't be so open minded that your brain falls out." Who would want to be that open minded?

And yet, many people believe in things that have no reasonable or plausible support. Some unreasonable ideas simply have high social status, or they are considered thrilling or mystical. That's one of the reasons that "New Age Spirituality" is so popular today. But if one feels it is important to be intellectually honest with oneself and with others, then irrelevant "reasons" like that will not play into one's judgments of an idea's truth or falsity.

Explanations and Ockham's Razor

There is a very old fundamental philosophical principle that advocates selecting the *simplest* of rival explanations for a given phenomenon. Let's take an example.

Suppose that one day I come home to my apartment and I notice that the window has been broken and the TV set is missing. There are any number of possible explanations for what I see, of which here are three:

1. An extraterrestrial descended to earth from a flying teapot, broke into my apartment, and made off with my TV.
2. A burglar broke into my apartment and made off with my TV.
3. My TV was teleported to another dimension by secret CIA agents, using techniques unknown to current-day science.

The simplest and most sensible of these three explanations, and thus the best of them, is #2 (unless we come across new pieces of information that give us reason to take alternative #1 or #3 seriously).

This same kind of thought should apply to phenomena that are claimed to be supernatural or paranormal. Consider, for example, the following two claims:

1. I believe that paranormal-seeming abilities in a human being are best understood through the postulation of a seventh sense, which conveys information and knowledge that would otherwise be inaccessible.
2. I believe that paranormal-seeming abilities in a human being are best understood through scientific explanations developed by researchers in the given domain. These explanations involve . . .[6]

What could possibly be humbler and more open minded than believing that scientific explanations are the simplest and the most plausible, and therefore the best?

The idea that good explanations are simple is a stance that is usually called "Ockham's razor."[7]

Sometimes one hears a statement like "Science will never be able to explain X!," where X might be the nature of consciousness, the origin of life, or the origin of the universe. But such cocksure pronouncements are both dogmatic and limited. How can anyone foretell what humanity will (or will not) eventually be able to explain? The fact that we *today* can't explain something doesn't mean that it will *never* be explicable.

Self-assured claims of this sort often are due to a confusion between what has been *explained* and what is, in principle, *explainable*. The nature of consciousness is a good example. There are many people who have pondered how consciousness arises in the brain and have offered plausible theories about it, but there is still no complete and totally accepted theory of consciousness that all scientists agree on; however, this doesn't imply anything about whether consciousness will one day be explained in the future.

One should simply be open minded and humble about future knowledge, and say something like this: "Today's science has not fully explained the nature of consciousness, but perhaps in the future we will be able to do so. We just don't know, today."

In the New Age movement, one often runs into claims like this: "I was present at an event that I couldn't explain, so it must be supernatural!" But why should the fact that one can't explain something on one's own mean that there is no scientific explanation for it, and never could be, never will be? A healthier outlook would be to think that if one can't explain something oneself, then perhaps someone else could do so—someone with greater knowledge or better qualifications than oneself. It is arrogant to assume that one's own inability to explain a given mysterious event is more reliable than long traditions of research, which might be able to point toward a genuine explanation. To dismiss the value of scientific findings and to think of one's own personal experience as the ultimate authority merely reveals a serious lack of humility.

To proceed solely on the basis of one's own limited life experience and to reject all the collective results of scientific research is an attitude that seldom leads to new insights. Someone who says, "I was there and I couldn't figure it out, so it just can't be figured out!" is merely confusing their own ability (or inability) to explain the given mystery with a more general, timeless meaning of "explainability." What they are really saying is this: "If *I* can't figure out what happened that time, then by God, *nobody* can." This is just cocky; one should be more humble.

Traps in Thinking

Socrates was very wise when he observed that one should not believe that one knows something, when in fact one *doesn't* know it. In his defense plea before the court that sentenced him to death, he said (or at least this is what Plato tells us), "I am wiser than this man; it is likely that neither of us knows anything worthwhile, but he thinks he knows something when he does not, whereas when I do not know, neither do I think I know. So I am likely to be wiser than he to this small extent, that I do not think what I do not know."[8]

To overestimate one's knowledge is not good. Nor is it good to let one's judgment be led astray. English philosopher Francis Bacon (1561–1626) was one of the first people to formulate some of the thinking traps that we all fall into so easily. Bacon believed that science should be organized so as to

11

be useful. Results of research should be collected and distributed, particularly through academies and journals, so that the same mistake is not made over and over again. In *Novum Organum* ("The New Tool"), published in 1620, he writes, "The mind of man is far from the nature of a clear and equal glass, wherein the beams of things should reflect according to their true incidence, nay, it is rather like an enchanted glass, full of superstition and imposture."[9]

Bacon wrote of four "idols" (and by this term he does not mean "false gods" but, instead, "false ideas") that often lead thought astray and that may cause the voyage toward knowledge to come to a dead halt in a blind alley. In the following list, I describe Bacon's four idols in my own words:

1. *Idola tribus* (the idol of the tribe): The first cognitive trap is that we humans tend to overgeneralize. We interpret facts on the basis of our habitual ways of interpreting situations and using our habitual notions. We ascribe to nature an order that may well not be there. We categorize and classify, and we believe that if some members of a category have a certain property, then so must all members. Through this cognitive trap are born prejudices against women, Germans, homosexuals, Africans, or whatever other category you can think of, which we have decided to generalize and to use in our further thoughts.

2. *Idola fori* (the idol of the marketplace): The second cognitive trap is to uncritically let one's thought be guided by faddish words, clichéd phrases, and widespread opinions. Human language is not precise but blurry. It is important to pin down what one is speaking about, for otherwise one may well cause totally unnecessary misunderstandings, and language may come to dominate over thought. An example is the term *openness*, discussed earlier in the chapter; it is often used in an erroneous way, and thus it guides our thinking down wrong avenues.

3. *Idola specus* (the idol of the cave): The third cognitive trap is to see the world solely from one's own perspective, thus letting wishful thinking get in the way of truth. In addition, own's own limitations, whether innate or the result of training, determine how we interpret things we observe. Here we run into what is called "cognitive bias," which is our tendency to *seek confirmation* of what we already believe, instead of looking for facts that might *cast doubt* on what we believe.

An example, discussed earlier in this chapter, is to say: "I can't explain it, so it must be unexplainable."

4. *Idola theatri* (the idol of the theater): The fourth cognitive trap is to have an uncritical belief in authorities or in famous teachings, doctrines, or traditions. One type of authority is the guru who has built up a theory about everything, and whose disciples blindly follow such theories. Another type is a fuzzy, fluffy theological system of thought, which, just like a play one sees in a theater, is dreamed up out of pure imagination. That some people claim it belongs to a venerated ancient tradition is hardly a reason to buy uncritically into it; in fact, that very fact is a good reason to give it especially critical scrutiny.

Rationality and Wisdom

What does it mean to be rational and wise? By "rational thinking," I mean coherent, contradiction-free reasoning. A rational argument consists, by definition, of plausible steps of reasoning, which is to say, the argument consists of a series of logical steps (but the premises may be either wise or silly). The opposite notion—that of "irrational reasoning"—denotes a confused and illogical chain of thoughts, which can lead one anywhere under the sun.

The advantage of rational reasoning is that the result is always correct, in a mechanical manner. That is, the output of a logical step of reasoning is always true, as long as the inputs are true. But the conclusion of a rational argument doesn't have to be true; in fact, if at least one of the premises is false, then the result of a rational argument will be false as well.

For example, if we take the following as our premises:

a. All types of animals, whether now living or now extinct, were created six thousand years ago;
b. Dinosaurs are extinct animals.
   then we can draw the following logical conclusion from them:
c. The dinosaurs were created six thousand years ago.

Even though the reasoning pattern is flawless, premise (a) is false, and therefore, the conclusion is false. False premises invariably lead to false conclusions!

Hmm . . . I am curious, reader, whether you happened to notice a serious logical error that I just committed in the last few sentences. If you did, my congratulations! If you didn't, then you might try to find it. Hint: Does a logical argument that uses at least one false premise really *have* to lead to a false conclusion?

The answer is: no. I was wrong in my claim. It was my translator, Douglas Hofstadter, who pointed out this serious flaw to me. I am grateful to him for having done so, and especially for providing the following four simple counterexamples, showing how wrong I was:

Premise 1: 1 + 1 = 1 (false)
Premise 2: 2 + 2 = 5 (false)
Conclusion (by addition): 3 + 3 = 6 (true)

Premise 1: Christer Sturmark was born in China (false)
Premise 2: Chinese people all speak fluent Swedish (false)
Conclusion: Christer Sturmark speaks fluent Swedish (true)

Premise 1: Barack Obama was born in Kenya (false)
Premise 2: Kenyan people all have belly buttons (true)
Conclusion: Barack Obama has a belly button (true)

Premise 1: Barack Obama was born in Hawaii (true)
Premise 2: Every Hawaiian-born person becomes U.S. president (false)
Conclusion: Barack Obama became U.S. president (true)

These four amusing examples clearly show that what I claimed earlier—"if at least one of the premises is false, then the result of a rational argument will be false as well"—was completely and utterly wrong. It was a tempting idea, to be sure, and maybe you thought it made perfect sense. Certainly I myself fell for it lock, stock, and barrel, but now, reader, you know that things are a bit subtler than that. The lesson for me is that we all live and learn, and by so doing, we can slowly but surely become better and clearer and more logical and more rational thinkers.

The conclusion of a rational argument doesn't have to be *morally* right or correct. A rational argument does not need to have any moral dimensions; however, this will be the case only if there are no moral values in the premises. Since moral values can vary from one person to another, the conclusions reached by a single rational method will not seem the same to all people, since different people will take different moral ideas as their premises. Moreover, what seems rational from one point of view will seem irrational from another. Rationality is thus context dependent.

If an adult were to write a letter to Santa Claus, the act could be described as irrational and silly. But a five-year-old child writing a letter to Santa Claus is fully reasonable. The child really believes that Santa Claus exists, and that the letter can have an effect on what presents Santa will leave under the Christmas tree.

For rationality to be morally fair, we need wisdom in addition. Rationality alone does not give wisdom. What, then, is meant by the notion of wisdom? And what distinguishes *wise* behavior from merely *rational* behavior? To be rational means to draw conclusions that are not in logical conflict with known facts. A rational conclusion must logically follow from the premises and must be contradiction free. But wisdom is more than that.

Human wisdom has to do with *ends*, not just with *means*. Wisdom has to do with what is morally good or bad. Wisdom requires rationality, but rationality does not require wisdom. A rational argument is a mechanical chain of logical steps, whereas wisdom includes moral judgments and a careful look at what the conclusions imply. Wisdom is more comprehensive than rationality, but for that reason it is far harder to capture in a precise description.

Wise reasoning is rational, but rational reasoning is not always wise.

Having Ideas and Forming a Philosophy of Life

What is the world like, and how should I live? What really matters? Sooner or later in life, we all ponder such questions. We all try to make for ourselves at least a fairly consistent picture of how the world is and what we ought to do in it. Each of us winds up constructing our own personal philosophy of life.

Having a philosophy of life means that one has a belief about the nature of reality and a sense of how one should live in it, and how one should treat oneself, other people, other living creatures, and the environment. Any philosophy of life has these two aspects, one of them being *descriptive* (trying to say how things *are*) and the other being *normative* (trying to say how we feel things *should be*). The descriptive aspect tries to reach an understanding of how the universe is constituted, while the normative aspect aims at establishing a system of values and a sense for human nature. All religions and other philosophies of life include these two dimensions.

The concept of "religion" is a manifestation of culture that is not easy to capture in a generally agreed-upon, all-encompassing description. A simple definition might be "the belief in a higher, supernatural power to which humans relate, and vice versa." Often, religions are based on the concept of one or several gods, or other supernatural beings. Many religious people speak of "a personal and conscious god," while others have a vaguer notion of what God is.[10]

A philosophy of life can of course be based on a vision of reality that does not involve any kind of god or supernatural power. A philosophy of life doesn't need magical beings; it just needs to provide a broad, clearly formulated, and coherent set of ideas, on both the descriptive and the normative side. Gods and superpowers are optional features. Thus, although every religion counts as a philosophy of life, not every philosophy of life is a religion.

As you know, holding the belief that no god exists is usually called *atheism*. Its opposite—*theism*—is the belief that there is a god (or gods). Does theism or atheism by itself constitute a philosophy of life? Well, the Swedish national encyclopedia defines the notion "philosophy of life" as follows: "A theoretical and value-laden set of ideas that gives rise to, or exerts a major influence on, a broad picture of humanity and the universe; it expresses a set of basic beliefs and includes a system of values."[11]

For this reason, one can't say that just any old set of ideas constitutes a philosophy of life on life. For example, if I believe that life exists on other planets, that in itself is not a philosophy of life. It's just an idea about one isolated issue. It has no moral implications (i.e., no normative side).

The same holds if I am an atheist—that is, if I believe that no god exists (or if you prefer, that God doesn't exist). Once again, such a stance is merely a point of view about one isolated question; it's not rich enough to constitute a whole philosophy of life, with all sorts of ethical implications and so forth. And much the same is true for theism. Theism—belief in God—doesn't necessarily imply a whole rich philosophy of life. It can be just a single isolated belief.

The desire to put a halt to painful tests carried out on animals, or to forbid abortion, or to have all jobs paid exactly the same—none of these is, in itself, a philosophy of life. Each of them might be a position coming *out of* a person's philosophy of life, however, and thus belonging to that person's set of moral views. In sum, a philosophy of life is a broad and reasonably consistent system of ideas.

The *descriptive* side of a philosophy of life contains not only a theory of what valid knowledge is and how one should best use it (in philosophy, this would be called "epistemology") but also a theory of what exists and what is real (this would be called "ontology").

The *normative* side of a philosophy of life includes basic ideas about human values and human rights, and usually, in addition, views about such things as racism or antiracism, feminism, animal rights, environmental issues, and so forth. All the values that we have belong to this side of our philosophy of life.

Most people, at some point in their lives, muse about the basic questions of existence, although some do so a great deal, and others not so much. But no matter how much or how little we ponder on such things, we all have *some* thoughts on these issues. Without a basic philosophy of life, we would have a hard time navigating around in the world and seeing our environment in a consistent manner.

What are the basic questions in a philosophy of life? Of course they can be described in many different ways. A list of the most central questions might include the following entries:

1. Questions about knowledge and knowing. Can we know something about reality? If so, what can we know, and how? What is knowledge, in the end? What methods can we use to reach knowledge? These are epistemological questions, belonging to the theory of knowledge.

2. Questions about what exists and about the nature of reality. These are ontological questions. They include questions about the possible existence of a god or gods, the implications of the "God" concept, and the compatibility or incompatibility of such a concept with the rest of our knowledge or beliefs about the world. These sorts of questions belong to theology, or the philosophy of religion.

3. Questions about morality—its nature, its existence, and its justification. What characterizes a good moral act? How can we decide whether a given act is morally right or not? Is there an objective morality, independent of one's ideas and desires? This aspect of philosophy is ethics.

4. Questions about the nature of humanity, involving good and evil, fate, consciousness, mortality, and immortality. Do we have free will? Is consciousness something over and above the chemistry of the brain? Does evil exist? If so, what is it? These questions belong to metaphysics.

5. Questions about how we should organize our social life and living conditions. What is the function or role of society? How should a person relate to authorities and laws? What rights and duties do we

have with respect to other people and other living creatures? These sorts of questions are mostly discussed in the philosophy of politics (and sometimes also in ethics).

Why should you (or anyone) consider such questions at all? Well, one good reason is plain old curiosity. You'd have to be pretty blasé to have no interest at all in how the world is, or ought to be. Another reason is that pondering such questions enriches your life. The more deeply you probe the world, the more fascinating it becomes.

Another reason can be to figure out if the ideas you have are really the result of independent thinking, or if they instead come from an unreflective acceptance of things that you were told by your parents, your friends, or representatives of the society you grew up in—your "cultural inheritance." If, at some point, you unquestioningly bought into a set of dogmas, values, and ideas about humanity and the world, then those are not truly your own ideas, and you run the risk of being a marionette easily manipulated by others. In such a case, it would behoove you, at least once in your life, to reflect about your value system and to develop your own philosophy of life. It is never too late.

My Point of View

Each of us has a point of view—a spot from which we look out at the world. I was born in 1964 in a country that was racially and economically quite homogeneous, and in which most people had pretty good lives, materially speaking. There was an outward sense of security—our country hadn't been in a war in two hundred years. When I was a kid, most women were housewives, although many were restless in that role; indeed, the women's liberation movement was soon going to be launched. Now, as I look back, I see a set of broad social changes that had to do with equality and gender roles. I also grew up in a democracy with a free-market economy, a blend of free enterprise with a strong public sector. Through the years, immigration, largely from Europe, Asia, and Africa, changed from a trickle into a flood. Last but not least, I was part of a major digital revolution in the 1990s; as a result of this revolution, Sweden became one of the leading nations in the world with computers and the internet. Today Sweden is among the most globalized of all countries, and it is now ethnically highly diverse, unlike when I was growing up. In short, in my country, I have witnessed vast changes in social life, economic life, and political life from the inside.

During this period of upheaval, I realized, when I was around twenty, that I had a *secular* outlook on the world, meaning first that I saw the concept of God as merely a human-created myth, and second, that I had a *humanistic* view of what it means to be a person. I was therefore a *secular humanist*—and that's what I still am today. This is the point of view that lies behind the arguments in this book.

We live in a world where people with highly varied philosophies of life, ideologies, and value systems have to be able to coexist and cooperate. For me, secular humanism is all about having a carefully considered set of attitudes about the world, not just concerning the nature of reality but also including a healthy and sensible set of values, which will help such coexistence to come about.

This is a point of view that is tenable, rationally and morally, both for the heart and for the head.

Secular Humanism

So what is secular humanism really about? The term "secular" originated, as I mentioned earlier, from the Latin adjective *saecularis*, meaning "worldly," in contrast to "ecclesiastical" (or "churchly"). In current usage, "secular" refers to the idea that human affairs should not be mixed in with (let alone controlled by) religious ideas or religious rules.

As for the word "humanism," in Swedish it actually has three quite different meanings:

1. Humanism as a *goal in education*. One can call oneself a humanist in Sweden if, at the university, one's major focus is or was the humanities, such as art history or comparative literature.
2. Humanism as a *general concern for the welfare of human beings*. One can say one is a humanist in Swedish if one has a strong commitment to people and human rights (which of course one can have as an atheist, as a Christian, as a Muslim, as a Buddhist, or as anything else).
3. Humanism as a nonreligious philosophy of life.

In English, by contrast, there are three different words corresponding to these different meanings. The first meaning is conveyed by "the humanities"; the second by "humanitarianism"; and the third by "humanism" (or sometimes "secular humanism").

The words "secular" and "humanism" thus have different meanings when they stand alone. An individual can be secular, yet at the same time believe in a god. Someone with such a stance considers their belief in God to be a purely private affair, and they feel society should be organized in a way that doesn't advocate or privilege any particular religious conviction.

When the two words are juxtaposed, as in "secular humanism," the meaning is a broad philosophy of life involving no gods or supernatural powers; instead, what matters are people and their concerns.

Secular humanism's way of construing reality builds on the notion that the world is natural, not supernatural. The world consists of matter, energy, and the laws of nature. There is no reason to believe that there are supernatural beings, like gods and spirits, or magical powers of any sort at all.

Secular humanism is therefore a philosophy of life, but it is not a religion. There are numerous other philosophies of life that don't see a need for godlike beings. Among this group of nontheistic, or atheistic, viewpoints, one can count the Chinese philosophy of life Confucianism, as well as Zen Buddhism, Theravada Buddhism, and Taoism.[12]

Secular humanism also deals with humans' attitudes toward each other and toward the world we all live in. It emphasizes each person's freedom, autonomy, value, and responsibility. It has faith in people's ability to seek knowledge and to learn from experience.[13]

I have found that many people conflate secular humanism and atheism. The latter concept will be carefully discussed later in the book, but here I wish to explain how it is related to secular humanism.

An atheist is a person who believes that there are no gods. This idea is limited to just one question: Is there a god? Since atheism deals only with this question, it does not constitute a full philosophy of life. Secular humanism, by contrast, is a rich and comprehensive view of the world, including values. Atheism is merely one facet of secular humanism's nature-based viewpoint, so a secular humanist is always an atheist, but an atheist isn't necessarily a secular humanist.

To put it another way, the relationship between atheism and secular humanism can be compared to the relationship between the intellectual belief "Jesus was not the Messiah" and the religions Judaism and Islam, both of which are broad philosophies of life that agree with that belief. Practicing Jews and Muslims believe that Jesus was not the Messiah, but obviously, holding that belief doesn't mean you have to be a Jew or a Muslim.

～

INTERLUDE: ON SPIRITUALITY AND FLOW

Have you ever experienced existential vertigo?

Once in a while I step out with my son Leo onto our patio in the evening, and we lie down side by side and stare up at the stars. Ideally, of course, one should do this in a place where there are no streetlights, no artificial light at all. And cold winter nights are the best since they are the clearest.

If Leo and I are in luck, we can see Jupiter, and with a pair of ordinary binoculars we can make out four of its moons. We see myriads of stars, and the center of the Milky Way forms a milk-white streak high up in the sky. (Incidentally, the Swedish name for the Milky Way is "Vintergatan," meaning "the Winter Street.") So here we are, Leo and I, two tiny earthlings, somewhere near the outer edge of our galaxy, which is just one among billions of other galaxies.

One time, it occurred to us that we might be looking down instead of up—after all, what direction is "up" in the universe? Who or what would determine that? In moments like that, I can occasionally be caught off guard, and I find myself spinning in "existential vertigo." I have the clearest feeling of being an infinitely tiny part of something that is infinitely vaster than I am. Leo and I then almost cease to exist, other than as two dust particles in the vast reaches of the cosmos. Could this possibly be called a spiritual experience?

My religious friends assure me that God is an indispensable ingredient in any genuinely spiritual experience. I, however, don't buy into that. The feeling of exaltation and awe when I plunge my whole self into a vast starry sky, into the universe's unfathomable depths, evokes in me a far more powerful spiritual reaction than any supernatural myth or legend could possibly do.

This kind of feeling can also be evoked in very different circumstances. When I sit at the piano and make music with other people, I usually have to sweat pretty hard just to hit the right notes and not to miss crucial harmonies, amateur that I am. But once in a while something very special transpires: I lose track of what I'm doing and suddenly I'm just part of the musical "flow." At such times, I become part of something bigger than myself—in this case, a musical experience that feels greater than the sum of its parts. This, too, is deeply spiritual for me, but it has no religious or mystical overtones.

Some people say that every human being has a need for religion. Can that really be the case? And who will decide what counts as religion?

What if what we really need is a philosophy of life without dogmas, without gods, without guilt, without shame, without a fear of hell, without the need to force other people to live according to "God's rules," and without endless bitter fighting over what is true and what is false? What if what we really need is a philosophy of life that affirms openness and questing, that encourages wonderment, fascination, and deep feelings for art, music, literature, nature, the cosmos, reality, and for other human beings?

For me, spirituality is not supernatural but has everything to do with nature itself. We are all just tiny pieces of something much greater. We are small pieces of humanity, tiny parts of the universe, minuscule slices of reality. This thought is dizzying enough for me; I feel no need to spice it up with divine beings or magical powers.

CHAPTER TWO

I BELIEVE THAT I KNOW

Concerning Reality, Knowledge, and Truth

It is not what the man of science believes that distinguishes him, but how and why he believes it. His beliefs are tentative, not dogmatic; they are based on evidence, not on authority or intuition.—Bertrand Russell

In order to be able to formulate your own viewpoint on reality, you will need some tools and a vocabulary of concepts. You'll need to understand concepts such as *truth*, *learning*, *faith*, and *knowledge*. You'll also need to decide for yourself which basic assumptions you will make. Is what *seems* real *really* real, or is it just a dream? Could everything be just an apparition inside my head? Or do genuine truths and falsities exist? And what good reason is there for being rational, anyway?

Someone with a rational attitude does not claim to be able to give definitive and exhaustive answers to all questions about the nature of the world. Contemporary science and knowledge are not yet at the point where they can explain all the phenomena on our planet and in our universe, and they almost surely never will be. Therefore, having a rational attitude means being intellectually humble. It means frequently engaging in self-critical examination of the ideas and concepts that one takes for granted—especially those that one is particularly strongly attached to. This contrasts strongly with traditional religious attitudes.

Many religious interpretations of reality are characterized by absolutist ways of phrasing things, or by claims about how things behave, based solely on religious writings and traditions. Things that today's science cannot yet explain are supposedly explained by reference to "God," which amounts to sweeping such issues under the rug. The following are examples: "We don't know how life originated, so there must have been a god that created it." Or, "We don't know how the universe arose, so a god must have created it." Or again, "We don't know what this consciousness inside our skulls is, so it must have come from God." And so on.

But one could instead have a totally different kind of attitude. Here's a simple parable to show what I mean:

I don't understand how it's possible to build a house like the one I live in. I myself am all thumbs, and I can barely screw in a lightbulb. But that doesn't mean that I believe that some god must have built my house. For the time being, I'll simply have to accept the fact that I don't how it was put together. The fact that something is a mystery to me right now doesn't necessarily mean that it will forever be a mystery for all people. Maybe I just don't understand it! Maybe I just don't yet have the right mental tools!

As in this parable, one simply has to accept the fact that many of today's big questions and mysteries are still unanswered. Perhaps one day they will be explained by science—or perhaps they won't.

What Is Knowledge?

What is knowledge? What can we know? What is truth? How can we determine whether something is true or false? The area of philosophy that tries to answer these questions is called, as was mentioned earlier, *epistemology* or *the theory of knowledge*. We will now familiarize ourselves with some basic tools so that we can make use of our faculty of reason in the best possible way.

I know the name of Sweden's capital. So I have *knowledge*, in this specific case. But might not knowledge in general be far more complex than this trivial example? What are the exact criteria that must be met so that we can say we *know* something? What does it mean, really, to *know* something?

Three criteria must be met if we're going to talk about knowledge—namely, belief, truth, and strong reasons. For me to know something (I'll call it "X"), the following must all hold:

1. I *believe* X.
2. X is *true*.
3. I have *strong reasons* for believing X.

It's easy to see that the first condition has to be met. I can't know something without believing it. If I don't *believe* that Paris is in France, then I certainly don't *know* that it is in France.

We can imagine situations where X is true and I have good reasons to believe X is true, but where the first criterion is not met: I simply refuse to believe that X is true. Take the case where a friend of mine has said mean things about me behind my back, and I've been told about this by several independent, reliable sources (thus I have good reasons to believe it). But I still don't *want* to believe it, and so I staunchly refuse to accept it. In such a case, I certainly don't *know* that I have been badmouthed; how could I *know* it if I don't even *believe* it? And yet the last two criteria are met. Such behavior on my part wouldn't be rational, but it would be psychologically understandable.

What about the second criterion? It's obvious that something has to be *true* if we are to *know* it. We can have an idea and have good reasons for believing it, but we can still be wrong about it, since the idea could simply be wrong.

Suppose that I have read several books about some topic in history and have talked with numerous knowledgeable people about it. Then I have good reasons for believing I am knowledgeable about the topic. But it could easily be that the books I read contained wrong pieces of information, and that the people I spoke with were misinformed or even were lying to me. Some of the ideas that I strongly believe in are simply not true. Thus I don't *know* them, even if I have good reasons to believe them.

Finally we come to the third condition: the strong grounds. Knowledge of X involves more than just X being true and my believing X. It's also crucial that I have good reasons for believing X. Let me give an example. Suppose that one day, out of the blue sky, a little inner voice whispers to me that there are 213 almonds in the bowl on my kitchen table. And suppose that just by chance, this is exactly right. The statement that I believe is true, but my hunch can't be counted as *knowledge*.

It all comes down to the fact that I don't have any good reason backing up my idea. I haven't counted the nuts, or even made a good estimate of their number using common sense; I simply happen to be right on the money. There actually *are* 213 almonds in the bowl. I'm just guessing, though, and it's merely a piece of luck that I'm right. I certainly don't *know* there are 213 almonds in the bowl. In sum, then, we need good reasons for a true belief to be counted as *knowledge*.

The words "faith" and "knowledge" are often placed in contrast to each other. But in everyday speech, "I know X" merely means that I have very good reasons for having faith in X. When I say that I "know" that the earth is round or that Paris is in France, all I mean is that I believe it, am convinced of it, have faith in it, and have lots of very good grounds for believing it. Of course I *could*, in principle, turn out to be wrong even in basic beliefs like these; maybe I've been systematically lied to on these topics since I was a baby.

From a knowledge-theoretic perspective, faith and knowledge are not opposite notions. The sense of "faith" we've just been speaking of is, however, in contrast to the *religious* sense of "faith," which means "I accept this as true without any proof or evidence." If you believe in something merely because you *want* to believe in it—because it makes you feel good or because it gives you hope for the future—then this is not a sufficient reason to call your belief *knowledge*. Nonetheless, people often believe things precisely because those things make them feel better or give them hope for the future.

Is What Seems Real Really Real?

Let us for a moment take a look at the most radical questions about belief. Anyone who has seen the movie *The Matrix* has at least considered the possibility that our entire world is just an illusion.

As far as we know, our brains evolved, through millions of years of natural selection, for the purpose of processing information that reached us via our senses. Our brains allow us to orient ourselves in the world, since evolution selected those brains that did the best in supplying information that aided survival. But what reason is there for believing that our brains give us *true* information about the world, rather than information that merely helps us survive in our environment? What evolutionary goal would be served by having the world faithfully mirrored in our brains? After all, evolution selected brains purely for their ability to survive, not for their ability to recognize truths.

And yet we humans, almost as a side effect, developed a unique capacity for finding and recognizing truths, even deeply hidden ones. This is clearest in the domain of mathematics. Much of mathematics was developed without even the slightest anchoring in the material world, and yet, much later, piece after piece of "unanchored" math turned out to be central in describing the physical laws governing the material world.[1]

For centuries, philosophers have pondered about whether humans can know anything at all about things outside their own bodies. Why should we rely on information that is fed into our brains by our sensory organs? In fact, we all know that our senses can trick us at times, so could they simply be tricking us *all* the time? Couldn't everything be a dream, a hallucination? Will the table in front of me continue to exist if I close my eyes? Can I even be sure that it exists when my eyes are open?

These dilemmas dreamed up by philosophers are quite legitimate. We can't be 100 percent sure that reality exists out there, independently of our minds. Nor can we be completely sure that the sun will come up tomorrow morning, nor even that we really have existed for longer than three minutes. Maybe the world was made from scratch just three minutes ago, with all our memories of childhood pre-implanted in our brains . . .

Nor can we be 100 percent sure that any consciousness other than our own exists. Could it be that all other people and beings are simply figments of our imagination? The philosophical term for such a notion is *solipsism.* Bertrand Russell, with a twinkle in his eye, gave the following argument against solipsism:

> Reading a book is a very different experience from composing one; yet, if I were a solipsist, I should have to suppose that I had composed the works of Shakespeare and Newton and Einstein, since they have entered into my experience. Seeing how much better they are than my own books, and how much less labor they have cost me, I have been foolish to spend so much time composing with the pen rather than with the eye.[2]

What, *in practice* (that is, in everyday life), is implied by the uncertainty that philosophers have about the nature of reality? Should it affect the daily actions and the daily thinking of thoughtful people? The only reasonable answer to this question has to be "no."

Even if, strictly speaking, we cannot *know* that reality exists, we can consider it extremely probable. Here are three simple arguments for this viewpoint:

1. Reality seems largely stable, largely the same, from moment to moment. The table stays there even after I've closed my eyes. My house looks about the same from one day to the next. All this suggests that we're not just encountering random things from moment to moment, but rather that we are plunged into the midst of a situation where there are reliable regularities—that is, laws of nature.
2. If I and my consciousness exist, then why shouldn't the same hold for others? Isn't that the simplest possible assumption?

3. Evolution shaped us to survive well. And for survival purposes, it's better to interpret one's environment correctly instead of wrongly. If I'm waiting to cross a street and spot a car coming fast in my direction, it's much more helpful to my survival if I believe that the car might run me over than if I don't.

It's always stimulating to entertain philosophical reflections, even if they involve radical skepticism. But when arguments of that sort are brought up in ordinary life ("How can you *know* she took the car? You can't know *anything* for certain!"), one should be on one's guard. Just because nothing is absolutely certain in a strict philosophical sense, that doesn't mean that every imaginable possibility is equally likely.

Armed with reason and a lively curiosity, we can set off on a voyage toward true knowledge about reality. We may never reach our destination, but we can come closer and closer with the help of clear thinking.

Matters of Fact and Matters of Taste

We can distinguish between two sorts of questions about reality, which should be treated a bit differently: matters of *fact* (objective questions) and matters of *taste* (subjective questions). There are clear answers to questions about matters of fact—answers that are either true or false. But for matters of taste, this isn't the case.

Here's a sample claim about a matter of fact: "There are three apples in the basket." The number of apples in the basket doesn't depend on the observer. Either there *are* three apples in the basket, in which case the claim is true, or there are *more or fewer* than three apples in it, in which case the claim is false. So in this case, we are dealing with a matter of fact, not one of taste.

Admittedly, there are borderline cases of apple-ness—for instance, rotten apples, immature apples, partially eaten apples—so that in certain circumstances, two reasonable people might disagree on the apple count. And there are also borderline cases of what "in the basket" might mean. (Suppose, for instance, that the basket is so narrow that the apples are stacked up in a vertical pile, with the highest one's center of gravity being located slightly higher than the basket's rim. Is it "in the basket"?) This second type of ambiguity could blur the question even more. If the reader prefers, then, we can replace this example of a matter of fact by the hopefully sharper example "3 is a prime number."

Another example of a matter of fact is the claim: "The earth is flat." This, as a matter of fact, is false.

And what about the claim "There is life on other planets"? Is this a matter of fact or one of taste? It seems to be a matter of fact, and it should be either true or false; however, we don't know which it is, and we may never find out. Nonetheless, either there *is* life somewhere out there, or there *isn't*. The fact that we don't know the answer, and even may never know it, doesn't make the claim less objective.

Such claims as "This house is haunted," "There is life after death," and "God exists" are also statements about fact, and each of them is either true or false. To be sure, this supposes that we have agreed on some definition of the concepts "haunted," "life," and "God" when we go to examine these claims' truth.

As far as matters of fact are concerned, there are established methods to distinguish what is true from what is false. This includes the scientific method, which we will look at carefully in a later chapter. By following such methods, we can make reasonable judgments about which claims are true and which are false.

With questions of taste, though, it's quite another matter. If I declare "Lady Gaga is better than Beyoncé" and you counter with "Beyoncé is better than Lady Gaga," which of us is telling the truth? Neither, of course! We simply have different tastes and preferences, that's all. So this is a matter of taste. I can like something that you don't like, and vice versa.

Even if there's no objective answer as to which of Lady Gaga and Beyoncé is better, there are many true statements about our exchange that can be made. For example, it's a *fact* that I like Lady Gaga better than I like Beyoncé. So if you were to say to me, "No, you don't like Lady Gaga any more than you like Beyoncé!," I would get annoyed at you, since what you're claiming is false.

Disagreements concerning questions of fact (or of taste) often turn out to be only *apparent* disagreements. Take, for instance, the question of God's existence. If you say "God exists" while I say "God doesn't exist," we'll only be genuinely disagreeing if we mean the same thing by the word "God." If it turns out that when you say "God," you mean "love," then we both agree that God *does* exist. But if what I mean by "God" is some very old scraggly bearded fellow who's sitting up on a cloud, and if you accept this definition, then we're in agreement that God *doesn't* exist.

Much the same holds for matters of taste. If we are going to be in genuine disagreement about which of Lady Gaga and Beyoncé is better, then we have to disagree in more than just a *linguistic* sense. For instance, we both have to mean the same thing by the word "better." But, in music, it's unclear what "better" means. When I say "better" regarding music, perhaps I mean "more musically original," whereas when you say it, perhaps you mean "more admired by the public." In such a case, we aren't really in disagreement, since your view and mine can both be true at the same time.

Many unnecessary quarrels could be avoided if we were taught how to distinguish between matters of fact and matters of taste. Too many discussions wind up on the rocks because the participants just don't mean the same thing by certain fairly common expressions they use.

What Is Truth?

Many theoretical attempts have been made at describing what truth is. The most reasonable such attempt is usually called the *correspondence theory*. The name comes from the idea that truth implies that there exists a correspondence, a link, between what is being said about the world and the world itself. The statement "There's an apple right there" is true if and only if there really *is* an apple right there. Aristotle (384–322 BCE) gave essentially this point of view in his book *Metaphysics*: "To say of what is that it is not, or of what is not that it is, is false, while to say of what is that it is, and of what is not that it is not, is true."[3]

The correspondence theory's notion of truth takes for granted the philosophical view of reality that is usually called *realism*. This means that reality exists and, in a certain sense, does so independently of our preconceptions of it. The opposite notion is called *antirealism*, which claims that there is no reality independent of our human notions.

A realistic viewpoint about the world is needed for us to be able to deal with our everyday environment, not to mention situations in scientific laboratories. Without a realistic viewpoint, we would simply be unable to relate to our environment.

Correspondence theory and philosophical realism turn out not to be so simple. Take the following assertion: There's a light-blue diamond on my dining room table.

Is it true? According to the correspondence theory of truth, this statement would be true if and only if there really *is* a light-blue diamond on my dining room table. But what does "diamond" mean? And what does "light-blue" mean? These are notions that humanity created. "Blue," just like all other color words, is a concept that has to do with the way our eyes and brains work as sensors of light, and the adverbial modifier "light" results in a variation on "blue" that certainly has no sharp cutoff, and that people can argue about forever. The adjective "light-blue" is thus fraught with all sorts of blurriness and ambiguity.

What about "diamond"? This, too, was a human invention. What stones in the world count as diamonds? There are all sorts of "diamonds in the rough" about which experts could squabble as to whether they actually *are* diamonds or are not. And what about the meaning of "diamond" that merely means "lozenge-shaped"? And what about the meaning that applies to cards (one of the four suits)?

And what does "on" mean? What if the table has a cloth mat on it? Are objects sitting on the mat also *on the table*? Does a candle whose ceramic candlestick has been placed on the cloth mat that is lying on the table also count as being *on the table*? And what about the flame atop the candle—is that, too, *on the table*?

We construct our concepts and words so that we can communicate about the world, so that others can understand what we have in mind when we are speaking with them—but that doesn't mean that sentences using familiar concepts are always precise in their meanings. As the foregoing shows, even sentences made up solely of common words can be anything but precise!

Another question has to do with how a given concept corresponds with the world. Consider the term "Santa Claus." Even though we adults know there is no Santa Claus, we see Santa Clauses all the time in shopping centers and in advertisements. So how do we use this word? What does it refer to? What sense does it make to *pluralize* it, when we all know that there is *only one* Santa Claus (and, to complicate matters, when we also know that this "unique" entity is in fact nonexistent)?

When I went to school, we learned that an atom was a tiny solar system with a nucleus in the middle and electrons in orbits around the nucleus, just like the planets orbiting around the sun.

In actual fact, this image doesn't have too much to do with reality. Yes, atoms consist of nuclei with electrons "in orbit" around them, but the concept of "orbit" is wildly different for atoms and for planets because, according to quantum mechanics, an electron in an atom is not localized to a specific point at a specific moment in time; rather, at each moment, each electron is blurrily located at *all possible* points in space, but with different *probabilities*. The solar-system model helps us envision atoms in a simple and helpful manner, especially when the concept of "atom" is first being introduced in schools. But we should never confuse such pedagogical models of reality with the actual reality that is "out there."

Quantum mechanics is a remarkable subject since its equations and calculations allow us to predict with astonishing precision the results of all sorts of physics experiments. In that sense, we can say that our quantum-mechanical models of reality are true, and they function nearly perfectly. But even so, we have no imagery for quantum-mechanical systems (such as atoms) that is compatible with our everyday imagery based on the ordinary objects surrounding us. The ability of quantum-mechanical particles to be in many places at the same time—in fact, in infinitely many places!—clashes violently with the day-to-day intuitions about the world that we have built up over years of life.

The same can be said (only it's even more counterintuitive) about the fact that two quantum-mechanical particles can be *entangled* with each other. What this means is that whenever one of two entangled particles is observed to be in a particular state (say, with its spin pointing "up"), then the state of the other particle is instantly determined to be in the *opposite* state (thus with its spin pointing "down"), no matter how far the two are from each other. It seems that the two particles are cosmically linked, even way across the universe from each other. This is a profoundly mystifying aspect of quantum mechanics. Albert Einstein was deeply suspicious of it, and he famously called it "spooky action-at-a-distance."[4]

Since quantum mechanics is filled with phenomena that seem mystical, it has become all the rage in New Age circles. Unfortunately, disciples of New Age thinking seldom have any inkling of what quantum physics is actually about, and in New Age writings I have never once run into any quantum-inspired idea that I could make head or tail of—they just toss about fancy-sounding words. If only quantum mechanics were that simple!

Does all this mean that we can never describe reality with genuine precision and certainty? Not at all; we can definitely do so. Not all descriptions of reality are equally good; some are better than others. If there are three apples on the table, then it's more accurate to say "there are three apples on the table" than to say "there are *four* apples on the table" or "there are three *oranges* on the table" or "there are three apples *under* the table" or "there are three apples on the *sofa*," and so forth. Even when we are dealing with concepts that we have completely invented ourselves, we can't just make up a bunch of random statements and think they are all equally true.

In science, we can always try to determine which statements about the world are better by carrying out experiments to test them. The models or descriptions that most accurately predict the experimental results are in practice the "truest."

Our models of reality can also be *applied* to construct technology—machines and tools. If those work, then they work, no matter what the nature of the theories behind them is. Take the invention of blue-colored LED lights, which was rewarded with the Nobel Prize in Physics in 2014. Thanks to this invention, along with already-existent red and green LED lights, people were able to produce *white* LED light (red + green + blue = white). This technological success didn't mean that our theory of light's nature constituted a *precise* and *complete* description of light. But the theory was accurate enough to allow us to produce white LED lights. Sometimes science's applications work beautifully, even when there are lingering questions about the scientific models that give rise to them.

Truth as Absolute, Truth as Relative

The attitude that there is a true reality whose nature we can learn more and more about, but probably will never reach in full, is usually called *critical realism*. In the last few decades, however, a completely different viewpoint about truth has arisen and become very popular in some academic circles. This is the idea that *truth is a relative notion*. What does that mean?

Well, take this common utterance: "What's true for me doesn't have to be true for you." It suggests that there aren't any objective descriptions or truths, but that everything depends on the person who's considering the situation. That is, a person's claims about reality don't have anything to do with how the world actually is; they are merely vehicles for phrasing one's personal thoughts, one's social connections, one's ideologies, and the various powers that one wields.

This radical view of truth can be traced back to a philosophical notion that in a way is the diametric opposite to the correspondence theory of truth. It is called the *coherence theory* of truth, and oddly enough, it originated in logic and mathematics. This theory maintains that a statement is *true in a certain framework* as long as it is compatible with a system of other statements, all of which deal with notions in that same framework, and are mutually compatible. In other words, statements belonging to a given framework can be called "true" as long as the framework has *internal* consistency—there's no need for the statements belonging to it to agree with the *external* world.

Let me illustrate this in a very personal way. I will never forget how deeply excited I was when, as a youngster, I learned about binary numbers, built on base *two* rather than on base *ten*. I learned, for instance, that "1111" stands for fifteen, as it is shorthand for $1 \cdot 2^3 + 1 \cdot 2^2 + 1 \cdot 2^1 + 1 \cdot 2^0$, or $8 + 4 + 2 + 1$. Likewise, "1001" stands for nine, as it is shorthand for $1 \cdot 2^3 + 0 \cdot 2^2 + 0 \cdot 2^1 + 1 \cdot 2^0$, or $8 + 1$. I learned how to add numbers in binary, and how to multiply them. Thus I discovered, for instance, that $10 \times 10 = 100$. Now you may find this discovery of mine rather uninteresting, since you already knew it! *Ten times ten equals one hundred!* But that's not what this equation means, in binary. In binary, it means *two times two equals four* (that is, $[2^1 + 0 \cdot 2^0] \times [2^1 + 0 \cdot 2^0] = 2^2 + 0 \cdot 2^1 + 0 \cdot 2^0$).

Both English sentences are true, but in the string of symbols "$10 \times 10 = 100$," which expresses both sentences, once using binary and once using decimal notation, some of the symbols stand for different things. Either choice is fine, so long as you remain self-consistent. Binary is one system; decimal is another system. Each one of them is internally consistent.

And then I also learned that the equation "$10 + 10 = 100$" is *true in binary*, although it is *false in decimal notation*. In binary, of course, it means "two plus two equals four," whereas in decimal notation, it means "ten plus ten equals one hundred." The first of these two claims couldn't be truer, while the second of them couldn't be falser!

For me, as a kid, all this was amazingly eye-opening. It showed me, in a far clearer way than ever before, how symbols correspond to ideas, and how the connection, though arbitrary, can be extended systematically so that there can be totally independent sets of ideas expressed using the same symbols but referring to different things. This was deeply exciting and liberating!

The coherence theory of truth first arose around 1830, when non-Euclidean geometry was discovered. One of this geometry's axioms was the *negation* of an axiom in the two-thousand-year-old Euclidean geometry. This axiom, often called Euclid's "parallel axiom," ran as follows: "Given a line *l* and a point *P* not on *l*, there exists one and only one line through *P* that is parallel to *l*." This statement was *true* in the framework known as Euclidean geometry, but it was *false* in the framework known as non-Euclidean geometry. Thus, to put it bluntly, a statement that was true in one system was false in the other, and vice versa. This sounds very much like truth as a relative notion!

Well, yes, but not so fast. . . . It turned out that the term "line," which was used in the parallel axiom (and its negation), didn't refer to the same notion in the two different worlds, so it wasn't really the case that *the same statement* was true for some folks and false for others. For instance, two *Euclidean* parallel lines are always at the same distance from each other, whereas the distance between two *non-Euclidean* parallel lines varies. In short, Euclidean lines and non-Euclidean lines are horses of very different colors, which is why the two geometries are so fascinating to compare. And this means that the term "line," which exists in both geometries, isn't referring to the same concept. In a non-Euclidean world, there don't exist any Euclidean lines, and in a Euclidean world, there don't exist any non-Euclidean lines!

All this is no more confusing than the fact that the equation "10 + 10 = 100" can be true in binary while also being false in decimal. Likewise, the sentence "Chicago is east of here" can be true for some people and false for others. The reason is quite simple: the word "here," when uttered by a San Franciscan, doesn't refer to the same place as when it is uttered by a Bostonian, just as "10" in binary doesn't refer to the same number as "10" in decimal.

Thanks to the discovery of non-Euclidean geometry (and later on, of other exotic geometries), mathematicians and logicians grew accustomed to the idea that expressions that looked identical could mean different things in different conceptual universes. This didn't mean that these thinkers thought that truth was relative; it meant that they understood that the *meaning* of some terms could shift according to context. A statement could thus be true in one axiomatic system and false in another—but the two statements weren't talking about the same *things*. This was a situation that superficially *resembled* truth-relativism, but it was not a situation where truth became a slippery notion that depended on the observer. Indeed, truth remained a central and fixed notion in mathematics.

By the way, the discovery of diverse "rival" geometries showed that all sorts of internally self-consistent mathematical systems can exist that don't have to correspond to the physical universe. The question then arose as to which of the "rival" geometries applied to the actual world. Eventually we learned that the vast cosmos we inhabit is described by a non-Euclidean geometry, although on a down-to-earth practical scale, we can treat our world as Euclidean.

Mathematicians are delighted when they can construct rival systems that appear to have "different truths"; however, they understand that this doesn't make truth relative. They understand that all this means is that corresponding entities in the rival theories, though called by the same name, are not the same things.

The coherence theory of truth, though it helps mathematicians understand "rival" axiomatic systems that are sealed off from one another, is not applicable to areas outside of mathematics—and the subtle lessons about truth and consistency that mathematicians struggled with and finally absorbed a couple of centuries ago should not be nonchalantly misapplied to the world at large, for the world is not an axiomatic system, let alone a collection of rival axiomatic systems.

To deny that truth is objective is not only naïve and wrong, but worse yet, it is morally problematic. In politics, it becomes very serious: one believes whatever one wishes when it agrees with one's personal goals, instead of believing what sense and reason would lead one to believe.

If you think there are trolls out there in the woods, and I don't believe there are any, then one of us is simply *right*, and the other is simply *wrong*. Even if we can't figure out which of us is right (without doing a lot of investigation), hopefully we agree that only one of us can be right. You might argue that your idea of trolls lurking in the woods is central to a coherent worldview (or "cosmology") that has a long tradition in your culture and that it therefore *must* be true. But that would be an absurd and hollow argument. Suppose someone slapped you in the face and then said to you, "No slap of your face occurred in *my* reality, so it doesn't matter." Would you accept such a weird relativism of truth?

If people were to accept the relativism of truth, then Hermann Göring could have been found innocent in the Nuremberg Trials by arguing that *his* reality was different from the *judge's* reality. If this kind of viewpoint were allowed in courts, then no one could ever be found guilty of any crime at all. Any old statement would have to be taken at face value, since no statement would be any righter or truer than any other one.

The viewpoint of truth-relativists—that "everything is equally correct"—could easily lead to a society just as immoral as the societies of religious fundamentalists throughout the world, who have cocksure and totally closed viewpoints, airtight and waterproof, about what truth is, and too often, their members don't mind killing people who don't agree with them.

Curiously, truth-relativism is also a self-defeating notion. If its ideas really were valid, then the claim "All truths are relative" would *itself* be true only for those people who liked it, and false for people who didn't. That is surely absurd. In short, truth-relativism is a self-contradictory philosophy, which, upon investigation, falls to pieces.

There are linguistic phenomena that on the surface resemble truth-relativism, such as the case mentioned earlier, "Chicago is east of here." A sentence can be true or false, depending on who says it. If the king of Sweden says, "I live in Stockholm," he is telling the truth, but if Alice Appletree, who lives in Ann Arbor, Michigan, says "I live in Stockholm," she is uttering a falsity. The very same statement is both true and false! Well, obviously this doesn't show that truth-relativism is valid. The two people are simply making different claims, despite using the same sets of words. When the Swedish king says "I," he is talking about one person, while Alice Appletree, when she says "I," is talking about someone else.

Or if I say, "The flowers are to the right of the piano," this can be true if I'm sitting facing the keyboard, ready to play, but if I spin around on the piano bench so that I'm facing the other way, then it becomes false. The phrase "to the right of" depends on a frame of reference that hasn't been specified, so the statement is imbued with ambiguity. But once the frame of reference has been specified, then the statement loses its ambiguity and becomes either true or false.

Movement and speed are also relative. We all are familiar with this from sitting in a train sitting still in a station and looking out the window at the train that is right next to ours. All at once the other train starts to move. Or is it my train that just started to move? It can be hard to tell! The speed of an object always has to be given relative to some fixed framework (i.e., frame of reference)—except for the case of light, which always moves at exactly the same speed relative to any observer, no matter what frame of reference the observer is in. This highly counterintuitive idea is the crux of (special) relativity, developed by Albert Einstein (1879–1955) in 1905.

Einstein was born in Ulm, Germany, and later he became Swiss and then American. His theory of relativity was a generalization of a far earlier "principle of relativity" originally posited by Galileo Galilei (1564–1642), which claimed that all frames of reference in physics are equally good (meaning that the laws of physics hold equally in all of them). Actually, Galileo's principle was limited to so-called "inertial" frames of reference, meaning frames moving at constant speeds with respect to each other, and to mechanical experiments (mechanics being the only branch of physics that was known in Galileo's day). Einstein, however, conjectured that Galileo's principle could be extended from *mechanical* phenomena to *electromagnetic* phenomena (meaning light in particular). This seemingly tiny mental step led to a gigantic revolution in physics early in the twentieth century.

A few years later, Einstein further generalized his generalization of Galileo's classic principle so that it would hold in *all* frames of reference, no matter how they were moving, and this super-generalization was called, appropriately enough, "general relativity," and the earlier "theory of relativity" was rebaptized as the "theory of special relativity." It is said that Einstein enjoyed joking about relativity theory when he was riding in a train, by asking the conductor, "Excuse me, but does Chicago stop at this train?"

In recent years, one of truth-relativism's foremost advocates in the academic world was the American philosopher Richard Rorty (1931–2007).[5] Rorty thought that all notions of truth are grounded in the basic axioms that we make about reality, and in the methods that we choose when we consider the world and experience it.

A classic example is when Galileo was accused of heresy in 1615. He had published astronomical results that were incompatible with the teachings of the Catholic Church, which asserted that the sun rotated around the earth. Cardinal Robert Bellarmine was the Vatican's representative in the trial. Galileo invited the cardinal to check out his claims, by looking out at the heavens through his telescope (which he himself had made). Bellarmine, however, declined to peer through the tube, saying that he could find better arguments in the Bible than anything that could be provided by any scientific instrument.[6] Galileo chose the methods of objective observation, while Bellarmine chose the gospels offered by holy scripture.

And so: Does the earth rotate around the sun, or does the sun rotate around the earth? According to truth-relativist Richard Rorty, it all depends on how you choose to relate yourself to reality. He believed that there is no fact of the matter, no objective way to say which of the two celestial bodies rotates around the other.

Another radical truth-relativist is the French philosopher Bruno Latour, born in 1947, who believes that objects do not exist until they are noticed.[7] He offers the example of bacteria, claiming that they didn't exist before their discoverers found them.

The mistake Rorty and Latour both make is to confuse the description of a phenomenon with the phenomenon itself. Of course people couldn't talk about bacteria and their properties before they were discovered. There were other theories of why people died of sicknesses. But this doesn't mean that bacteria only began to exist at the moment that we humans became aware of them.[8]

An objective reality exists, which includes atoms, even though science's models are always provisional and subject to revision, thus not "true" in the absolute sense of "perfectly flawless and everlasting." Scientists are not dogmatists but are open to new evidence. (Incidentally, this intellectually honest refusal, on the part of scientists, to declare that they have reached the absolute and final truth, even when there is enormous evidence for a set of ideas, demonstrates the profound humility of the scientific attitude.) But the most extreme of truth-relativists go even further, flat-out denying the very concept of "truth," as well as the existence of reality. Unfortunately this stance is often accompanied by deep scientific ignorance and even contempt for science.

The notion of truth cannot be relativized. On the other hand, it's quite reasonable to believe that various factors, such as political and economic power, affect our priorities and our ideas in such disciplines as sociology and other social sciences, and perhaps also some ideas in the natural sciences. In particular, such matters can have huge effects on what types of research will be financed by governments and other funding agencies. But this doesn't allow one to conclude that the very concept of truth is unreliable. If a theory works, then it does so independently of the factors that might lead people to believe in it.

Social Constructions

The theory of *social constructivism*, though highly dubious for many reasons, is nonetheless rooted in a very reasonable premise—namely, that the conditions in which people live have effects on the perspectives they adopt about the world around them and, in particular, their perspectives about research and scientific analysis. Factors such as race, gender, class, and political ideology exert influences on what type of research is pursued by a society and what results are taken as important.

A classic example of this effect is the Soviet geneticist Trofim Lysenko (1898–1976), who, during the 1930s, criticized Darwin's theory of evolution and launched an alternative, and ideologically colored, theory of the origin of species. He claimed, among other things, that heredity is not determined by genes or chromosomes, but can be modified by environmental factors. His theories were mostly erroneous, but they were novel and were ideologically aligned with Marxism-Leninism, and thus they appealed to Stalin. By contrast, Mendelian genetics, the scientific theory accepted in the West, was labeled "bourgeois pseudoscience."[9]

Social constructivism, inspired by such thinkers as French philosopher Michel Foucault (1926–1984), claims that the scientific quest for knowledge is not what it seems to be; indeed, Foucault saw science as an evil tool designed expressly to help white, heterosexual, bourgeois, middle-aged, and economically well-off males to marginalize poor people, women, homosexuals, ethnic minorities, and non-European cultures in general.

History shows, unfortunately, that science has at times been exploited in just this way. During the nineteenth century and for quite a while during the twentieth, biology and anthropology were used in order to justify racist oppression. Also, medical science was used to justify the locking-up of, and the performing of lobotomies on, individuals exhibiting deviant forms of behavior. (Lobotomy is a surgical operation on the brain, introduced in the 1930s by Portuguese neurologist António Egas Moniz (1874–1955). It involves cutting the nerve pathways that link the frontal lobes to the rest of the brain. The idea behind it was to treat patients who had grave anxiety problems or psychotic behavior, to make them calmer; however, the procedure also destroyed important parts of the patient's personality and emotional life, and it is no longer used anywhere.) Women who were perceived as deviating from society's moral norms were forced to undergo sterilization. Such examples show how science and "social engineering" have been misused in our history, and how they are often cited as support for the oppression and the separation of people into different classes, always following the whims of the currently reigning authorities.

Social-constructivist analysis can be helpful and justifiable in many cases. Insight into social constructions is important when we are trying to understand the context in which we live. Here such theories have much to contribute. We are in fact surrounded by social constructions all the time: for instance, when we use coins in a grocery store, we are using a social construction. After all, a coin is just a little stamped piece of metal, whose worth and function as a tool in everyday transactions is not intrinsic to the object but is merely *constructed* by our society.

To take all of reality as a mere social construct is problematic, however. The traditional social constructs of "masculinity" and "femininity" provide a clear example. Since time immemorial, women have been associated with feelings, and men with reason and intellect. These ancient social constructs are still largely accepted today, but they not only are false but also extremely limiting for the potential and the chances of an individual person.

Other harmful social constructs are norms concerning sexual orientations, sexual relationships, and the nature of the family, as well as myriad assumptions about consumerism, success, and status.[10]

Postmodernism and Education

Social constructivism and truth-relativism are usually considered to belong to the overarching trend of ideas called *postmodernism*.[11] The phrase "socially constructed" is but one part of a system of contemporary sociological jargon that belongs to a postmodernist syndrome that also includes *cultural relativism* and the closely associated notion of *value-relativism*.

Postmodernism gives rise to deep epistemological, moral, and political problems. The most extreme versions of postmodernism claim that the natural sciences are just a social construction—a random "tale" no more valid than any other random tale that anyone might choose to make up out of whole cloth and to propagate.

In 1996, American physicist Alan Sokal (1955–) jumped into the fray as a powerful critic of postmodernism when he tricked the editors of the trendy academic journal *Social Text* into publishing a completely nonsensical article he had written, which he submitted to the journal ostensibly as a social-constructivist criticism of attempts to combine Einstein's theory of general relativity with quantum mechanics. His sarcastic article, with the wonderfully obscure title "Transgressing the Boundaries: Towards a Transformative Hermeneutics of Quantum Gravity," was an absurd spoof of the "deconstructionist" style of writing. In it, Sokal donned a haughty intellectual persona who virtuosically attacked "the so-called scientific method," using rafts of standard postmodernist jargon terms.

Thus, for instance, Sokal's postmodernist persona wrote that the concept of "an external world whose properties are independent of any individual human being" was merely "dogma imposed by the long post-Enlightenment hegemony over the Western intellectual outlook." This truth-relativist persona went on to insist that scientific research "cannot assert a privileged epistemological status with respect to counterhegemonic narratives emanating from dissident or marginalized communities." (In simpler words, science has no valid claim to truth; any person has their own truths.) On and on he went, did Sokal, spewing jargon and using pseudoscientific arguments left and right, even borrowing New Age ideas from believers in ESP and other paranormal phenomena, all in order to give the appearance of making an ardently male-bashing attack on what the journal's editors would have called "phallocentric science." And just as Sokal had hoped, the editors fell for his hoax—lock, stock, and barrel.[12]

Another fine example of postmodernist nonsense is furnished by my own land of Sweden. A few years ago, I participated in a debate about schools, in which an education professor was part of a panel discussion. The professor claimed that the scientific theory of evolution and the biblical account of creation are simply two different "tales," or paradigms, about the story of humanity. There was no discussion about whether one of them was "truer" than the other. To my mind, such a stance is intellectually dishonest. If everything is just a matter of opinion, then what sense does it make to debate about anything?

Another prime example of such an attitude came from Swedish education professor Moira von Wright. In 1998, she wrote a report for the Swedish Ministry of Education about physics teaching in schools. In it, she claimed that the "scientific content" in physics courses should be modified "for the sake of equality," and in the following terms she rejects the process of scientific thinking:

> The notion of the superiority of scientific thinking is incompatible with the ideals of equality and democracy. . . . In the scientific community, some ways of thinking and reasoning are rewarded more than others. . . . If one fails to notice this, one runs the risk of drawing misleading conclusions—for example, mindlessly jumping from the idea that scientific thought is more rational to the idea that it should replace everyday thinking.[13]

Later in the same report, von Wright writes:

> What does it imply for equality that physics texts consider it crucial to convey to students a mechanical and deterministic picture of the world, and that they stress its superiority? To impose such narrow knowledge having a fixed interpretation is incompatible with our schools' goal of equality—and yet, this is exactly what most physics texts do, thereby contributing to the maintenance of the asymmetrical and hierarchical relationship between the masculine and the feminine in the scientific world. When physics is uncritically put forth as the only truth, it takes on a scientistic stance and thus exercises a (negatively) symbolic power over students' acquisition of knowledge. . . .[14]

A gender-conscious and gender-sensitive physics would involve a relational point of view about physics; moreover, a great deal of the standard scientific content of physics would have to be removed.[15]

The conclusion of von Wright's report is as follows:

Perhaps, to start off, we should ask ourselves in a new way how to get girls to be more interested in careers in the natural sciences. For instance, in place of the old stock question about how we might try to interest girls more in physics, we could turn things around and pose the novel question: How can we get physics to be more interested in gender, and in the feminine perspective?[16]

The attitude of von Wright's report is not just anti-intellectual but also anti-scientific. And in addition, it is condescending toward women to assume that they are less able than men to think in a rational manner. And yet, from 2010 through 2016, Moira von Wright served as president of Södertörn University in Stockholm.

Relativism and Politics

Truth-relative thinking grows even more worrisome when it worms its way into politics. Suppose, for example, that statistical studies show that Swedish women have a lower average salary than Swedish men do (which is in fact the case). Then imagine a debate in the Swedish parliament in which a politician takes the podium and states: "Well, that study may be true for *you*, but it isn't true for me! *My* truth is that women earn more than men do!"

Any politician who doesn't accept the idea that there are objective facts out there is in big trouble. Obviously it's often hard to arrive at the truth in debates about society and politics, and politicians often choose whichever aspect of reality they want to bring out. They carefully choose what to focus on and what to ignore, especially when making presentations involving numbers and statistics. If the subject matter is sufficiently complex, it becomes almost impossible to decide if the statistic in question was or was not "cherry-picked" so as to lend support to a particular interpretation.

Statistics do not lie, as many people believe (at least as long as they haven't been intentionally falsified), but one can carefully select pieces of information in order to place stress on just one aspect of reality. In this way, statistics can be used to manipulate people. Politicians and other people with important roles in society are also skilled at presenting facts mixed in with values. Something that is presented as purely factual can subliminally convey a great deal of hidden ideology.

Statistics is a way of describing reality, but it doesn't tell anything about the way that complex links work—and that's what politicians often play with and distort. One party can claim that employment has gone up, and the opposing party can claim that joblessness has risen. Both claims are supported by statistics, and in fact both can be right in the same time period. To understand how this can be the case requires careful analysis—and even careful analysis may leave out other important factors. In other words, combating relativism in politics is a very difficult battle, but it is vital that it be fought, because to claim that truth does not exist is both dishonest and immoral. To make such a claim is to plunge into profoundly treacherous waters.

～

INTERLUDE: ON THREE GREAT MYSTERIES

Will we ever come to know how everything began?

On my nightstand I always place three books about what I consider to be the three greatest mysteries of existence. The first one deals with the origin of the universe; the second one deals with how life began, about the transition from inanimate to animate matter; and the third one deals with the mystery of consciousness. The books themselves, and their authors, change from time to time, but at all times you will find books on my nightstand about these three mysteries.

These three riddles have fascinated me ever since I was small, and in this respect I am hardly unique, for these questions are often considered the greatest mysteries we know, and we humans have wondered about them as long as we have been able to wonder.

All through history, insightful researchers, philosophers, and thinkers have dreamed up various theories to answer questions about the origin of the universe, the origin of life, and the nature of consciousness, but at the present, there is no uniform consensus on the answers. Our innate curiosity is always pushing us further along the pathway to try to answer these questions.

Some people claim that scientific answers will never be found for these questions. They lie beyond the realm of science, so they say; all these riddles are unsolvable. It is religion's sacred duty to answer them, not science's.

Unfortunately, I cannot for the life of me comprehend this line of argument. It strikes me as an utterly alien style of thinking to dismiss in advance the notion that science will one day be able to answer questions that we today consider to be mysteries. Why should anyone deny the idea of new pathways to knowledge? Why should one shut one's eyes, when one can face the world with one's eyes wide open?

Such negative attitudes shut out possibilities. It feels so limiting to say, "This will never be done; we will never come to know this; it's impossible!" Down through the centuries, science has solved innumerable mysteries that for people at an earlier time had seemed hopelessly unsolvable. We learned how to fly; we learned how to cure diseases; we learned what the inconceivably distant stars are made of and what their source of energy is; we learned how to see new phenomena on incredibly tiny scales (literally microscopic!), and on incredibly vast scales (literally astronomical!). Thanks to the theory of natural selection, we have explained nature's enormous variety of complex and beautiful creatures. Phenomena that once were totally inexplicable are today no longer mysteries. Why shouldn't this sort of thing happen over and over again?

I wish we all had a more open and humbler attitude about what we don't yet understand. Maybe someday we will all easily understand how consciousness arises as an emergent result of the activities of the billions upon billions of neurons (and trillions upon trillions of synapses) that constitute our brains.

And perhaps one day we will explain how life originated from inanimate matter, through chemical processes and a great warmth coming from far away from our planet, a warmth that we call the sun. Perhaps in a remote time we will be able to understand how the universe's creation (something coming miraculously out of nothing) can be accounted for through the remarkable leaps of quantum physics—or perhaps it will take some other kind of cosmological theory of which we don't yet have even a glimmer.

Or, perhaps, we just won't ever find any explanation for these mysteries. But that is unknown right now. I surely wouldn't want to discount anything in advance. After all, our limited knowledge is in constant development. Humility and openness require me to say, "Maybe one day we will finally fathom all these mysteries."

Until then, I will continue to read those three books on my nightstand—always ones by new writers, always ones with new ideas. Reading such books gives me the feeling that my thoughts are soaring, very fast and very high. It is a dizzying feeling. And perhaps, one lovely day, there will be just one single book on my nightstand—a book that will give plausible and well-supported answers to all these questions.

CHAPTER THREE

BELIEFS BASED ON GOOD REASONS

Concerning the Grounds that Underlie One's Convictions

The fundamental cause of the trouble is that in the modern world the stupid are cocksure while the intelligent are full of doubt.—Bertrand Russell

Now that we've begun exploring the nature of reality and truth, numerous questions arise: What should I believe? Why should I believe it? What does it really mean to believe in something?

The word "belief" has numerous meanings. When one is talking about the time of day or what the capital of Kenya is, one often says things like, "I think it's about 5 o'clock" or "I believe it is Nairobi." This is one type of belief. By contrast, the phrase "I believe in Henry" means not only that one is convinced that Henry exists but also that one has confidence and faith in Henry. One can also say, "I believe in justice," meaning that one wishes to see justice everywhere.

The phrase "I have faith" often suggests a religious belief concerning one or more gods, but it can also be used without any religious implications, as in "I have faith in you."

What, then, does believing in something involve? It makes sense to distinguish between two sorts of belief.

The first could be called "cognitive belief" (from the Latin *cognitio*, meaning "knowledge"). This can be described as the belief that something behaves in a particular manner or has a certain property. A typical example is "I believe that it's five o'clock." Similarly for "I believe it'll rain tomorrow morning" or "I believe there is life on other planets." A cognitive belief is a claim of truth.

The second sort of belief is faith-like belief. This kind of belief is all about confidence, faith, trust, and reliance. It's more about *believing in* something or someone than about *believing that* something is the case. "I believe in you" doesn't mean "I believe you exist" (although of course this is implied) but, rather, "I know that I can always rely on and trust you."

We all have many things that we believe in, both in the cognitive sense and in the faith-like sense.

Most of our ideas about the world are cognitive beliefs. We believe that the table in front of us exists. We believe that we ourselves exist. We believe that the world existed last week and will exist next week. We believe that the sun will come up tomorrow morning (well, maybe not, if we're way up north in Sweden around December 21!). We believe that the earth rotates around the sun. None of these things can be definitively and absolutely *proven* in the epistemological sense of total and eternal certainty, but nonetheless, we clearly have excellent reasons for making these claims of truth.

In order to think clearly about the nature of beliefs, one has to distinguish between *evidence* and *proof.* We often hear it said that one cannot prove that scientific theories—for example, the theory of evolution—are true. This is certainly the case. The term "proof," in its purest meaning, applies only to mathematical arguments.[1]

Outside of math, strict proofs of this sort don't exist; therefore, from a purely epistemological point of view, we can never logically *prove* that Einstein's theory of gravity is correct, nor can we even prove that the earth is round. The fact that statements outside of mathematics cannot be proven in this strict and narrow sense, however, doesn't in the least imply that all cognitive beliefs outside of math are equally well (or equally poorly) grounded. Although science can't and doesn't give us *absolute* and *infallible* proofs, its power is that it gives us *pieces of evidence* (some stronger and some weaker) for various theories.

Humans believe in all sorts of things. We all build up, whether consciously or not, our own image of reality by accepting certain ideas and rejecting others.

When we try to describe how the world around us is constituted (in other words, when we engage in natural science), we formulate *theories* (or ideas, or models) about how things behave. For the reasons given earlier, we can't make a *watertight proof* of a theory, but we can nonetheless find pieces of evidence for it by carefully checking out the truth of various ideas that would support the theory, and of various conclusions that follow from the theory. Each time we verify a support or a conclusion, we add credibility to the theory. We can also *falsify* theories, (i.e., show that they do not work); this happens when experiments to check our plausible guesses disagree with what we had predicted.

To believe in a theory means to believe that certain things behave in a certain manner, or have certain properties. Of course we can have rather half-hearted beliefs, such as "I'm not too sure, but I would suspect that . . .," and we can have extremely strong beliefs, such as "I am convinced beyond any doubt that . . .," but wherever we may be located along this continuum, we are still dealing only with likelihoods, never with absolute certainties. When there are several rival theories about the same phenomenon, we generally believe the one that we consider to have the strongest total evidence.

When we say that we *know* something, we mean not only that we believe in that thing but also that we have strong reasons for doing so.

Aside from having cognitive beliefs, most of us also have faith-like beliefs. Thus, we believe in our friends and in our loved ones. We have confidence in their abilities. We may well also declare that we believe in love, or in a political ideal such as democracy—meaning that we have confidence in the power and strength of those abstractions.

We constantly rely, unconsciously, on our senses. This kind of reliance is a faith-like belief. If our eyes tell us there is a tree in the garden, then we believe that it is there. Of course, if there remains some doubt about the tree's existence, we can also walk out into the garden and touch it, but to draw any conclusions from that act, we would still have to rely, in a faith-like manner, on our sense of touch. Without having faith in our senses, it would be very difficult to survive and move about in this world. If we had no such faith, it would be very hard to live.

"But What Do You Believe in, then?"

I have often been confronted with the question, "But what *do* you believe in, then? Surely you must believe in *something*!," usually expressed in a frustrated tone of voice.

I've never fully understood what lies behind this question since, of course, I—like every sane adult human being—have myriads of beliefs: I just don't happen to believe in any god. But that one disbelief doesn't mean that I am beliefless! I am, after all, human. So all right—what sorts of beliefs do I have?

Well, on the *cognitive* side of the ledger, here are a few typical examples. I believe that 10 + 10 = 100 (in binary!); I believe that rooks move along a chessboard's rows and columns, but not diagonally; I believe that my car has a steering wheel but not a keel; I believe that matter is made of atoms; I believe that living creatures are made of cells; I believe that heredity is transmitted by genes made of DNA; I believe that unicorns are found only in fairy tale books; I believe that there is life on other planets; I believe that Darwin's theory of evolution correctly describes the origin of species. I also believe that "God" is a concept that was created by humans and that the word doesn't refer to anything, or correspond to any kind of reality in the external world.

On the *faith-like* side of the ledger, I also have a myriad beliefs. Of course I have no faith-like belief in any god, since that would first require me to have a *cognitive* belief in a god (one can't rely on something that one totally disbelieves in!). But, to give some positive examples of my faith-like beliefs, I believe in humanity's great potential, in people's ability to reason, and in our willingness to take responsibility; I also believe in equality, nonracism, nonsexism, and human rights.

Those are a few of my answers to the question forming this section's title.

Is Belief in Science a Kind of Faith?

A widespread misconception is the idea that when someone *believes in science*, that is comparable to other people *believing in God*. The problem is that the phrase "believe in science" is misleading. It is *reasonable* to believe in science—that is, to have confidence in the methods that give rise to scientific theories. A scientific belief is very different from a religious claim that one merely accepts on faith, for such claims can be erroneous or even fraudulent. When it comes to questions about the world, science is built on ways of *reasoning* and *testing* that, throughout the centuries, have proven themselves to be of far greater reliability than claims coming from prophecies, revelations, dreams, or holy scriptures.

A consequence of this is that most well-supported scientific ideas give us practical knowledge about the world around us. If one evening I look out the window and see that the sky is clear, and if I then hear a weather forecast predicting sunny skies the next morning, I can feel pretty confident that it'll be sunny when I wake up. Of course this is not a rigid, fixed, dogmatic belief. The sky might grow overcast before I go to bed, or I might hear an updated weather forecast that says something different, and then my expectations will change. My belief about the next day's weather is thus grounded in the familiarity with the world that I have built up over many years, and also in my sensory information sources. And my belief can flexibly change as a function of new evidence that I receive. All this is similar to how a science builds up its results, whether it's physics, chemistry, biology, or any other science.

Most people believe in science, in the sense that we normally all rely on the findings of established and serious researchers. We thus have a faith-like belief in science, although that doesn't mean that we dogmatically believe every scientific claim that we read. Plenty of scientific ideas turn out, in the long run, to be wrong; the beauty of science, however, is that through the collective act of thousands of people engaging actively in the search for truth, the false ideas eventually get weeded out in a steady and reliable fashion, and true ideas are left as their legacy.

Throughout the past several centuries, the methods of science have proven themselves to be highly successful in describing the world, explaining the world, and predicting events in the world. The same cannot be said about religious beliefs. In fact, religious scriptures and revelations have, throughout history, often provided conceptions of the world that are in direct conflict with what we know today.

In contrast with religion, science progresses by constantly testing its hypotheses; it is therefore a self-correcting way of knowing. And yet today, there are many people who look upon science with great skepticism. In theological circles, there are even some people who claim that *science needs theology* in order to be self-sufficient.[2] A frustrated researcher with a sense of humor once said, "Science is not a religion. If only it were one! For in that case, it would be far easier to get money to support research."

Intellectual Honesty

In today's information-rich society, there is a furious competition among ideas that we are exposed to. In newspapers we can read astrological horoscopes, on late-night television we can see ads for magical cures to dreaded diseases, on the web we are fed claims of miracle diets, and on and on. Fortunetellers and psychics claim to be able to reveal our future or to read our minds. Tabloid newspapers overflow with articles about reincarnation or Indian gurus who can make jewels appear out of thin air. In street fairs, hawkers offer to sell us magnetic bracelets that will give us energy or crystals that will restore harmony to our lives. And various kinds of evangelists suggest (or demand) that we ought to believe in this or that god. Given this relentless onslaught of competing claims, how on earth can a sane person figure out what they ought to believe?

Claims, ideas, and products are often paraded in front of us without any serious knowledge or research backing them up. Instead, what lies behind the scenes is simply the desire to make money or to propagate an ideology. We desperately need a sane and verifiable way of checking out the veracity of so many claims.

It is reasonable to believe in a scientific finding if it has been reasonably well confirmed in experiments. At least we can rely on a given scientific finding until a new and more reliably confirmed finding comes along to replace it. This, of course, means that we need to be open at all times to the idea of testing and revising our beliefs.

We won't, alas, ever find any ironclad mathematical *proof* that the heavens aren't filled with randomly flying teapots, or that we aren't constantly walking through invisible, odor-free, perfectly silent, intangible elephants. But the fact that we can't disprove these very silly ideas in an absolutely watertight fashion doesn't mean that we should *believe* in them! The same goes for trying to disprove the idea of life after death; we certainly can't disprove it in a logically indefeasible fashion, but on the other hand, what reason do we have for *believing* it?

How do you know that the next time you walk in front of a mirror you'll be able to see yourself? Maybe you won't be there at all! And maybe the next time you open your hand, the pebble you're holding in it won't fall to the ground but will just sit there, suspended in mid-air! Or maybe that little hummingbird hovering outside the kitchen window will suddenly fly right through the glass pane without breaking it, then turn into a flying dinosaur, and bite your head off! But we all doubt these things, and for very good reasons. Nonetheless, we can't *prove* them. Proof is limited to mathematics.

But if we can't be sure of anything, how can we manage in life? When we ask ourselves such questions, it is good to keep in mind Ingemar Hedenius's principle of intellectual honesty: "Believe in an idea if and only if you have good reasons to think it is true." Though trivial sounding, this is an important principle that can help us navigate around in the everyday world. Ideas should be tested, and even strange claims should be given a chance. Where could you ever find a more open-minded attitude than that? This is why Hedenius's principle is important.

Whenever you run into a new claim about the world's nature, ask yourself such questions as these: "Are there good reasons for taking this idea seriously? Are there good reasons for dismissing it? Could there be other ways of explaining what it claims? What would be the best way to explain what it claims? How strong is the evidence behind it? Does this claim mesh well with other things that we know are true? Is it improbable or plausible? Could there be hidden reasons that someone is proposing this idea?"

In many situations in life, we must rely on the ideas and the knowledge of other people. We have to believe what we find in physics textbooks, knowing that they were written by professional experts in physics. Contrariwise, we place little or no faith in the juicy and lurid headlines that we encounter in tabloid newspapers, such as the *Weekly World News* (now defunct, alas) in the United States, the *Daily Mirror* in Britain, or *Aftonbladet* in Sweden. Consider the following delightful sampler of *Weekly World News* headlines:

"World's Ugliest Woman Dies after Looking in Mirror"
"Dolphin Grows Human Arms"
"Adolf Hitler Was a Woman!"
"Duck Hunter Shoots Angel"
"Plane Missing since 1939 Lands!"
"Satan's Skull Found in New Mexico!"
"Heaven Photographed by Hubble Telescope"
"Horse Born with Human Face!"
"Barack Obama Is a Robot!"
"Teenage Girl Eats 4,000 Sponges!"
"Hillary Clinton Adopts Alien Baby!"
"*Titanic* Survivors Found on Board Life Raft!"
"NASA Takes Photo of Ghosts in Space"
"Mini-Mermaid Found in Tuna Sandwich!"
"5,000-Year-Old UFO Found off the Coast of Chad"
"Farmer Shoots 23-Pound Grasshopper"
"Bald Man Fries Eggs on His Head"

What makes us read these headlines skeptically? What clues put us on our guard?

Well, sometimes it is their sheer implausibility. Sometimes it is their outrageousness. Sometimes it is their grotesqueness. Sometimes it is their childishness. Sometimes it is their similarity to magic. Sometimes it is their humorousness. Sometimes it is because we have seen "similar" things before. In other words, after being exposed to enough tabloid headlines, we start to develop an unconscious sixth sense for "this kind of nonsense."

Another way of stating this is that in reading tabloid newspaper headlines over a long period of time, we gradually learn to recognize certain recurrent types of patterns. When we were little, we were intrigued and amused by such headlines, but throughout time, we learned to be suspicious.

You obviously can't check out every single possible claim on your own, and so you have to rely on others to help you find your way. We all therefore need to develop a sense for which types of authorities we can trust, and which types of "authorities" are not actual authorities at all. A physics professor presumably knows more physics than I do. The *New York Times* is presumably far more reliable than the *Weekly World News*. But that doesn't mean that there are certain sources that we should blindly rely on at all times and in all circumstances. We can't just go around believing everything that every single physics professor utters, nor should we accept uncritically every single word that we read in the *New York Times*. In the final analysis, after enough training and experience, you yourself have to be able to judge the plausibility of the claims you encounter.

The elegant lady sitting next to you at a fancy banquet tells you that she was kidnapped by space aliens and was held in their spaceship for two weeks. She adds that in her house she has some objects that she took from the spaceship and that she'd be more than happy to show them to you. You, presumably, are skeptical. But *why* are you skeptical? Essentially, it's because her claim runs against all your prior experiences and all the lessons you have learned about the world throughout your whole life. Additionally, though a bit less important, if such alien abductions really took place, then surely they would be widely reported in all the newspapers.

You conclude that your tablemate is talking nonsense—but can you *prove* that it's nonsense? No, all you can do is take her tale as being extremely unlikely.

It would make sense for you to believe her tale if the *simplest* and *most reasonable* explanation for her tale is that she really was abducted by space aliens. But in this case, that is *not* the simplest and most reasonable explanation. Much simpler and more likely are such possibilities as that she is just pulling your leg, or that she wants to trick you, or that she is desperate for attention, or that she is drunk, or that she is seriously mentally ill—or even that she just wants to lure you back to her house. Or maybe she's just confused or tired. All of these explanations for her far-fetched story are far more reasonable than for you (or anyone else) to swallow the idea that the alleged alien abduction really took place.

"Extraordinary Claims Require Extraordinary Evidence"

An excellent rule of thumb, when one is told a surprising tale, is to ask one-self the following: Is the most likely explanation of this claim that it really is true? Another excellent rule of thumb is Carl Sagan's dictum "Ordinary claims only need ordinary evidence, but extraordinary claims require extraor-dinary evidence." Carl Sagan (1934–1996) was a highly influential American astrophysicist and author, and his catchy phrase echoes earlier statements by the great French astronomer and mathematician Pierre Simon de Laplace ("The weight of evidence for an extraordinary claim must be proportioned to its strangeness") and British philosopher David Hume ("A wise man propor-tions his belief to the evidence").[3]

You have a friend who tells you that it's going to rain tomorrow. You believe her because there's no particular reason to be suspicious of her utter-ance. She adds that she just watched the weather report on television, which you did not do. Now the next day it may turn out that her statement was wrong, but even so, it was extremely reasonable of you to have believed her.

Another good friend tells you that Tibetan monks can float in the air when they meditate. After a bit of thought, you conclude that his claim runs headlong against the entire modern scientific worldview. It contradicts all our experience involving gravity and the classical laws that govern the mo-tion of objects that are subject to forces. Now this doesn't mean that you can *prove*, with watertight mathematical rigor, that your friend's claim is wrong. But after a few more minutes of thought, it occurs to you that if Tibetan monks really could hover in the air, this would constitute a revolutionary scientific fact, and it would have spread like wildfire in the news media. The first person to have observed a hovering monk and to have described the phenomenon and given strong evidence for it would surely have received a Nobel Prize in Physics. But no such thing has taken place.

Could it nonetheless be a totally new phenomenon that simply hasn't yet hit the newspapers? Well, your friend adds that meditating monks who hover in mid-air are nothing new; in fact, this has been going on *for thou-sands of years* in Tibet. He also casually states that he himself hasn't witnessed it, but he heard it from a friend of his, who knew someone very smart who had lived for many years in Tibet.

All at once, out of the blue, the little slogan "Believe in an idea if and only if you have good reasons to think it is true" pops into your head. Either there exist hovering monks, or the meditating monks are boringly stuck on the ground. After open-mindedly considering your friend's statement, you decide that there are good reasons for believing that *all* the monks in the world, including those in Tibet, are subject to the laws of gravity, even when they are meditating, and so they are very unlikely—fantastically unlikely—to be hovering in mid-air.

The fact that you draw this conclusion doesn't show that you are overly skeptical or dogmatically closed minded. Quite the contrary! It shows that you are highly *open* to thinking about new things that come your way, that you consider claims with care, and extraordinary claims with extraordinary amounts of care. In this case, your careful reflection led you to the conclusion that your friend most likely was hoodwinked by his own friend, or perhaps that in the chain of witnesses and tale-tellers there was one weak link, who was unwittingly taken in, or who even was lying.

Unmasking What Is Implausible

Another excellent method for finding a reasonable way to think about unusual claims is the principle called *reductio ad absurdum*. Basically, the idea is to carry out a thought experiment. First you suppose that the idea you're considering is true, and then you do your best to see what sorts of consequences its truth would lead to. Could it lead to absurdities? If so, it makes good sense to conclude that the tentative assumption is false.

Let's think about a classic example of such a thought experiment in physics.[4]

Some 2,300 years ago, Aristotle claimed that objects having different masses fall at different speeds. A stone weighing one pound falls more slowly than a stone weighing ten pounds, so he thought. But nearly 2,000 years later, Galileo believed that Aristotle had been wrong, and that all objects actually fall equally fast (in a vacuum).[5] Galileo was able to reason his way to this conclusion. Here is how his *reductio ad absurdum* line of thought ran.

Let's suppose that Aristotle was *right*, and that a ten-pound stone falls much faster than a one-pound stone falls. So now let's imagine tying the two stones together with a very light string before we release them. The ten-pound stone will be held back by the slower-falling one-pound stone. Conversely, the lighter stone will be pulled along faster by the heavier one. So the light stone will fall faster than it would have if it were not tied to anything, while the heavier stone will fall more slowly than it would have otherwise.

The conclusion is that the two tied-together stones, which together weigh *eleven* pounds, will fall a bit more slowly than the *ten*-pound stone would fall when on its own. But this contradicts Aristotle's claim, which is that a heavier object (the eleven-pound "compound stone") will fall *more quickly* than a lighter object (the ten-pound stone).

And so, by using pure imaginative reasoning, we have reached the conclusion that the compound stone will fall both *more slowly* and also *more quickly* than the ten-pound stone. This is *absurdum* (two opposite conclusions cannot both hold at the same time, obviously), and as a result, we have to revoke the initial assumption (namely, Aristotle's claim) that led us into this box canyon.

The principle of *reductio ad absurdum*, then, is that one first "tries on" a claim by tentatively assuming it's true, and then determining if this idea leads, via reasoning, to a self-contradiction (or to completely unreasonable consequences). If this happens, then one should go back and reject the claim that was tentatively assumed true. In short, here's the *reductio ad absurdum* recipe:

1. Assume that Claim X, which you've been told, is true.
2. By reasoning, check out the consequences that this would lead to.
3. If it leads to crazy or self-contradictory consequences, reject Claim X.

Here's another example of how *reductio ad absurdum* works. An itinerant inventor knocks at your door and tries to sell you a little crystal. He tells you that this crystal has fantastic energies trapped inside it, and that you will feel incredibly good if you wear it on your body. Moreover, if you place the crystal inside your car, your fuel consumption will go down by 20 percent. The crystal costs only $99, so quick as a wink you'll make up for its cost in your fuel savings. So . . . should you believe this?

Let's see if we can make use of the principle of *reductio ad absurdum*. So let's assume the inventor is telling the truth. The crystal thus works exactly as advertised. Well, in that case, why wouldn't all auto manufacturers have already built such crystals into their vehicles? Any of them would be overjoyed to have a device that would make their cars 20 percent more efficient than their competitors' cars! What an incredible advantage that would be in the marketplace!

Well, couldn't it simply be the case that the inventor hasn't yet told any car manufacturer about this little device? Hmm. . . . Now why would an inventor hold back incredibly valuable information from all auto manufacturers? Suppose that General Motors (GM) could purchase exclusive rights to this crystal and could also get a patent for the general technique. If all GM cars were 20 percent more efficient than their rivals, GM would gain a huge advantage over all rival car makers.

Now in all likelihood, GM would be willing to pay the inventor a very high sum for this marvelous opportunity. So why hasn't the inventor chosen to become hugely rich in this very simple manner? Is he perhaps a highly idealistic person who merely wishes to help the world, and who doesn't give a hoot about money matters? But in that case, why is he trying to sell you his crystal for $99? Why isn't he just giving his crystals away to you (and others)? As you can see, the whole scenario is growing more and more unreasonable, more and more absurd, more and more dumb, more and more *absurdum*. And so, bottom line, we are dealing with a case of *reductio ad absurdum*.

We first presumed that the wandering inventor was telling the truth, and then we showed that this assumption led to absurd consequences. The upshot is that the inventor's claim about his crystal has been shown to be almost certainly false.

No method of reasoning is infallible in the real world, and even inside mathematics, truly brilliant thinkers can make serious mistakes in their thought processes. Let us take a look at the curious case of Giovanni Girolamo Saccheri (1667–1733), a very original Italian mathematician and Jesuit priest whose life's highest goal was to defend his hero Euclid by rigorously proving that Euclid's parallel postulate (mentioned earlier) was true.

Before Saccheri's day, many people had tried to do this (though always in vain), but none had ventured down the *reductio ad absurdum* pathway. Saccheri decided that this was the ideal route to follow, so he took a wild bull by the horns by assuming that the *negation* of Euclid's parallel postulate was true. This was certainly a daring approach. The bold Jesuit then proceeded to prove one brand-new theorem after another after another, all of them based on this strange new postulate, the postulate that he despised but had nonetheless decided to explore. Although each new theorem ran deeply against his intuitions, Saccheri just gritted his teeth and kept on going for a couple of hundred pages. Finally one day, he hit a result that struck him as so extremely strange and implausible (namely, that there is a maximum possible area for triangles) that he effectively threw up his hands and famously proclaimed, "This result is *repugnant to the nature of the straight line!*"

This meant to him that he had hit up against total absurdity. Just what he wanted! Hallelujah! Following the *reductio ad absurdum* recipe discussed earlier, Saccheri joyously went back and rejected the initial assumption that was responsible for the "repugnant" result—and this assumption, of course, was the negation of Euclid's sacred parallel postulate. Well, if the *negation* of Euclid's postulate had to be *rejected*, then Euclid's postulate itself had to be *accepted*. Goal achieved! Hurrah!

Near the end of his life, Saccheri published a book called *Euclides ab omni naevo vindicatus* (Euclid Freed of Every Flaw) in which he set forth his findings, and thanks to this book and its self-confident title, it was believed for a while, at least by some, that Euclid's parallel postulate had finally been rigorously proven, and that Euclid had been vindicated. What a happy day for Saccheri and for Euclid!

But unfortunately, the "repugnance" that Saccheri felt so viscerally was not a true contradiction or paradox. It was just an emotional reaction, just a gut feeling, just an intuition. And roughly one hundred years after Saccheri's death, a few other courageous mathematicians (most famously, the Hungarian János Bolyai and the Russian Nikolai Lobachevsky) followed in his footsteps by rejecting Euclid's parallel postulate and accepting a rival postulate in its place—but these hardy souls, instead of simply declaring their results absurd or repugnant, slowly began to see a new kind of geometry taking shape before their eyes, one whose "straight lines" weren't straight in the same way that Euclidean lines are straight, but one whose inherent internal logic, though surprising, was nonetheless impeccable. Indeed, it was perfectly reasonable—in *this* geometry—that there *was* a maximum possible area for triangles. And thus was born non-Euclidean geometry—indeed, doubly born—in both Hungary and Russia at the same time.

Because he was too rigid in his preconceptions, poor Padre Saccheri missed his chance at founding what would surely have been called "Saccherian geometry." Though courageous, he was not courageous enough. He misused the principle of *reductio ad absurdum* by claiming to have found the longed-for absurdity a bit too early, a bit too easily. "Repugnance to the nature of the straight line" was not a good enough reason to throw all his strange theorems out the window. And so the lesson is that one has to be extremely careful in one's thinking, and not let oneself be too easily led (or misled) by one's prejudices. This is easier said than done, of course. After all, in many ways, being reasonable and sensible is an art, not a science.

Many people are willing to attribute magical powers to exotic Tibetan monks, or healing powers to mysterious crystals, or paranormal powers to rare human beings. If one then expresses skepticism about these attributions, one often runs into the objection, "You can't *prove* that the monks aren't hovering in mid-air when they're meditating! Why are you so limited? You should be more open minded! Not everything can be explained through science! Show a bit more humility!"

We find ourselves once again among the widespread and irrational prejudices against rationality. In such a challenging context, it becomes all the more crucial to repeat the fact that it is hardly possible to be more open minded than a person who is willing to entertain ideas of all sorts, who will consider their consequences, and who will accept whichever ideas turn out to have the strongest support. If, by contrast, someone is willing to buy into ideas that lack any strong evidential support, and if they are not prepared to change their mind even when strong evidence appears for rival ideas, then that person can hardly be called "open minded." In fact, such an inflexible, dogmatic stance is the hallmark of closed-minded prejudice.

A special case in these discussions is when the question concerns whether or not something exists. It's not possible to give an ironclad proof of something's *nonexistence* (think of Bigfoot or the Loch Ness monster, for instance), but in principle it is quite possible to give strong evidence of something's *existence*.

So suppose you have a friend who claims that there exist green swans.

"No, no, no!" you protest. "There aren't any such things! Green swans don't exist!"

"Give me evidence for that claim!" retorts your friend.

Well, now you're in a fix. If you want to give powerful evidence that there aren't any green swans, you'll have to travel throughout the earth and check out every last swan that you come across. But even that won't suffice to prove your claim. You also have to definitively prove that you didn't miss a single swan in your search. That's too much!

Your belief that there don't exist green swans is built upon the lack of green swans both in your own observations and in those of other people. But you can't give irrefutable evidence for the nonexistence of green swans—only partial evidence. (And by the way, assuming that green swans *do* exist will not lead to an absurdity, so the *reductio* approach won't work in this case.) On the other hand, if your friend wants to prove you wrong, all she needs to do is to produce one single green swan! It would seem, then, that the burden of proof lies on your friend, not on you. You can calmly lean back in your armchair and suggest to her that she should set out on her green swan chase.

Much the same holds for the case of the hovering Tibetan monks. If monks can fly while meditating, then it should be easy as pie to show that they exist; all you need is that one of them should do so under controlled conditions, and have someone take a video of them. But to show they *don't* exist, all you can do is present partial evidence. As we said earlier, the evidence for something's *non*existence is necessarily of a weaker sort—scattered pieces of evidence that, taken together, add up to a plausible trend. In the case of the allegedly hovering monks, however, such partial evidence of their nonexistence still amounts to a strong argument (as long as no concrete evidence turns up in favor of their hovering).

In all situations where there is a dispute about the existence or nonexistence of some hypothesized entity (hovering monks, the Himalayan Yeti, ESP, the Higgs boson, a green swan, the devil, ghosts, dark matter, the continent of Atlantis, God), the burden of proof lies on those who claim the entity in question exists.

~

INTERLUDE: ON EXPLORING RIVAL MOVES IN CHESS AND RIVAL MOVES IN THE GAME OF LIFE

Are you a chess player? I hope so—at least an occasional player.

The better a chess player you are, the smaller the set of possible moves you will consider before making your next move. Does that sound odd?

Well, it shouldn't. The goal in chess is to find the best move, and you have to do so in a very limited amount of time, since the chess clock is always ticking away (at least in serious competitive play); therefore, it's key not to waste precious time checking out a whole bunch of terrible moves. Instead, you want your mind to pinpoint and focus down on just those few moves that are promising, and among those, of course, your goal is to pick the very best one.

Much the same holds when we find ourselves facing questions about the complex world we live in: with the help of knowledge accumulated over years of experience, we can—and we must—do our very best to quickly sort out promising from unpromising potential futures and thus prioritize our mental explorations of what path to take next (that is, which "move" to make next in the game of life). Only in this way can we reach deeper levels of understanding of the world.

One extremely central function of the vast, deep, and intimate knowledge that comes out of long experience is that it can help keep one's explorations within reasonable limits—and by the way, the negative connotations usually attached to the word "limit" don't apply here. Quite the contrary, I mean limitations that have highly positive effects.

To show concretely what I mean, I will use chess as a metaphor for life. In most situations that arise on a chessboard, each player has about thirty-five different moves that could be made in principle. (It depends a bit on whether one is in an opening stage of the game, in the middlegame, or in the endgame, but thirty-five is a pretty accurate estimate in general.) A skilled chess player knows, however, that nearly all of those thirty-five-odd moves are not very good—in fact, most of them are downright terrible. And so good players quickly focus down on just a handful of moves—three, four, or five at most. And how do they pick those select few out of the thirty-five? It's all thanks to years of experience!

In a complex situation in a chess game, a Grandmaster, trying to rapidly sort out the wise possibilities from the wild ones, spots familiar patterns and makes subtle analogies to situations previously encountered throughout years and years of play. This intimate knowledge of chess situations drastically reduces the Grandmaster's search for what to do next. People who have less chess experience simply won't see those patterns and won't spot those analogies. Because of their lack of experience, they will be open to all sorts of moves that the Grandmaster would discard in a flash, and they won't even have an inkling of what's deeply wrong with them.

A Grandmaster should obviously not listen to the random musings of a child who's just learned the rules, nor even to the "advice" offered by an enthusiastic amateur who's played a hundred games. Listening to such novices would be an utter waste of the Grandmaster's crucial time, with that chess clock ticking away. It's vitally important not to be that open minded! After all, chess games, like lives, are finite!

But at what point does one draw the line? Should a Grandmaster listen to the intuitions of a twelve-year-old chess prodigy who has won a few local tournaments? Or suppose that this young chess prodigy had recently played against Magnus Carlsen (the current world champion as of 2021) and given him a run for his money? At what point should one open one's mind up and entertain ideas that might otherwise seem to be coming out of left field? Well, these are subtle judgment calls, and obviously there can be no fixed recipe that will flawlessly give the "right" answer.

In everyday life, we simply cannot constantly be open minded about everything. We can't explore all well-meaning suggestions coming at us from all directions. We have to rely on our own personal experience (and the vicarious experiences we have had) to sort out the wise from the wild, allowing us to quickly pare down our options.

Thus you don't have to be open minded about trying liver yet again, if every time you tasted it in the past you always hated it. And by the same token, I can't afford to keep on trying to read new books by an author whom I've always hated in the past, even if a bunch of back-cover blurbs praise the books to the skies. And Birgitta, despite the well-meaning urgings of her friend Gunilla, doesn't have to keep on trying the same online dating service that she has already tried several times, always with utterly disastrous results. Be self-confident enough to rely on your experience! (As long as you have lots of it, that is.)

If you have much knowledge and experience in a given aspect of life, you can and you should be sufficiently self-confident to rapidly dismiss most of the random ideas and well-meant suggestions that come your way, just like the Grandmaster who rapidly dismisses suggestions from greenhorns. Of course, the occasional outside suggestion is good, and hopefully you (or the Grandmaster) will have had enough experience to recognize that that suggestion is different in quality from the others. But, of course, there's no ironclad guarantee. By rapidly dismissing most outside suggestions, you naturally run the risk of throwing out something precious—throwing out the baby with the bathwater—but that's life. Life, like chess, is always a matter of risk taking—but hopefully it's well-considered risk taking, thoughtful risk taking, rooted in long years of experience.

Suppose that mathematician A has been exploring a certain difficult problem in number theory for many years and tells mathematician B, who is not a number theorist, about the problem. B, who has never heard the problem before, blurts out, "Hey, I think it probably has to do with the distribution of the prime numbers. You should check it out." Should A just suddenly drop everything and try to explore B's hunch? Well, there are of course all sorts of unknown factors in such a situation, but all other things being equal, it wouldn't make any sense at all for the far more knowledgeable A to just switch tracks instantly, simply because B, a greenhorn outsider, reacted that way. Most likely what B said is naïve and wrong. Of course, it would be very different were B a world expert in the area of number theory that A is working in, because then B, too, would be drawing on deep and vast knowledge.

Although one should not confuse life with a chess game, one can certainly gain insights from the wonderful analogies to life that chess playing affords.

CHAPTER FOUR

WHAT IS SCIENCE?

Concerning Theories, Experiments, Conclusions, and Science's Essence

The whole of science is nothing more than a refinement of everyday thinking.—Albert Einstein, *Physics and Reality*, 1936

At every moment we are building up beliefs about how things behave in our world. Without such beliefs about the world's nature, we would be lost: we wouldn't have any idea what to do in any situation we faced. Each human brain is therefore constantly hard at work, trying to make as much sense as it possibly can of what is going on in its surroundings.[1]

The more we humans come to understand the world, the more inspired we are to go hunting for new knowledge. This snowballing process starts early in life; just think how much knowledge a small child has to have in order to survive and not hurt itself—and the curious child acquires such knowledge by playful experimentation, just like a curious scientist carrying out careful experiments. When exploring its environment, the child proceeds by trial and error, putting all sorts of objects in its mouth to taste them, twisting and turning things about, trying to put one thing into another thing, building towers and watching them fall down, climbing up on tables and chairs, sliding under beds, jumping on and off of sofas—in short, incessantly checking everything out.

Children, of course, wind up getting bruises, cuts, scratches, splinters, and shocks—but bit by bit, all of this pain due to endless curiosity leads the child to an ever vaster, ever deeper, ever more intimate understanding of how the world is made and how it works. This includes a sense for other people, for all sorts of animals, for trees and plants, for water, weather, and wind, for what will float and what will sink, for what tastes good and what tastes bad. The child learns about wagons, tricycles, bicycles, and cars, about electricity, wires, plugs and outlets, about toasters, stoves, refrigerators, and microwave ovens . . .

A small part of a young child's world knowledge is acquired from adults who describe things and explain things, but the majority comes from the child's own poking around and exploring, from repeated trials and errors. Eventually, though, as children grow, the knowledge they are acquiring starts to come more and more from other people, rather than from their own personal experiences. The direct tests that once were so frequent become more and more spaced out, and instead information comes more and more often from such activities as reading books, magazines, and newspapers; watching television, movies, and videos; carefully searching for desired things on the web; accidentally bumping into random things on the web; having countless unpredictable conversations with friends; absorbing the vicarious insights that are thus made available; listening to and pondering the myriad clashing opinions that one hears in the wide arena of public discourse . . .

Many of these clashing opinions are just opinions, of course—not facts. And most of them play almost no role in our lives. To be sure, we may constantly read and register what is called "news," but as long as it doesn't touch us directly, we don't need to take a position on its truth or lack thereof (although of course we may on occasion be led, by interest or curiosity, to reflect more carefully on certain claims that we've encountered). Yet the moment a claim or an opinion becomes important to us *personally*, for one reason or another—the moment we want to use it as a basis for our own actions and let it guide us in our own vital decisions—that's when we ought to return to the youngster's mode of personal trial and error.

And this is exactly where the scientific approach comes in. In truth, science is nothing but *a more systematic quest* for knowledge, where one uses basically the same methods as little children do: one takes nothing for granted, one checks everything out oneself, by twisting it and turning it around and around, by touching it and tasting it and smelling it, by intimately interacting with concrete reality.

At the Very Heart of Science

Science is a highly structured and systematic method of exploring the wide world by means of trial and error. As early as 2000 BCE, the Babylonians (and not too much later, the ancient Greek philosophers) developed methods of investigation and ways of thinking that can be said to have laid the groundwork for science.

Islamic philosopher Muhammad ibn al-Hasan (965–1038) is often described as the founder of the scientific method. He lived in Basra and Cairo, and wrote "Kitāb al-Manāẓir" (A Treatise on Optics), which contained, among other things, a scientific description of optics and of vision. He devised the idea of carrying out systematic and repeated experiments in order to support or refute hypotheses and theories. He combined observations, experiments, and logical reasoning in his theory of vision, according to which light rays bounce off objects and into one's eyes, rather than the reverse.[2]

In science, a distinction is made among facts, hypotheses, and theories. A *fact* is usually a very specific observation, or the result of an experiment. A *theory* is a larger set of ideas whose purpose is to predict or explain the results of experiments. The term "theory" is used to describe a set of interrelated ideas that explain a specific event or a general set of phenomena. Theories are therefore more complex than individual statements. An uncertain claim presented on its own is usually called a *hypothesis* (a Greek word meaning "guess").

When doing science, one proposes testable hypotheses that come from a theory that one wishes to check out. The hypotheses are then checked against various observations and experiments. A hypothesis is, in this sense, a first crude attempt at explaining a narrow class of observed phenomena, while a theory is a more ambitious explanation of a wide class of observed phenomena.

A scientific theory not only has to be compatible with already-established facts; it also has to be able to *predict* the results of at least some experiments. A theory that cannot make any predictions is vacuous. Hypotheses and theories therefore have to be *testable*—meaning that there must be some manner in which they can be *falsified* (that is, shown to be wrong). If a theory or hypothesis isn't testable, then it is scientifically useless.

Let's take a look at an episode in the history of science that clearly shows the difference between theories and hypotheses, and that also shows how science really works. This is the story of the discovery of the planet Neptune.

At the start of the seventeenth century, German mathematician Johannes Kepler (1571–1630) published three laws describing the motions of planets in their orbits around the sun. He showed, among other things, that the planets' motions were neither circular in shape nor uniform in speed, thus going against beliefs (or dogmas, one might say) that had reigned for thousands of years.

Several decades later, Isaac Newton (1643–1727) made Kepler's laws of motion clearer and more precise by proposing his own theory of gravitation. In various ways his theory was counterintuitive. After all, how could a heavenly body such as the sun give rise to a mysterious force that pervaded all of space and that affected objects that were immensely remote from it? Yet this was part and parcel of Newton's new theory. Many critics saw the theory as supernatural and mystical. But by the start of the nineteenth century, Newton's theory of gravitation and Kepler's laws describing the planetary orbits were firmly established.

French astronomer Urbain Le Verrier (1811–1877) was a professor at the Sorbonne in Paris, and he carried out wide-ranging investigations of the planetary motions. His careful observations revealed that there were very slight irregularities in the motion of the planet Uranus as it crossed the starry vault. These aberrancies could not, at first sight, be reconciled with Newton's and Kepler's laws. For this reason, Le Verrier came up with a hypothesis: the tiny deviations in Uranus's pathway must be caused by the gravitational force due to some new, unknown planet lying somewhere out beyond the orbit of Uranus.

Guided by this guess, he then carried out theoretical calculations using his old friends, the reliable theories of Newton and Kepler. This led him to a concrete and testable prediction: A heretofore unseen planet having mass m and following orbit o should be observable at time t and position p in the heavens (where of course these were given as precise figures). He then sent a letter to astronomer Johann Galle at the Berlin Observatory, saying exactly when and where the hypothesized planet should be visible. The letter arrived on September 23, 1846, and that very evening Galle confirmed Le Verrier's predictions, spotting a new planet exactly where it had been predicted.

This story clearly shows the relationship among theories, hypotheses, observations, and testability. Many people, nonetheless, think that theories are weaker than facts. They will cry out, "But that's just a *theory!*" This is simply a misunderstanding of the word "theory" as it is used in science.

In ordinary parlance, the word "theory" usually denotes an idea proffered by someone without too much evidence. We often hear such things as, "Oh, so *that's* your theory as to why the jewel disappeared from the locked desk drawer!" or "The police theorized that two robbers were involved, not just one." In such cases, the term "theory" means a plausible guess that has been put out there for people to consider as a possibility, but without yet having any strong support.

But in scientific circles, the word "theory" has quite a different meaning. It describes a set of tightly interrelated ideas that not only are logically coherent but also provide an explanation for many diverse phenomena and, moreover, can give rise to new predictions. A set of ideas has to *earn* the respected label of "theory" by being confirmed in various ways; until then, it's just random guesswork.

Of course there isn't a watertight boundary between everyday thinking and scientific thinking, and the term "theory" therefore has a certain degree of ambiguity to it. Sometimes it denotes an idea having only the flimsiest of support, while other times it denotes a tight cluster of ideas having enormous support, and anything in between these two extremes is of course also possible.

In science, it's crucial that any theory should always be open to modifications and improvements, but at the same time, for a set of ideas to deserve the label of "theory," it must already have passed numerous tests of credibility, so that it has numerous pieces of evidence supporting it. Some of these pieces of evidence may be extremely strong, and others may be relatively weak. But scientific theories always have *some* degree of evidence and support.

Einstein's theory of general relativity is an extremely strongly supported theory, which has survived well over one hundred years of experimental tests. Darwin's theory of evolution is also very strongly supported by observations throughout the world. On the other hand, there are theories of so-called "dark matter" and "dark energy" that are still being worked out, and that explain some phenomena but leave other phenomena totally unexplained.

There is a curious body of knowledge in theoretical physics known as "string theory," which is filled to the brim with perfectly self-consistent mathematical ideas, but which so far has no experimental support at all. For this reason, string theory is a pretty anomalous case of the term "theory," but what makes string theory count as a *theory* (at least in the eyes of some) is that its ideas and equations all cohere very tightly (for an analogy, think of the many internally consistent theorems of non-Euclidean geometry), even if they haven't yet been experimentally supported (and even if they never receive the tiniest trace of experimental support).

In short, there is a whole gamut of possibilities for scientific theories, ranging from very strongly confirmed to quite rickety. Scientists are used to the fact that their word "theory" covers a wide range of degrees of credibility; however, lay people often seize upon this vagueness in the term and use it to heap scorn on theories that make them uncomfortable. This is why we occasionally hear such annoyed outbursts as "But evolution is only a *theory*!"

People who say such things don't realize that evolution is a deeply corroborated *scientific* theory, and as such, it is accepted as truth, not as speculation.

What, then, makes a theory count as being scientific? Within science, there is a standard distinction between the *natural* sciences and the *social* sciences. These two types of science are quite different from each other, even if the differences are often exaggerated by outsiders. The primary goal of the natural sciences is to find general laws that describe how nature works. A theory of particle physics doesn't merely describe how the particles located in South Dakota behaved in the year 1912, or, more generally, how the particles in South Dakota behave, but how *all* particles of a specific type behave at all times and under any given circumstances.

In the social sciences, one is often more interested in the individual or the specific case. The goal might be to describe a particular culture's customs and traditions, or to describe the view of humanity that playwright August Strindberg expresses in his writings. On the other hand, in some of the social sciences, especially psychology, the goal is to formulate very general theories, such as how visual perception works in the brain, or why people generally have foreign accents if they learn a language after a certain age, or what makes people remember some things and forget others.

One reason that it is important to define the word "science" is that we need to be able to distinguish genuine science from *pseudoscience*, which is an activity that adopts the external trappings of science but that doesn't follow the methods that characterize genuine science. Thus pseudoscientists may create large bodies of technical-sounding jargon, in imitation of what real scientists do; or they may take up the habit of writing down long equations filled with Greek letters and subscripts, because that *looks like* what physicists do; or they may use lots of graphs and complex-looking diagrams with circles and arrows and so forth, trying to imitate the rigor and precision that are characteristic of genuine science. But behind the scenes, pseudoscience does not respect the central and indispensable principle that makes science what it is—namely, trial and error, theory and experiment, confirmation and refutation—in short, a rough-and tumble competition among all sorts of theories, where only the fittest survive.

Science is not just a set of statements or findings or answers to questions; it is an entire set of ways of exploring the world. The defining feature of science is that it subjects all sorts of hypotheses to reliable tests, and it mercilessly drops any hypothesis that doesn't hold up under serious scrutiny.

The constant application of this method could be said to be the criterion that distinguishes between genuine science and fake science. Claims in the natural sciences that don't rely on this method—claims that have no way of being tested—are therefore not scientific. Such a claim might be: "I can walk on water as long as no one is observing me in any way whatsoever." Another might be: "There are billions of little tiny invisible negative-mass elephants that live in purple caves on the surface of antiprotons on the dark side of the moon." Sometimes a claim can, on first glance, give the impression of being scientific, especially if it is couched in highly technical terms. But if, on careful scrutiny, it cannot be put to a rigorous test, then it must be relegated to the category of *pseudoscience.*

Sometimes science is criticized for being dogmatic, and for being too sure of its "truths." This type of criticism, however, is based on a profound misunderstanding of how science actually works. The truth of the matter is that built into the scientific method is the process of merciless self-criticism, which means that scientists are constantly trying to check and recheck and improve their own results and, of course, also the results of their colleagues.

To be sure, some scientists are emotionally committed to verifying and confirming and defending their own ideas, and they are thus prejudiced and somewhat blinded, but of course other scientists are not committed to those ideas, and they will therefore make up for these people's prejudices and blindnesses. As a result, any theory that is flawed will not last long in the harsh world of science.

Experimentation, Rejection, and Revision: The Very Core of What Science Is

We use the methods of science—namely, testing and modifying our theories—to gain knowledge. In this manner, we gradually build up better and better descriptions of reality.

Of course throughout the centuries there have been major scientific mistakes, such as the idea that one can read someone's personality traits from the shape of their skull (an idea called "phrenology"). This notion was proposed by Franz Joseph Gall (1758–1828), a physician in Vienna. The thought was based on the idea that the brain consists of a number of organs that correspond to various mental functions, abilities, and character traits, such as the faculty of color vision, the faculty of speech, mathematical ability, musical ability, sexual desire, timidity, gregariousness, self-confidence, honesty, irascibility, generosity, altruism, religiousness, reliability, devotion, self-sacrifice, and so forth. The idea of phrenology enjoyed quite a long life, but it was nonetheless a pseudoscience.

Another example is given by the so-called "N-rays," announced in 1903 by the highly respected French physicist Prosper-René Blondlot (1849–1930), very shortly after German physicist Wilhelm Röntgen (1845–1923) had discovered X-rays, which subsequently acquired the name "Röntgen rays" in certain languages (especially German, of course, but also Swedish). Blondlot, after experimenting with X-rays, was convinced that he had discovered a new form of radiation, and he named it after the University of Nancy, where he was a professor. Numerous other researchers rapidly repeated his experiment and jubilantly confirmed the existence of the N-rays. In the year following the original announcement, Blondlot and more than a hundred colleagues claimed to have detected N-rays emanating from all sorts of substances, including the human body; oddly enough, however, the rays were apparently not emitted by green wood or by certain sorts of treated metals. One researcher, named Gustave Le Bon (1841–1931), was so eager for credit that he claimed that he, not Blondlot, was the original discoverer. (Undoubtedly Le Bon was hoping to receive the Nobel Prize in Physics for his work, and it's clear that a Le Bon Nobel would have delighted palindrome lovers. But sad to say, Le Bon and palindrome lovers were disappointed.) In the end, despite all these initial confirmations, it turned out there weren't any such rays at all. After a while, all the experiments started failing, and soon the whole thing had gone up in smoke. How could all these researchers have made the same mistake? Well, they simply fooled themselves, because they all *wanted* to see something. This is one of the most famous tales of scientific self-deception.

Perhaps the most notorious such tale in modern times is that of "cold fusion," defined as the spontaneous fusing together of two atomic nuclei at or near room temperature, releasing a huge burst of energy (usually called "binding energy"). If this phenomenon were real, it would mean that unlimited amounts of energy would be trivially available—an indescribable boon for humanity. Such nuclear-fusion processes do in fact occur at extremely high temperatures, such as within stars, but the idea that they could also occur at everyday temperatures was highly speculative. Nonetheless, in 1989, experiments were announced by two respected chemists in Utah, and for a whole year the scientific world hotly debated whether the claimed coldfusion miracle was real or not. In the end, since it was never reliably reproduced anywhere, the finding was discounted, and since then nothing more has been heard of cold fusion, except in a few isolated laboratories whose work is not supported by reputable sources and whose findings are not published in mainstream journals, because they do not meet the requisite criteria of rigor. Nonetheless, there are still scientists (or rather, fringe-scientists) who *want* cold fusion to be true, and who are still trying to realize it.

Luckily, cases like these are a tiny minority, and as the tales we have told reveal, sooner or later every bubble of pseudoscience will pop (although, sad to say, new bubbles are constantly being blown by new generations of pseudoscientists). By contrast, the usual pathway in science is quite different: an initial finding is confirmed, refined, developed, and extended, and the ideas grow deeper and deeper and richer and richer as the discoveries multiply.

Critics of science sometimes state: "It doesn't make sense to rely on science, because every day new facts turn up that previously had been taken to be false." Often the case of Isaac Newton and Albert Einstein is trotted out as a linchpin example. The argument runs as follows: "For two centuries, everyone believed in the holy Newtonian worldview, but then along came Einstein and showed that the great Newton had been mistaken. Newton then went straight out the window, just like that! Poof! And sooner or later, the very same thing will happen to Einstein. And on and on it'll go. So you just can't rely on anything or anyone in science!"

This glib-sounding argument is completely baseless because it wildly misrepresents the situation. Newton did his work at the end of the seventeenth century and the beginning of the eighteenth century. He created a theory of mechanics at whose core lay his three laws of motion and his law of gravitation. This was the so-called Newtonian worldview, and it perfectly explained the motions of the planets and of terrestrial mechanical systems for two hundred years. It also supplied perfect explanations for the behavior of gases (thermodynamics), of liquids (hydrodynamics), of cars, of airplanes, and of untold thousands of other machines.

Some two hundred years later, at the beginning of the twentieth century, Albert Einstein developed his theory of general relativity, a description of space and time that—in a certain sense—replaced Newton's model. Einstein's theory was *a tiny bit more accurate* than Newton's, but that doesn't mean that he showed that Newton's theory was way off base or useless. How could he possibly have done such a thing? After all, Newton's theory had *worked sensationally* for all those years!

The truth is that Newton's theory was not invalidated at all; it was simply cast in a new light. It emerged as a *limiting case* of Einstein's theory—namely, when velocities are not super-fast, approaching the speed of light (and this condition holds for all macroscopic objects, like hockey pucks, people, Porsches, and planets), and when the force of gravity is not extremely powerful (and this condition holds for the earth's gravitational field, of course, as well as for the sun's much stronger gravitational force, which gives rise to the solar system). In other words, Einstein's theory is almost exactly equivalent to Newton's theory, except in the most unusual of physical circumstances, involving fantastically high velocities or fantastically strong gravitational fields.

It is fine to say that Einstein's theory of gravitation gives a better approximation to reality than Newton's theory does. But it is also crucial to keep in mind that as long as we are in everyday situations, Newton's theory gives us extremely accurate, nearly perfect predictions. As was stated earlier, Newton's theory itself is a rigorous consequence of Einstein's theory, at "low" velocities (much faster than the speed of the fastest rockets ever built) and "low" gravitational fields (like those of ordinary stars and, of course, those of planets, which are far, far feebler). And Newton's theory is still super-reliable, in the sense that engineers use it all the time to build their bridges, their rockets, their satellites, their cars, their boats, and so forth. Only when it comes to extremely precise measurements, such as are needed in devices like GPS systems, do engineers occasionally need to make use of Einstein's theory, and also quantum theory, developed somewhat later.

Scientific theories are good approximations of reality, and they are always getting better. Einstein's relativity theory and the subsequent quantum theory are better approximations to reality than Newton's theory, and it's quite possible that at some future time they, too, will be "supplanted" by some more accurate theory, but even if that happens, they will certainly *not* go "out the window"—no more than Newton went out the window one hundred years ago. That idea is just misguided thinking.

For another case study, consider that, in the nineteenth century, it was proven beyond all reasonable doubt that light is a type of electromagnetic wave. This was a direct consequence of the wonderful equations discovered by Scottish physicist James Clerk Maxwell (1831–1879), which united the formerly independent theories of electricity and of magnetism into one single, beautiful set of four equations. Out of this marvelous unification came radio, television, telephones, electrical motors, movies, microwave ovens, and countless other technological advances of the twentieth century.

But in 1905, Albert Einstein, exploiting a deep analogy that he had guessed at, suggested that maybe light did *not* consist of electromagnetic radiation after all but, instead, was made of *particles*! For many years, no physicists gave this idea the time of day. In this belief, Einstein was a renegade, an outlier, a loner. But eighteen years later, in the early 1920s, experimental discoveries about light (especially the Compton effect) were made that simply could not be reconciled with Maxwell's equations, and at that point, it started dawning on physicists that maybe, way back in 1905, Einstein had been right, after all, about light's nature. This didn't mean, however, that Maxwell had been *wrong* when he described light as electromagnetic waves; what it meant was that there was another, deeper layer to the picture, and that under certain very unusual circumstances, light exhibited behaviors that could only be explained in terms of a particle picture. But in all the situations where Maxwell's theory of light *had* worked, it *continued* to work, and always *will continue* to work—which is to say, in almost all situations that we on earth normally come across.

Once again, we see that the meanings of "true" and "false" in science are subtler than might be expected. A theory can be *strictly speaking* false, but *for all practical purposes* perfectly true! This holds both for the Newtonian worldview, as described earlier, and the Maxwellian picture of light as waves. Each of them was once considered to be universally, perfectly, totally valid, but now we understand that each theory has its limits of validity; however, that doesn't invalidate the theories for normal situations. In fact, the old theories are still the *best* way to understand most ordinary situations, which is why they continue to be taught as the core of all physics curricula in high schools and universities, and continue to be profoundly relied on in all engineering schools.

At the present time, our picture of light is as a kind of hybrid entity, sometimes wavelike and sometimes particle-like. It is an uneasy balance, since nothing that we humans know "in real life" behaves in this mixed manner, but we have to get used to it, since it's the best picture we have for this mysterious phenomenon. Even Einstein, who was the originator of our modern picture of light, remained baffled by his own beast, the "light quantum" (or *photon*, as we say these days), all the way until the end of his life. In 1951, only four years before his death, he laments in a letter to his lifelong friend Michele Besso:

> All these fifty years of conscious brooding have not brought me any closer to answering the question "What are light quanta?" Today, of course, every rascal thinks he knows the answer, but he is deluding himself.[3]

Perhaps someday in the future we will arrive at a deeper and more satisfying picture of the nature of light than we have today, but that is only a vague hope.

But whether our current picture of light is totally satisfying to us or not, it is extraordinarily accurate and extraordinarily fruitful in terms of its applicability. For example, our current understanding of light and of atoms, even if not the last word on the subject, is adequate to allow us to build wonderful devices that can give us very detailed pictures of the goings-on inside our bodies. In particular, we routinely use MRI scanners—standing for "(nuclear) magnetic resonance imagery"—in hospitals, to discover tumors, as well as in laboratories, to study the processes in various organs of the body.

MRI scanners use strong magnetic fields and radio waves, and they exploit certain properties of the hydrogen atom, whose nucleus consists of a single proton. A water molecule (H_2O) contains two hydrogen atoms and one oxygen atom, and roughly three-quarters of the human body consists of water molecules. The nucleus of a hydrogen atom—that is, a proton—can be thought of as a very tiny magnet that can face in various directions, as a function of the magnetic field that it is immersed in (one can think of a magnetic field as consisting of lines of force, along which the proton "wants" to align itself). When the MRI scanner produces a strong magnetic field, myriads of tiny proton-magnets rapidly line up along it. At that point, radio waves are directed at the sample, and the protons absorb the waves' energy, which changes their orientation; then, when the radio waves are turned off, the protons return back to their original orientation, and in so doing they release the energy they just absorbed, in the form of new outgoing radio waves. But these radio waves now contain useful information about the positions of the hydrogen atoms, and when the waves are received, they can be processed by fast computers and thus can yield very accurate pictures of the interior of the body.

Suppose that Mary is suffering from severe chronic headaches, and she goes to the doctor, who orders an MRI. The result shows that Mary has what seems to be a benign tumor in a specific spot in her brain. Soon an operation is done, and when the surgeons open up Mary's skull, they find the tumor in the exact spot that they expected, and it also has the exact shape and size that they expected from the pictures provided by the MRI scanner. The surgeons cut out the tumor, and very soon Mary is once again healthy, with no more headaches. And so the picture the surgeons saw was true. What conclusions can we draw from such a story?

The MRI scanner was based on theories of electromagnetic radiation, of magnetic fields, and of the particles making up atomic nuclei (protons, in this case), as well as of the forces through which all these entities interact with one another. The surgeons found the tumor just where it should have been. Does this stunning success (along with the millions of other successes of these theories) mean that these theories are absolutely perfect? Certainly not.

Suppose, for instance, that string theory is someday shown, through various experiments and cosmological observations, to yield a far deeper view of the nature of particles than we ever had before.[4] In particular, suppose that we learn that protons and other particles are best thought of as infinitesimal "strings" that "vibrate" in ten or eleven dimensions. If this happened, although it would clearly be a huge triumph for string theory, it would not imply in any way, shape, or form that Mary's tumor, discovered using the MRI scanner (which was based on earlier theories, not on string theory), was an illusion.

We can draw an important conclusion from the little story we've just considered. Many theories belonging to current-day physics and chemistry have technological applications—measuring instruments, computers, and so forth—and these devices work exactly as their inventors intended them to do (showing the insides of bodies, for instance). Now these theories will never turn out to clash, in any important sense, with more accurate or "truer" theories that might arise sometime down the line—in, say, fifty or one hundred years. (If they did clash, that would mean that the devices based on them would have been tricks, or worse yet, could never have been invented or worked at all.) The theories that will arise tomorrow may be both broader and deeper and thus "truer" than today's theories, but even so, they will not *invalidate* today's theories. Today's theories will never be shown to be *false*; they will just come to be seen as less accurate approximations than the newer theories.

What all this teaches us is that there can be *good*, *better*, and *even better* scientific descriptions of reality. All these "rival" descriptions can possess diverse levels of accuracy and quality, but they all will have their usefulness and their range of applicability. They are therefore more friends than they are rivals, since one of them will take over in circumstances where the other loses some of its accuracy. This is how science is structured.

How does the scientific method work? To explain this, I will have to speak about two ways of drawing conclusions from known facts: deduction and induction.

Deductive Processes of Inference

The act of doing science includes rules for logical reasoning—that is, how to validly draw conclusions from premises. These rules have a name: *rules of inference*. Their purpose is to *preserve truth*. This means that if we apply a rule of inference to one or more premises (or "inputs") that are true, then we can be assured that the "output" (that is, the conclusion) will also be true. This is the key defining property of any rule of inference: its conclusion *must* be true, as long as the premises are true.

There are numerous valid rules of inference, and here I'll give a few examples of them. The first rule is called *modus ponens*, and it works as shown in the following:

> *Premise 1: If a burner on a stove is set to "high," then the water in a pot placed on the burner will start to boil after a while.*
> *Premise 2: The front-right burner on my stove is set to "high."*
> *Conclusion: The water in a pot placed on my stove's front-right burner will start to boil after a while.*

This conclusion follows logically from the two premises, and so it has to be true, as long as the two premises are true.

The second inference rule I'll exhibit is called *modus tollens*, and here's how it works:

> *Premise 1: If a burner on a stove is set to "high," then the water in a pot placed on the burner will start to boil after a while.*
> *Premise 2: The water in a pot that I placed on my stove's front-right burner didn't ever start to boil.*
> *Conclusion: The front-right burner on my stove wasn't set to "high."*

Just as before, this conclusion follows logically from the two premises, and so it has to be true, as long as the two premises are true.

The first rule of inference is used to *confirm* something: If A implies B, and we know A, then we know B as well. The second rule of inference is used to *refute* something: If A implies B and we know that B isn't the case, then we conclude that A cannot be the case.

The third rule of inference I'll show is one that is called *instantiation*. Here is how you can schematize it:

Premise 1: All X's have property P.
Premise 2: A is an X.
Conclusion: A has property P.

Suppose, for example, that we know that all ripe strawberries (that is, all X's) have the property of redness. Suppose also that we know that a certain object A is a ripe strawberry. Then we can conclude with absolute certainty that object A is red.

Later, we'll see how these rules of inference are used in science. It's important to remember that a conclusion reached with a rule of inference doesn't have to agree with reality—after all, the premises themselves might not agree with reality! The following is such an example:

Premise 1: All dogs have eight legs.
Premise 2: Fido is a dog.
Conclusion: Fido has eight legs.

We've used *modus ponens* correctly, and the conclusion follows with ironclad logic from the premises; however, the conclusion just happens not to be true. This is because the first premise is not true. But logic is not concerned with the truth of the premises; it is concerned solely with the rules of inference being applied in the proper way—and here the *modus ponens* rule *was* correctly applied. We can safely say that *had the two premises both been true*, then the conclusion reached via *modus ponens* would also have been true.

Another classical rule of inference is known as *tertium non datur* (sometimes called the "rule of the excluded middle"). It states the obvious fact that a statement and its negation cannot *both* be true. It can't be the case, for instance, that my king *is* in check and also that it is *not* in check. That doesn't make sense. The earth cannot simultaneously be both *round* and *non-round* (i.e., flat). That also makes no sense. Similarly, the Loch Ness monster can't both exist and *not* exist at the same time! How much sense would that make? Not too much.

As we will see, this rule of inference plays a significant role in reasoning about whether God exists, as well as in many other ideas about the nature of reality.

Inductive Processes of Inference

Earlier I described rules of inference that are central to the deductive manner of reaching conclusions. But there are also *inductive* ways of deriving conclusions validly. This involves moving from a statement about a single case (or a few cases) to a much wider generalization. The following premise is an example: *Every time I release a stone from my hand, it falls to the ground.* This is an experience I've had when I opened my hand that was holding a stone. Using induction, I draw this more general conclusion: *Stones fall to the ground whenever one lets go of them.*

In contrast to deductive rules, an inductive rule isn't guaranteed to lead from true premises to true conclusions. The premises can be true, and yet the conclusion can be false. A classic example due to Bertrand Russell runs as follows:

Imagine a turkey in the barnyard that gets fed every day at noon. At first it is skittish and suspicious. But after sufficiently many repetitions, the turkey starts to get used to the idea. It notices that every day when the church bells ring (which they do only once a day, exactly at noon), it gets fed. So the premise is: "So far, whenever the church bells have rung, I've gotten food." The inductive conclusion is a leap to greater generality: "Every time the church bells ring, I will get food." After a while, the turkey starts going right up to the farmer every time the church bells ring. The day before Thanksgiving, however, things don't go exactly according to the turkey's expectations. In fact, when the church bells ring at noon, the turkey gets picked up right then and there and carried off to be slaughtered. As the turkey learns (though for a regretfully short time), induction is not always a reliable way to draw conclusions. But the conclusion is often very believable, as was the case with the falling stones.

It's important to keep in mind that scientific conclusions are always, in principle, tentative. Even if the premises in an inductive inference are true, the conclusion can turn out false, as I've just pointed out with the turkey. The conclusion has to be adjusted when new facts arrive. We can *never*, in a strict sense, show that *every* stone will fall to the ground when we release it, but we can consider the conclusion to be very likely. We have lots of experience under our belt, so we can assume without worrying that the *stone-dropping-to-the-ground* scenario will always take place. The fact that it is not a totally watertight conclusion reached by perfect, ironclad logic doesn't mean that one should feel skeptical about it.

Scientific research is often based on a theoretically formulated set of testable hypotheses, and it then proceeds from these hypotheses, using both inductive and deductive rules. Sometimes this way of proceeding is referred to as "the hypothetico-deductive method."

Conclusions reached via *deduction* are always true, as long as the premises are true and the deductive steps are carried out correctly. But when we are seeking to understand the world more deeply by doing science, we often use *inductive* rules of inference. Since the results of induction cannot be guaranteed to be true, we can't be sure that all our conclusions, even though they were all reached by the hypothetico-deductive method, will always jibe exactly with reality.

A simple way of describing science is this: We get an idea about how things are connected, or how they work. We make a theory about their behavior. Then we generate a set of testable hypotheses. (If a hypothesis is to have any explanatory value, it has to be testable in some fashion or other.) We then start to investigate what happens in reality. Do the results of experiments agree with what we expected? Either our hypotheses are confirmed, or they are refuted.

This is often an iterative, or repeated, process. One goes back and tests one's ideas again, refines them, and tests them again. The hypothetico-deductive method can be summarized as follows:

1. Start out with a given theory, or formulate a new theory.
2. Devise one or more testable hypotheses.
3. Test these hypotheses through observation or experiment.
4. Depending on the results of the tests, the theory's believability will be either strengthened or weakened.

What role does induction play in all this? We usually use induction in step 1, when we formulate a theory. For example, if all the ravens we've so far seen are black, then we could conjecture, on this basis, that all ravens in the world are black.

Case Study: Louis Pasteur and Spontaneous Generation

A classic application of the hypothetico-deductive method is an experiment that was carried out in the middle of the nineteenth century by the great French chemist and biologist Louis Pasteur (1822–1895). He devised this experiment to test the then-popular theory of "spontaneous generation."

People have always been puzzled by the mystery of the origin of life. How did life on earth come to exist? How can inanimate matter become living? Aristotle had the idea of "spontaneous generation," meaning that life could just spring up on its own, and for many centuries this was the reigning theory. Fruit flies swarming in a compost pile and worms growing in rotting flesh were thought to be evidence of the spontaneous generation of small animals in a suitable environment. At least some thinkers argued that this was what was happening.

Pasteur wasn't so sure. He placed two pieces of meat on a table. Then he moved one of them into a closed drawer, leaving the other one out in the open air, exposed to the environment. To test the theory of spontaneous generation, Pasteur created a testable hypothesis: if spontaneous generation is true, then worms should soon come to exist in both pieces of meat. What actually happened was that in the piece of meat that was left out in the open, worms appeared after a few days. The other piece of meat, in the sealed drawer, remained worm free. Pasteur concluded that his experiment had shown the theory of spontaneous generation to be wrong. He applied the deductive rule of inference *modus tollens* as follows:

> *Premise 1: If spontaneous generation is true, then worms should turn up in both pieces of meat.*
> *Premise 2: No worms turned up in one of the two pieces of meat.*
> *Conclusion: The theory of spontaneous generation is not true.*

For a serious researcher, one single experiment is not sufficient to draw a strong conclusion. Pasteur thus devised a series of new conjectures and tested them with new experiments. Eventually, his experiments with the pieces of meat led him to reject Aristotle's theory of spontaneous generation.

Pasteur later formulated a theory of his own, which was that bacteria are floating in the air, and that these bacteria cause putrefaction. This was the first step in our current understanding of bacteria and infections, and the theory has major practical implications—both for how we handle food today, and for how we try to avoid letting wounds get infected. Pasteur's subsequent scientific research was also of great import; for example, he was able to immunize a young child who had come down with rabies. This was a historical first in medicine.

This is an example that lies at the very heart of science's quest for knowledge. Science tries to give us accurate descriptions of reality. Theories that fail to pass tests are dropped. Theories that *do* pass tests live longer but have to be put to further tests. This, at least, is how science *ideally* functions.

When is it reasonable to conclude that a theory describes reality to an adequate degree? The answer is, it's when the theory can explain one or more phenomena better than earlier theories, and when repeated tests give the same results over and over again. Assuming that there aren't other tests that cast doubt on the given theory, the theory is gradually accepted by the scientific community. If the theory can furthermore be applied to the building of new pieces of experimental apparatus and the devising of new technical devices, then one can say that the theory has demonstrated its viability.

Of course there is a certain resistance in science to the acceptance of new theories. The more pathbreaking and surprising a theory is, the more evidence is required before the scientific world will accept it. This is why, for instance, Albert Einstein's radical suggestion, in 1905, of light as being composed of particles was treated with such disdain—it was just too radical, and there wasn't sufficient experimental evidence for it. This was a somewhat sad episode in the history of ideas, but fortunately, it was quite exceptional. Indeed, we should in general be very relieved that scientists are not just instant pushovers for any new idea that comes along. Science would be nothing but a huge random mishmosh of true and false teachings if its practitioners were to warmly welcome any old new idea willy-nilly into the fold without seriously vetting it first.

Experience and experimentation are central in the practice of science. But can we always rely on experiments? Maybe the experimenter forgot about some factor, or maybe the desire to find a certain result affected the experimenter's objectivity? It's perfectly reasonable to be dubious about just one isolated experimental result, especially if it's very different from previous results.

It is therefore a key principle that tests and experiments should be repeatable. Other researchers throughout the world should be able to repeat a given experiment in their own labs, and of course they should get the same result. If a theory holds water, then it should work no matter who is testing it. And if a theory is widely accepted, that means that no one—so far—has been able to find any holes in it. For this reason, it seems reasonable to assume that the theory is a good description of reality, which can be applied in diverse ways, at least until another description of reality comes along that is even better.

Falsifiability

Viennese philosopher Karl Popper (1902–1994) was among the twentieth century's most famous philosophers of science. Popper introduced the term "falsifiability" and claimed that this notion, in actuality, is what lies at the core of scientific progress.

For a theory to be falsifiable means that there must some way to devise an experiment or a situation that could, in principle, show that the theory is wrong. Otherwise put, there must be some way to state what would suffice to definitively prove the theory wrong. If this can't be done, then the theory lacks scientific content and has no scientific worth.

In this way, falsification is an example of the deductive rule of inference *modus tollens*. We have a theory A that gives rise to an experiment that, if the theory is correct, should give result B. We carry out the experiment, and it turns out that it doesn't give result B but, instead, gives not-B. Then we can deductively conclude that not-A is the case, which is to say, we have falsified theory A.

According to Popper, this idea is the true core of science. No scientific theory can ever be considered *definitively* true; at best, it is *tentatively* true, waiting until the day that it is falsified. As he puts it:

> Progress consisted in moving towards theories which tell us more and more—theories of ever greater content. But the more a theory says, the more it excludes or forbids, and the greater are the opportunities for falsifying it. So a theory with greater content is one which can be more severely tested.[5]

The concept of falsifiability can also be illustrated by the well-known game of Twenty Questions. One poses a series of yes/no questions—twenty at most—to try to figure out what sort of object (or animal or person) another person is thinking of. Of course one chooses yes/no questions whose answers provide as much information as possible. Thus ideally, each new question cleaves the set of possible objects (or animals or people) into two nearly equal-sized groups—one group corresponding to a "yes" answer and the other to a "no" answer. This is the most efficient way to try to figure out the answer to the riddle. (One doesn't ever ask a question to which one almost surely knows the answer beforehand, since that would simply be wasting one of the twenty allowed questions.)

Doing an experiment to test a scientific theory is like posing a question about the nature of reality. If there is no possible way to devise a question to which the answer might be "no," then the theory is not a scientific theory.

A theory that only leads to "yes" answers, no matter what experimental question is being asked, is not falsifiable, and thus it tells us nothing at all about the world. The goal of science is to say things about the world, and a "yes"-only theory is thus not scientific but merely pseudoscientific.[6]

Falsifiability is thus a way to distinguish science from pseudoscience. In his autobiography, Popper describes how, before he became a teacher, he was pondering the nature of science:

> Early during this period I developed further my ideas about the demarcation between scientific theories (like Einstein's) and pseudoscientific theories (like Marx's, Freud's, and Adler's). It became clear to me that what made a theory, or a statement, scientific was its power to rule out, or exclude, the occurrence of some possible events—to prohibit, or forbid, the occurrence of these events. Thus the more a theory forbids, the more it tells us.[7]

If you offer a scientific theory, then you have to be able to say under what sorts of circumstances you would concede that your theory is untenable. This is Popper's famous "demarcation criterion" for distinguishing science from pseudoscience. (More generally, the "demarcation problem" in the philosophy of science concerns how and where one can draw a line defining what counts as genuine science. Such lines can be drawn between science and *pseudoscience*, or between science and *religion*, or between science and *superstition*.)

Mathematics and Its Absolute Truths

Do there exist any scientific findings that are *absolute* truths—facts that can never be overturned when new discoveries come to light? Yes, but only in one domain: mathematics (and related topics like logic). The world of math is truly an *a priori* science. (Knowledge that is *a priori*—Latin for "from the outset"—is knowledge whose validity doesn't depend on sensory experiences or on contact with the physical world.)

A conclusion that we can reach *a priori* is one that we can draw without ever leaving the armchair in which we're so comfortably ensconced. In other words, we don't need to poke about in the world, making this or that observation, in order to draw such a conclusion. Mathematics does not depend on any type of external observations. All conclusions are reached via logical derivations. (Or at least from the published articles of math proofs, that's how it would seem. In real life, though, mathematicians almost always first reach their conclusions intuitively, via analogies that skip over logic. Mathematicians use just as much guesswork as the rest of us do, but then they slyly cover their tracks by writing articles in which no traces of the original guesswork are visible, everything being couched in a purely logical language, a bit as if the author were a robot rather than a human being.) What this implies is that a mathematical theorem, once it has been rigorously proven, can never be invalidated by any kind of external observation, measurement, or experience. (As was mentioned earlier, mathematical theorems are statements that are derived via purely logical steps from axioms. A set of formal symbols, plus a set of axioms, plus a set of rules of inference constitute a *formal system* or *axiomatic system*. A theorem that is derived inside such a system is always true inside that system.)

Of course there should be some degree of humility even among mathematicians. No less than other beings, mathematicians can make errors. Even if no experiment can reverse a mathematical result, one can nonetheless have made logical errors in one's (alleged) proof of a theorem. Mathematics thus has a special status in the world of science. If one has not committed any errors in one's derivation of a theorem, then the theorem is enshrined forever in the Museum of Theorems, so to speak. (Caveat: the proven theorem is enshrined only in the Museum belonging to the formal system one has used. There can be *other* Museums for *other* formal systems in which the given theorem doesn't appear at all. In fact, in 1931, the young Austrian logician Kurt Gödel showed that, in any given axiomatic system of sufficient complexity, there are certain statements that one would definitely *want* to be theorems, because they express *truths*, but that have no proof inside the given system. In other words, there is a serious gulf between what is *true* and what is *provable* in a given axiomatic system. Gödel showed that there are infinitely many *true* but *unprovable* statements in all sufficiently rich axiomatic systems. This astonishing result, which took decades to be absorbed by the mathematical world, goes by the name of "Gödel's incompleteness theorem.")

Blind Experiments and Placebos

How can one avoid making mistakes when exploring the world scientifically? Of course, there is no magical recipe for avoiding mistakes. Sloppiness, wishful thinking, and the quest for glory and fame have led many a researcher astray, down through the centuries. But despite human fallibility, science as a *collective* human activity has certain built-in properties that serve as excellent devices to keep one honest and on the straight and narrow path. (Of course, this takes for granted that the scientist in question is an honest person.) Among the most important of these methods, especially important in the realm of medical science, are what are known as *blind* and *double-blind* tests.

The goal of such tests is to eliminate any unconscious biases or wishful thinking on the part of the researcher and, whenever human subjects are involved, to eliminate such tendencies on the part of experimental subjects as well. Today we realize that a person's desires and beliefs can have both positive and negative effects in such situations as tests of new types of medicines.

A patient who has insomnia may wish to take a sleeping pill every evening. The patient can obtain such pills from a doctor, but the doctor might, without telling the patient, give the patient just some sugar pills instead of true sleeping pills. The patient takes these pills and has no problem sleeping through the night. This effect is called the *placebo effect* ("placebo" being Latin for "I will please [you]"). It comes down to the idea that a person's psychological expectations can sometimes have a medical effect that is just as efficacious as a genuine medicine.

The opposite of the placebo effect is called the *nocebo effect*, and this is when having positive expectations leads to a negative result. A good example of this phenomenon shows up in certain African nature-based religions. In a tiny Nigerian hamlet, a gas-driven generator disappears one day from one of the huts. Someone stole it. The village's witch doctor calls a meeting of all the village's residents and announces that he has cast a curse that will take effect immediately on the person who stole the generator. The spell will, so he claims, make the person have trouble eating, have intense stomachaches, and gradually become very sick. A few days later, the generator suddenly reappears. A young boy had felt sick, and then he heard about the spell. He hadn't been able to eat or to drink normally, and in the end, he brought back the generator. This is a clear example of the nocebo effect. The belief in magical powers causes the symptoms of the spell to appear. Both the placebo effect and the nocebo effect have been carefully researched and documented.

If we wish to investigate the effect of a new medicine, we have to avoid the placebo effect by using what is called a *blind test*. What this means is that one group of subjects receives the actual medicine, while another group of subjects receives a preparation that *looks* exactly like the actual medicine but it is actually just a sugar pill that doesn't contain any genuine medicine at all. Neither of the groups knows which one is getting the actual medicine and which one is getting the sugar pills. In other words, the subjects are "blind" with respect to this piece of knowledge. Only under such controlled circumstances can the test show whether the medicine actually has an effect, and that it isn't merely thanks to a placebo effect or some other kind of lucky interference.

In some tests of medicines, it often turns out that the group that received sugar pills shows a strong positive effect. In this case, it is clearly a placebo effect. Hopefully, the group that received the *genuine* medicine will show a *stronger* positive effect. The difference between the two positive effects should thus show how powerful the genuine medicine is. If there is no difference at all between the two positive effects, then the medicine is of no value.

What if the researcher has some biases or engages in wishful thinking? Can't that have an effect on the results of a blind experiment? Of course! It's for this reason that people have devised *double-blind* experiments. In such situations, neither the researchers nor the subjects know in advance which group received the true medicine, and which one got the sugar pills. Only when the statistical analysis of the experiment has been carried out can it be revealed which group received what.

Blind and double-blind experiments are crucial to scientific research that involves effects on human beings. These ideas are particularly important in dealing with the plethora of pseudoscientific treatments and medications that are so rife today, especially in New Age circles and in practitioners of alternative medicine. These kinds of experiments, and fake medicines, unfortunately lure and deceive many people; the victims wind up getting tricked not only monetarily but also emotionally. The methods and medicines that are usually associated with New Age medicine (and other alternative types of medicine) almost always fall short when they are properly and seriously tested in a blind or double-blind manner.

And so, what should one call an alternative medicine that actually *does* work? Well, it is then a genuine *medicine*.

The Darker Side of Science

There are times when we encounter representatives of science who exhibit great rigidity, perhaps most commonly when we deal with physicians. We've all had the experience of going to the doctor and feeling that we weren't being listened to carefully. Whatever's running through the doctor's head would seem to be far more important than whatever our little souls might be trying to express.

If you feel frustrated about this haughty attitude on the part of your doctor, then keep in mind that it's not *science* that's at fault here, but just this particular *representative* of science, whose mind was made up in advance and who therefore doesn't feel any need to listen to you.

The fact that someone is a practitioner of science is not any surefire guarantee that the person will not cheat. Obviously the culture of scientists, just like any other human culture, is filled with people who have their own emotional needs and psychological agendas. Thus there are researchers who publish falsified results simply to gain fame and glory.

There are also researchers who persistently balk at accepting new facts and results, because were they to accept them, they would have to abandon beliefs they had long had, and on which their careers were built. There can also be financial interests behind the scenes, which lead people to do their best to cast doubt on new results, since if the results were accepted, they would hurt the sales of a long-established medication.

An example of this concerns research on ulcers. In 2005, Barry J. Marshall and J. Robin Warren were awarded the Nobel Prize in Medicine for their discovery of a bacterium that was given the name *Helicobacter pylori*. (This bacterium was first discovered in 1982 in cultures taken from patients suffering from chronic inflammation of the stomach.) For many years, ulcers were seen as the result of stress and bad nutrition. But Marshall and Warren had a different theory, which was that ulcers were caused by a bacterium. After some time, Marshall was able to prove his theory. He decided that he himself would drink a *Helicobacter* bacterial culture, and soon thereafter he was stricken by a severe stomach inflammation. The causal link was confirmed, and Marshall and Warren then showed that patients with ulcers can be permanently cured with the help of antibiotics. Before the connection between bacterial infection and ulcers had been confirmed, however, the illness was often chronic, and patients often suffered relapses.

At first, Marshall and Warren's discovery met nothing but skepticism. Very significant financial profits were at stake. Since the pharmaceutical industry earned a great deal of money from various ulcer medications, it had no interest in having the reigning theories of ulcers cast in doubt. After all, it's always better for the pharmaceutical industry when patients suffer frequent relapses, rather than winding up being cured for once and for all. Marshall and Warren's research results were held back from publication at least in part because of the high financial stakes if their findings were taken seriously.

As a general rule, the quest for scientific knowledge is the most successful of all approaches in the end, but it can take much more time than it ought to. Attempts to hold back progress because of vanity, money, or stubbornness don't generally succeed for very long. Sometimes scientific researchers are accused of patting each other on the back and of hiding each other's mistakes. But in truth, the world of scientific research could scarcely be described as protective of unsuccessful theories. Any researcher who succeeds in blowing an established theory to bits will probably be both rich and famous—and may even win a Nobel Prize. That's a pretty strong motivation to try to falsify theories.

Are Religion and Science Mutually Compatible?

A question that recurs time and again in public discussions is whether science and religion are compatible. This question is often turned into an excuse for lengthy and supposedly profound meditations.

But in truth, the question is rather easy to answer: Religion and science are compatible just so long as the religion in question doesn't make any claims that run against scientific knowledge. More simply put: As long as religion doesn't say anything about the world, then there's no conflict. Take, for example, the question about belief in a creator who made the universe out of nothing, and thereafter played no role in it. A belief in this kind of god is, in practice, quite meaningless. It doesn't tell us anything about why we are here, where we are headed, how we should live, or the nature of reality.

On the other hand, if one believes in a god that has various desires and powers, things get messier. A god that listens to prayers and carries out miracles is definitely *not* compatible with science. Nor is the notion that the soul survives after the body dies, nor is the idea of reincarnation, or other similar supernatural views. It's not that these kinds of ideas, in and of themselves, are *necessarily* wrong; it's just that science's findings unambiguously indicate that they are wrong.

Another frequent question is whether theology can in some way contribute to science these days or, conversely, whether science can contribute to theology. To the first question I can't see any possible answer other than "no." Whether or not its claims are correct, theology is all about questions that lie beyond the domain of science (except for those theological claims that directly conflict with science). Science, however, can help theology abandon certain preconceived notions, such as the thesis that humans stand apart from, and are not an outcome of, natural evolution.

Has religion ever done anything useful for science? Well, yes—if one looks at the history of science, one has to say that it has. The notion that the world is comprehensible has historically been connected with the idea that there is an intelligent creator. If one *didn't* believe in the world's comprehensibility, then it would be pointless for one to devote oneself to science. In this sense, religion paved the way for science's progress. This fact does not, however, increase the likelihood of the existence of a god or an intelligent creator; but the idea of comprehensibility helped science make its first inroads.

~

INTERLUDE: ON THE USEFULNESS OF SCIENCE

Can one believe too strongly in science? After all, history has taught us that science is not always right. It is easy to forget that science is not simply the business of producing huge batches of truths about the world. Science is far more than that; science is a way of investigating reality—a way of forging models, pictures, and theories about how the universe works. Science's pictures of reality often have to be revised, and we don't know any better way to attain knowledge of reality than by following the scientific path. In that sense, it's not possible to believe too strongly in science.

What we call science should actually be judged on the basis of how effective its findings are, when it gives us tools and medicines to defeat diseases like cancer or cholera, or when it gives us clever ways to overcome the familiar constraints of nature, such as gravity, allowing us to send humans to the moon, or perhaps soon to Mars.

Is science always a force for good? Well, consider the fact that nitroglycerin—the very same thing that can save the life of a patient with a weak heart—can also blow people to bits. And so it is with all knowledge. Whether science is moral or not comes down to how its findings are used. The knowledge in itself is neutral.

What meaning does science have for us in our daily lives? When we hear a news item about a scientific finding, it often concerns something extremely abstract and remote, such as the discovery, in 2012, of the Higgs boson (hardly something that I need to know when I show up at my office), or the landing of the space probe Rosetta on the surface of a comet in 2014, some ten years after its launching. But we shouldn't lose sight of the fact that the 2014 Nobel Prize in Physics was given for the development of LED lights. This outcome of many years of research and development improved the lives of hundreds of millions of people all around the globe.

For me personally, science has almost certainly extended my life expectancy and given me the chance to have many more experiences. I have quite decent chances of living twice as long as my ancestors did, just a few hundred years ago. I can spend twice as much time with my precious loved ones. I can watch my son as he grows up, reads books, plays games, and gains bits and pieces of knowledge. And it's thanks to science that I have all this good luck. I am constantly thankful to science for these astonishing gifts.

CHAPTER FIVE

GHOSTS IN THE HEAD

Concerning Our Wonderful but Easily Fooled Brains

When I think, it's not me who's thinking, but my brain that's thinking.—
Four-year-old Leo Sturmark

Why do we need to follow the methods of science in order to learn about the world? Aren't our wonderful brains good enough by themselves, without imposing some kind of formal discipline on them?

The human brain is often described by scientists as "the most complex phenomenon that we know in the universe." And indeed, the capabilities of our brains are vast and truly stunning. But even so, a perfectly healthy brain can go off track in all sorts of ways.

Our Brains Pull the Wool over Our Eyes

The human brain has quite a few salient shortcomings, some or all of which may well be repaired by evolution throughout the next million years or so. But for the time being, we'll just have to put up with them.

Optical Illusions

The simplest example of what I mean is optical illusions. We can easily be convinced that two lines of exactly the same length have very different lengths. Look, for example, at the two tables shown in figure 5.1 (this illusion is just one of many designed by Stanford University's brilliant psychology professor Roger Shepard in the early 1980s). Would you say that the diagonal straight line constituting the piano bench's left side is the same length as the horizontal line constituting the squarish table's far edge? Probably not. The former is far longer, isn't it?

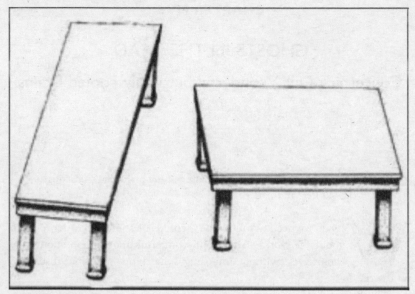

Figure 5.1.　Two Tables

Well, actually, no. The two line segments are exactly the same length. Your eye correctly tells you that if these two pieces of furniture were sitting side by side in a room, the left one (the piano bench) would be far longer than the right one (the squarish table) is wide, but your brain then converts that correct piece of information into the false idea that the two lines *in the image* are of vastly different lengths. The false inference is deeply compelling, isn't it? Every time I look at this amazing image, I *still* can't believe it, and I have to actually *measure* them both. And then I simply gasp at the trick my brain has played on me.

For all these reasons, we need *tools* to help us understand the world as it really is—tools that are more reliable than our unaided senses and our gullible brains.

Synesthesia

On occasion, our brains put together pieces of information and draw wonderfully creative conclusions from them that have no rational basis. A classic case is the drawing of two shapes called "Kiki" and "Muma." Take a look at the two shapes in figure 5.2 and decide which one is "Kiki" and which one is "Muma."

Figure 5.2. What are these figures called?

Virtually everyone who looks at these two shapes is convinced that the one on the bottom is called "Muma" and the one on the top is called "Kiki." Now what possible reason would there be for this? The pictures don't show anything that you deal with in everyday life, and the two names are merely nonsense words. So what on earth would lead you to the confident conclusion that one of the words has to be the name for one of the pictures, and the other for the other?

There is no rational reason. It simply comes down to the fact that we all hear "Muma" as a softer and rounder sound than "Kiki," which tends to sound pricklier and more angular. And we tend to see the image on the bottom as soft and round, while the one on the top looks pricklier and more angular.

This kind of connection between sounds and shapes is a form of *synesthesia*, which means a conflation of sensations coming from different sensory organs. Another example of synesthesia is the feeling many people have that certain digits are associated with certain colors. Thus the digit "7" might seem blue to you, while "3" might seem yellow, and so forth.

A well-known example of a person with synesthesia is Daniel Tammet, who can recite more than twenty thousand digits of π (3.14159265358979323846264 . . .) by heart. In his book *Born on a Blue Day* (2006), he explains that to memorize so many digits, he didn't think of the digits themselves; rather, he saw them in his mind's eye, as a series of colors and undulating landscapes.

Our brains play many such tricks on us, without our being in the least aware of them. And illusions don't always involve sensory perception. Our poor brains are susceptible in many different ways. This is one reason, among many, that it is crucial that we have developed scientific procedures and double-blind tests.

The Anchoring Effect

Israeli American psychologist Daniel Kahneman has investigated how we make decisions and how we can be influenced in our decision-making processes.[1] One phenomenon that he has identified and studied is the so-called *anchoring effect*. The idea is that whenever we try to estimate a numerical value of any sort, we can be unconsciously biased by numbers that we have just seen, whether or not they have anything to do with the question at hand.

One of Kahneman's examples that bring this bias out in the open is the classic "wheel-of-fortune experiment."[2] In this experiment, a subject is asked a question having a numerical answer between 1 and 100, such as "What percentage of the world's nations are located in Africa?" (It's crucial that the subject should not know the answer and, therefore, can only guess at it.) Before subjects receive the question, they are first asked if they would like to do a little gambling using a lottery wheel. (Most say yes.) Such wheels can normally stop on any number at all between 1 and 100, but this wheel happens to be rigged in such a way that it stops only on 16 or on 45. The experiment is done many times with many different subjects (say, one thousand). This means that roughly five hundred subjects spin the wheel and get a "16" before answering the numerical question; the other subjects, also roughly five hundred in number, get a "45" before answering.

It turns out that on the average, the subjects who got the "16" make lower guesses at the percentage of the world's countries located in Africa than those who got the "45." This bias occurs even though the lottery wheel obviously has nothing whatsoever to do with the question they received.

The subjects' estimates are thus unconsciously biased by the number they saw just before making their guess. This shows how easily our brains can be fooled, and how they don't by any means always make well-grounded guesses. It's important to be aware of such vulnerabilities when we are following our intuitions or our "gut feelings." This kind of unconscious bias is particularly troublesome in court cases or other situations in which numerical estimates by witnesses can have major consequences for the parties concerned.

Bloated Self-Images

We humans also suffer from a tendency to overestimate ourselves and to have overly positive self-images. A clear example of this is an experiment in which a subject is shown one hundred small photos of randomly chosen faces scattered here and there on a computer screen.[3] One of the many photos on the screen "just happens" to be of the actual subject of the experiment. The experimenters then measure the time it takes for the subject to find and point out their own face among the many on the screen. The experiment is then repeated a large number of times, each time with the faces scrambled, so that they appear in different locations on the screen.

What subjects do not know, however, is that after the first run or two, their own photo gets replaced by a manipulated photo, which makes the subject look more attractive (at least by conventional standards) than they actually are. It turns out that subjects tend to locate their *enhanced* portrait among the random faces faster than they locate their *unenhanced* portrait.

What this tells us about our self-image is, needless to say, somewhat troubling. Perhaps it once was useful for people to have a slightly bloated sense of their own appearance; perhaps people who had that sort of self-image would tend to survive better. But is this a good thing for us today? It's often said that optimists are happier than realists, but we should nonetheless be aware of this bias on the part of our brains.

Focus and Blind Spots

We are also not quite so observant as we might wish. When I give talks about philosophy and tools for clear thinking, I often show a short video taken from a famous experiment in which subjects watch a video of six people walking about and tossing basketballs back and forth, and they are supposed to count how many passes are made by the three players who are wearing white. The result is very surprising. The estimates of how many passes take place varies greatly among people who watch the video clip. But much more striking effects occur as well. I don't wish to reveal too much; instead, I strongly recommend that you check out the video yourself,[4] and that you read further only after having done so. Just try to count accurately how many passes are made by the three players in white.

Many people who watch this video are oblivious to other interesting things that occur in it at the same time as the ball passings, and they are quite amazed when they watch the video a second time around.

The Likelihood of a Zombie Virus

Another example of our brains' deficiencies is how poorly most of us gauge risks and probabilities. Our brains are not particularly well constructed for such judgments, and this fact can give rise to very serious misjudgments. Test your own ability by carrying out the following thought experiment:

1. Right now there is a zombie virus that is rapidly spreading throughout the country. It's known that one out of five hundred people will catch the virus and become infected. All these people will turn into zombies within a month, which of course is not so great for them.
2. There is a test that one can take at the doctor's office to check if one is carrying the virus. But the test isn't totally reliable.
3. It has been established that for anyone who *has* the virus, the test is 100 percent accurate. In other words, anyone who has the virus will learn from the test that they are infected.
4. It has also been established that the test is only 95 percent accurate for those who are *not* carrying the virus. This implies that 5 percent of all healthy people who are tested will have a false positive, meaning that they will be told they are infected, even though they aren't.

Now you don't want to become a zombie. Therefore, you are very anxious to find out if you're infected or not, so you go to your doctor and get tested. After a few days, you receive the result: the test says that you are carrying the virus!

Of course you're very upset with this result. But you know that the test isn't totally reliable, so now comes the question: How great are the chances that you really *are* carrying the virus? I'm not looking for an exact answer, mind you—just a rough estimate. Are your chances of being infected greater than or less than 50 percent? Think about this for a little while before you check out the answer.[5]

Three Doors and One Car

We can also fall for probability illusions that clash violently with our intuitions. The following is a classic example. You're a contestant on stage in a TV quiz show, and a big curtain is pulled back, revealing three large closed doors. The host tells you that behind *one* of the doors there is a brand-new sports car. If you open *that* door, you'll get the car! But behind the other two doors are *goats*, and that's what you'll get if you open either of those ones. You'd love to win the car, but you're not all that excited about taking a goat home.

The host knows what's behind each of the three doors. The problem, of course, is that *you* don't know. Now the rules of the game are simple:

1. You are first to pick a door and then to go stand right next to it, but *not to open it*. As soon as you've made your choice, the host steps forward and opens a different door, and behind that door you see a goat.

    So now you are standing in front of one of just *two* closed doors, with the third door being open and a goat conspicuously standing behind it. The host now offers you a new option that you can take or reject:

2. Would you now like to switch to the *other* closed door, or would you prefer to simply remain with your *original* choice of door?

What do you think? Will the chances of winning the car *increase* if you make the switch, or will they stay exactly the same? Think this over for a moment before reading any further.

Most people say that it doesn't matter one hoot whether you switch doors at this stage or not. When, at the outset, you chose a door, you had a one-in-three chance of winning the car. And now you have a new opportunity to choose one of just *two* doors. It would seem obvious that there is a 50 percent chance that you are right now standing at the "good" door, and a 50 percent chance that you are standing at the "bad" door. So what earthly purpose would it serve to switch doors at this point? None whatsoever, right?

The problem, however, is that this reasoning, although seemingly perfect, is dead wrong. In truth, making the switch that you have just been offered would *double* your chances of winning the car: they would jump from *one* in three to *two* in three! This claim plays havoc with our intuition, and yet it is completely correct.[6]

Let us pause for a moment to put some perspective on what is under discussion. We are actually in quite curious cognitive territory here because it *seems* to us that concerning this door-choosing TV game, we used ironclad reasoning to reach our conclusion (namely, that it doesn't matter in the least if we switch or not—our chance of winning the car won't be increased). Isn't that what this book is all about—using reasoning, not just relying on prejudices or holy scriptures or random guesswork, to reach conclusions? But now, unfortunately, we're being told that our reasoning *wasn't* reasoning but, instead, was faulty! We are being told that our reasoning processes misled us! So doesn't the preceding discussion lead us to *doubt* our reasoning processes, rather than to *believe* in them? And so, isn't this discussion undermining the book's key thesis—namely, that we should always use reasoning to reach conclusions?

Well, whether we like it or not, reaching the truth is sometimes quite tricky, and reasoning itself is certainly not mechanical or trivial. What my discussion of the three-doors game is trying to bring out is that, just as there exist *perceptual* illusions, there also exist *reasoning* illusions. We humans are susceptible not only to being optically tricked, but also to being *logically* tricked—to being led down the garden path by falling for smooth-sounding words and flawless-seeming arguments.

There is a wonderful book by American logician Raymond Smullyan, titled *The Chess Mysteries of Sherlock Holmes*.[7] Many of the "chess mysteries" involve chessboards that Sherlock Holmes and his faithful friend Dr. Watson randomly stumble across in a London chess club that they frequent, while others are encountered in aristocratic country estates or even aboard steamships sailing to exotic destinations. The questions asked are of this sort: "Who moved last—White or Black?" Or "What piece was moved last?" Or perhaps "What piece was taken last?" Such puzzles are called *retrograde chess-analysis puzzles*, meaning that they involve looking at a snapshot of a chess situation and reasoning backward to where this situation could have come from.

Figure 5.3 is a sample problem, as narrated by Dr. Watson.

Holmes was the first to speak. "The first retrograde problem I ever solved was a 'can't castle.'"

Figure 5.3. A Retrograde Chess Problem

"Do you remember what it was?" inquired Sir Reginald with interest.

"Oh yes," replied Holmes, "only I think it is too simple to interest you— a mere bagatelle, you know."

"Why don't you show it to us anyhow? It might be amusing to learn what got you started on these problems, and to test ourselves against your beginning skill."

"Very well." Holmes set up the following position.

"It is Black's move," said Holmes. "Can Black castle?"

Since Holmes had described this problem as "simple," I thought I might have a chance of solving it, and so exerted myself to the utmost to do so. And, I cannot help but be proud to say I was the first to get it. I made a couple of mistakes in exposition, but the mistakes were more in the nature of omissions than of faulty reasoning, hence were not serious. Here is the analysis I gave (with all the gaps filled).

White's last move was clearly with the pawn. Black's last move must have been to capture the White piece which moved before that. This piece would have to have been a knight, since the rooks could not have got out onto the board. Obviously none of the Black pawns captured the knight, and the Black queen's rook couldn't have captured the knight because there is no square that the knight could have moved from to get to that position. Likewise the bishop couldn't have captured it, since the only square the knight could have come from is d6, where it would have been checking the king. Hence either the king or the king's rook has made the capture. So, Black can't castle.[8]

This puzzle may seem hard enough to break anyone's brains, and yet it is among the easiest of all the problems in the book. This shocking realization may make readers of *The Chess Mysteries of Sherlock Holmes* feel daunted, possibly even feeble minded. And to add insult to injury, Smullyan plays masterful tricks on his readers in many problems. What typically happens is this: Watson examines the given situation and speaks very sensibly about it. As we read Watson's analysis and consider his reasoning, we find no flaw whatsoever in it. We are convinced that he has revealed the true answer to the puzzle. And in some of the puzzles (such as the one described earlier), Watson is indeed correct, and that's all there is to it. But on occasion, to our surprise, Sherlock Holmes (as channeled by author Smullyan) very gently points out a subtle flaw in what Watson has just said.

It turns out that Watson has overlooked one unusual possibility, or has not taken into account a certain subtle fact about the board, or has forgotten about some obscure rule of chess. All at once, everything swivels! Watson suddenly realizes, and we readers also realize, that the situation is not as he analyzed it after all, totally persuasive though his reasoning was. And yes, Watson was indeed using reasoning; he certainly wasn't using just random guesswork or following baseless superstitions or holy scriptures. Even so, Watson's reasoning, although tempting—and possibly even utterly convincing—was *flawed* reasoning.

What makes *The Chess Mysteries of Sherlock Holmes* even more provocative is that, on occasion, Watson will good-naturedly bounce back from such a cognitive shock, and then he will reason his way to a totally different conclusion, a subtler one that takes into account the tricky feature that Holmes pointed out—and then Holmes will take Watson's *new* conclusion and once again poke holes in it!

It turns out that Watson has overlooked something yet again, but this time the overlooked idea is even subtler. Smullyan obviously takes enormous delight in tricking his readers, leading them to overlook fatal holes in Watson's reasoning not just once, but two or three times in a row!

Let's take a look at some Smullyanian Sherlockian shenanigans of this sort. The following extract is taken from the chess mystery that Smullyan calls "A Matter of Direction":

Together, Holmes and I sauntered over to the club. It was empty except for two occupants: Colonel Marston, whom we knew fairly well, and a distinguished, intelligent-looking gentleman with a very pleasant and humorous manner.

"Why, Holmes," said Marston, rising from his place at the chessboard, "let me introduce you to a very dear friend of mine, Sir Reginald Owen. We have just finished a most delightfully bizarre and eccentric game. The playing was utterly wild on both sides, though perfectly legal, of course."

"So I see," remarked Holmes, looking at the board.

Figure 5.4 shows the position of the chessboard.

"Why is it, Marston," said Holmes, "that every time I see you at chess, you always seem to be playing White?"

Marston laughed at this, but suddenly his face dropped. "Why, Holmes," he said, "how on earth did you know I was playing White? I'm sure the last move was made some time *before* you or Watson entered the room. So how could you possibly know?"[9]

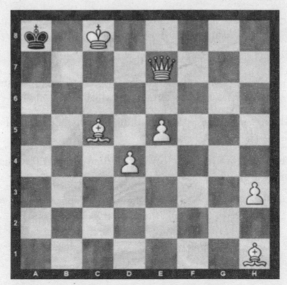

Figure 5.4. Position of the Chessboard

At this point, the story continues with amusing repartee, although the answer is not revealed. We'll thus skip ahead to the evening, when Holmes and Watson have returned to Holmes's domicile at 221b Baker Street. After donning his favorite dressing gown and stoking his well-loved pipe, Holmes turns to Watson and asks this question:

"Have you yet figured out how I knew Marston was playing White?"

"Why, no," I replied. "I used all the methods you have taught me, I examined the whole room thoroughly, but could not discover a single clue!"

At this, Holmes burst into a roar of laughter. "The whole room, Watson, the whole *room!*" Did you examine the rest of the building as well?"

"I never thought of that," I admitted meekly.

"My dear Watson," Holmes said, laughing harder than ever, "I was only jesting, you know. It was hardly necessary to examine the whole building, nor even the whole room, nor the table, nor the players, but merely the chessboard."

"The chessboard? What was peculiar about the chessboard?"

"Why, the position, Watson, the *position*. Don't you recall anything peculiar about the position?"

"Yes, I do recall that at the time I regarded the position as highly unusual, but I cannot see how one can deduce that Marston played White from it!"

At this Sherlock rose. "Let us set up the position again . . ."

"Now there," said Holmes after reconstructing the setup of the afternoon, "can't you deduce which side is White and which side is Black?"

I looked long and carefully, but could discover no clue at all. "Is this an example of what you call 'retrograde analysis'?" I asked.

"A perfect example," replied Holmes, "albeit it an extremely elementary one. But come now, you see no clue whatsoever?"

"None at all," I said sadly. "Superficially it would appear that White is on the South side. But this is really quite superficial. The game is clearly in the end-game stage, where it is not too uncommon for one of the kings to be driven to the opposite end of the board. So it seems that White could really be on either side."

"There is *nothing* in the situation which arrests your attention, Watson?" asked Holmes despairingly.

I looked again at the board. "Well, Holmes, I suppose there is one feature which would probably arrest anybody's attention, namely that the Black king is now in check from the White bishop. But I can't see that this has any bearing on which side is White."

Holmes smiled triumphantly. "All the bearing in the world, Watson. And here is where retrograde analysis comes in! In retro-analysis, one must delve into the past. Yes, the *past*, Watson! Since Black is now in check, what could have been White's last move?"

I suggest that this is an excellent opportunity for you, dear reader, to examine the board and see what comes to mind. Perhaps it will coincide with what Watson noticed, or perhaps not. In any case, please do give it a try, and then read on.

I looked again at the board and replied, "Why, it could easily have been the White pawn on e5 just having moved from e4 and discovering check from the bishop. This, of course, assumes that White is South. But on the other hand, it could also be that White is North, in which case his last move was the pawn on d4 from d5. I see no basis for deciding between these two possibilities."

"Very good, Watson, but if it is really true, as you have said, that White's last move was with one of the pawns on e5 or d4, then what could have been Black's move immediately before that?"

I looked again and replied, "Obviously by the Black king, since it is the only Black piece on the board. He couldn't have moved from b8 or b7, hence he must have moved out of check from a7."

"Impossible!" cried Holmes. "Had he been on a7, he would have been simultaneously in check from the White queen and the other White bishop on c5. If the queen had moved last to administer check, Black would have already been in check from the bishop. Had the bishop moved last, Black would have already been in check from the queen. Such an impossible check is technically known in retrograde analysis as an 'imaginary check.'"

I thought for a moment, and realized that Holmes was right. "Then," I exclaimed, "the position is simply impossible!"

"Not at all," laughed Holmes. "You have simply not considered all the possibilities."

"Now look, Holmes, you yourself have just proved that Black had no possible last move!"

"I proved nothing of the sort, Watson."

At this point I was getting a bit impatient. "Oh, come now, Holmes, you just proved to my entire satisfaction that the Black king had no possible last move."

"True enough, Watson, I proved that the Black *king* had no possible last move, but this hardly proves that *Black* had no possible last move."

"But," I cried, "the king is the *only* Black piece on the board!"

"The only Black piece on the board *now*," corrected Holmes, "but that does not mean that the king was the only Black piece on the board immediately prior to White's last move!"

"Of course," I replied, "how stupid of me! White on his last move could have *captured* a Black piece. But," I exclaimed, more puzzled than ever, "whichever of the pawns on e5 or d4 moved last made no capture!"

"Which only proves," laughed Holmes, "that your original conjecture that White's last move was with one of those two pawns is simply incorrect."

"Incorrect!" I cried, bewildered. "How can that be?"

Before we continue with this chess mystery, dear reader, let me suggest that you take a little break, and give your best try at figuring out what Holmes means by this. How can Watson's watertight-seeming analysis possibly have been wrong?

Well, now that you've presumably given it some thought, let's go on . . .

Then it dawned on me! "Of course!" I exclaimed triumphantly. "I see the whole thing now! How silly I did not see it all along! White's last move was with the pawn from g2 capturing a Black piece on h3. This capture simultaneously checked the Black king and captured a Black piece, and it was this piece—whatever it was—which made the preceding Black move!"

"Nice try, Watson, but I'm afraid it won't do! If a White pawn had just been on g2, then how on earth could the bishop on h1 ever have gotten to that square?"

Here was a new puzzler! At this point I said, "Really, Holmes, I am now thoroughly convinced that the position is simply impossible!"

"Really now? Well, well! This only affords another example of what I have often remarked: Conviction, no matter how firm, is not always a guarantor of truth."

"But we have exhausted *every* possibility!" I exclaimed.

"All but one, Watson—the *correct* one, as it happens."

"It seems to me that we have really covered every possibility. I am certain that we have *proved* this position to be impossible!"

Holmes' expression grew grave. "Logic," he replied, "is a most delicate—a most fragile—thing. Powerful as it is when used correctly, the least deviation from strict reasoning can produce the most disastrous consequences. You say you can 'prove' the position impossible. I should like you to try to give me a *completely rigorous* proof of this fact. I think that in so doing, you may yourself discover your own fallacy."

"Very well, then," I agreed, "let us review the possibilities one by one. We—or rather you—have certainly proved that neither pawn on d4 or e5 could have moved last. Correct?"

"Absolutely," said Holmes.

"Likewise the pawn on h3?"

"Right," said Holmes.

"Surely the bishop on h1 didn't move last!"

"Right again," said Holmes.

"And certainly the other bishop on c5, and the White queen, could not have moved last. And *surely* the White king didn't move last!"

"I'm completely with you so far," remarked Holmes.

"Well then," I said, "the proof is complete! No White piece could have moved last!"

"Wrong!" exclaimed Holmes triumphantly. "That is a complete *non sequitur*."

"Just a minute now," I cried, a bit beside myself. "I have accounted for *every* White piece on the board!"

"Yes," said Holmes, highly amused at my consternation, "but not for pieces *off* the board."

At this point, I was beginning to doubt my sanity. "Really now, Holmes," I cried in utter desperation, "since White moved last, the piece he just moved must be on the board, since Black has not yet moved to capture it. Pieces don't just move off the board by themselves, you know!"

"Wrong," said Holmes, "and therein lies your whole fallacy!"

At this point, I blinked my eyes and shook myself to convince myself that I was really awake. With the utmost control, I calmly and slowly said, "You honestly mean to tell me, Holmes, that in chess a piece may leave the board without being captured?"

"Yes," replied Holmes. "There is one and only one type of piece which can do that."

"A pawn!" I said, with a profound sigh of relief. "Of course, a pawn on reaching the eighth square promotes. But," I continued, "I do not see how that can help us in the present situation, since the White queen is not now on the eighth square—regardless of which direction White is going."

Holmes replied: "Is there any rule in chess which demands that a pawn, when it promotes, must promote to a queen?"

"No," I replied. "It can promote to a queen, rook, bishop, or knight. But how does that help us here? . . . Hallo," I said, "of course! It may have promoted to the bishop on h1—which of course means that White is North. But how does that leave a last move for Black? Ah, I've got it! The promoting White pawn was on g2 and *captured* a Black piece on h1; this Black piece made the move right before that! So indeed, White must be North!"

"Very good, Watson," said Holmes with a calm smile.

"One thing, though, that bothers me, Holmes: Why on earth should White have promoted to a bishop when he could have had another queen?"

"Watson," Holmes very carefully replied, "that question belongs to psychology and probability, and certainly not to retrograde analysis, which deals not with probabilities, but only with absolute certainties. We never assume that a player has played *well*, but only that he has played *legally*. So however improbable it is that a given move was made, if no other move was possible, then that must have been the move which in fact *was* made. As I have told you many times, when one has eliminated the impossible, then whatever remains, however improbable, must be the truth."[10]

In this chess mystery, Watson learns a lesson in humility that we humans all need to learn, as we become ever more aware of the nature of reasoning— namely, there are mental traps into which one can fall, even when one is convinced that one is reasoning with extreme care. This is a humbling lesson, but it's also a wonderful and central lesson about life. Repeated exposure to all sorts of mental traps is *helpful* to people who want to deal with the world as it really is, and not simply with a wishful-thinking world.

It is important to be exposed to many different sorts of situations in which the wool can be easily pulled over one's eyes without one's suspecting anything at all. After a while, one starts to build up a feeling for where traps for the unsuspecting may lurk. But there are no ironclad guarantees, no surefire recipes, no perfect procedures for correct thinking. Even the most reliably scientific of thinkers will fall into cognitive traps now and then.

Newton made mistakes; Maxwell made mistakes; Einstein made mistakes; Feynman made mistakes; even Raymond Smullyan made mistakes in his dearly beloved subject of retrograde chess analysis (there is in fact one mistaken answer in the book, and when he realized his error, Smullyan was crestfallen, though he soon rebounded with amusement). *Errare humanum est.* This is why I have stated, over and over again in this book, that even science does not yield absolute, permanent truths. And that "weakness" of science is actually the glory and the power of science: it is always willing and able to change its mind, given enough evidence.

Choice Blindness

One of the most astonishing phenomena that have been revealed by psychological experiments is called *choice blindness*, which is often accompanied by a deeply felt need to rationalize one's decision after the fact. In a typical experiment, a questionnaire is given to subjects, in which each question is quite complicated and nuanced. To answer each question, the subject is given several choices from which to pick just one. After the subjects have made all their choices, they are asked to explain why they made these particular choices; however, little do they realize that they have been tricked. What seems to be a carbon copy underneath the actual questionnaire is, in fact, not identical to the copy that they marked up; the answer to one of the questions has actually been changed to the *opposite* of what the subject chose.

Here is a typical kind of statement that a subject might choose as an answer to a question:

> *It is more important in a society to protect citizens'*
> *personal integrity than to develop their welfare.*

If the subject chooses this answer, the "carbon copy" instead shows that they chose the following answer:

> *It is more important in a society to develop citizens'*
> *welfare than to protect their personal integrity.*

Another typical example of an answer might be the following:

> *Even if an action has a chance of hurting someone innocent,*
> *it can be morally defensible to take that action.*

If this is the answer chosen by the subject, the "carbon copy" instead shows that the following answer was chosen:

> *If an action has a chance of hurting someone innocent,*
> *it is morally indefensible to take that action.*

The interesting part of this experiment is when the experimenter and the subject sit down together to talk about the reasons behind the subject's answers. The actual filled-out questionnaire is handed to someone outside the room, who then tears off and hands back the "carbon copy" underneath the original. At this point, the subject is asked to explain the reasoning behind each of their chosen answers. Many subjects completely fail to notice that some of their answers have been altered—and of these people, a large fraction will even try hard to justify an answer that is the *opposite* of the answer they actually chose.

Such experiments naturally lead people to do some serious soul searching. These findings can't help but make us wonder: how often do we first decide in favor of certain values and ideas, and only then, after the fact, try to find a rational basis for them? In the best of all worlds, obviously all value judgments should come out of firm reasoning and convincing arguments. But quite clearly, that's not the kind of world we live in.

Social Thinking and Abstract Thinking

The type of problem solving that our brains are not so good at often involves abilities that were developed late in evolution. As a species, we simply haven't been able to refine those abilities yet. Perhaps in a million years or so, we'll be a lot better in these kinds of thinking, but that won't help us right now.

Evolutionary psychologists believe, for example, that our ability to *think abstractly* arose quite late in our brains' developmental process. Our ability to *think socially*, on the other hand, is considerably older. This is why we often have a hard time in reasoning logically about abstract phenomena but find it quite a bit easier to make judgments about social situations. Here I will present a beautifully conceived experiment that unambiguously demonstrates that we have a rougher time thinking about abstract situations than thinking about social situations.

On a table in front of you there are four cards. Each card has a letter of the alphabet on one side and a number on the other side. And you are told that your goal is to figure out whether all four cards obey the following rule:

> *If there is a "D" on one side, then there will be a "3" on the flip side.*

Here are the cards as you see them on the table:

D F 3 7

We can describe the situation in words, as follows:

> *The first card has the letter "D" on its front side.*
> *The second card has the letter "F" on its front side.*
> *The third card has the number "3" on its front side.*
> *The fourth card has the number "7" on its front side.*

Which card, or cards, should you turn over in order to see if the rule was always followed or not? Take a moment to figure out what you think.

For many intelligent people, this puzzle is quite tricky to solve. But before you give up, please consider an alternate way of posing the same puzzle.[11] You're a police officer who has to check whether the law concerning alcohol consumption by minors is being followed in a certain bar. The law runs as follows:

> *No one under eighteen may drink alcohol in this bar.*

So now you enter the establishment and find four people standing at the bar.

> *a beer drinker*
> *a juice drinker*
> *a woman aged twenty-five*
> *a boy aged sixteen*

Once again, we can translate the situation into words, as follows:

> *The first person is drinking a beer.*
> *The second person is drinking a glass of juice.*
> *The third person you recognize, and you know she is twenty-five years old.*
> *The fourth person you also recognize, and you know he is sixteen years old.*

You thus know what two of the people are drinking, but not their ages; you also know how old the other two are, but not what they're drinking. So the question is: Which person, or people, do you need to examine further, in order to see if the drinking-age law is being faithfully respected by the bar owner?

Do you think this puzzle is simpler than the previous one? Most people are sure it is. The earlier, more abstractly formulated puzzle is felt to be far harder to solve than the puzzle involving a social situation. And yet, these two problems are deeply identical (or "isomorphic," as mathematicians would put it, meaning that they share *exactly the same abstract structure*, when the surface-level details are stripped away). It's just that the first puzzle takes the basic idea and embeds it in an *abstract* context, whereas the second puzzle takes exactly the same basic idea and embeds it in a *social* context.[12]

Since logical and abstract thinking is not as easy for us as social thinking is, there is a good reason for trying to practice and improve one's abilities in abstract thinking. This book is in fact trying to give you some tools for such practicing.

Here is a riddle that deals with a remarkable island called Enlightenment Island:

You are traveling to the Enlightenment Island. There are two families who live there: the Troolies and the Trixters. The Troolies always tell the truth, and the Trixters always lie.

This time you bump into Eve and Dan. They say strikingly little. In fact, Eve says only one sentence, which consists of three words. Based on her statement, you conclude that they both belong to the same family. You aren't sure which family it is, but from her statement, you are certain that they come from the same family. What was it that Eve said?

Think about it for a moment. Don't give up too easily![13]

The art of thinking logically requires training and practice. A classic riddle that challenges logical thinking runs as follows: A man is staring at a portrait on a wall, which shows a young man. Another person walks by and asks, "Who's that, in the painting you're looking at?"

The man says, "I have no siblings, but this young man's father is my father's son." What do you think? Most people conclude, after a while, that the man is looking at a self-portrait. But that's not correct.[14]

Moral Thinking

To add insult to injury, it turns out that our intuitive moral judgments are also quite inconsistent. Here is a classic example, due to British philosopher Philippa Foot (1920–2010) back in 1967, called the *trolley problem*. Imagine that you are standing near trolley tracks, and you see a trolley approaching. Some ways down the track there are five people whom you don't know, all of them tied down to the track. Now it happens that just where you're standing, there is a two-way fork in the tracks. If the trolley goes down one of the two possible ways, it will continue straight to where the five people are tied down, and if it goes the other way, it will veer off onto a siding. Unfortunately, however, a little way farther down the siding, there is yet another person tied down to the tracks. You are standing right by a lever that can determine which side of the fork the trolley will take. You can throw the switch at this very moment. If you don't do anything, then five people will die in a few seconds. If you throw the switch, you'll save those five, but the person on the siding will die.

What should you do? What is morally permissible here? What is your moral duty in this situation?

The most standard reply is that one should throw the switch, but if we consider a variant of this thought experiment, things seem quite different. Suppose that you are standing on a bridge right above the trolley tracks. As before, there is a trolley barreling down the tracks toward five quivering, tied-down people; however, there is no fork in the tracks and no lever to pull. Instead, there happens to be a very fat and heavy man who is nonchalantly sitting on the railing, with his legs dangling down. If you just give him a gentle little shove, he will go tumbling downward right in front of the oncoming trolley, and this will bring the trolley to a halt, and thus the lives of the five imminent victims will be saved. On the other hand, the fat man will lose his life instead.

So now, what will you choose to do? The most typical answer is that one should *not* push the man off of the bridge down to his doom. But what's the difference? In each case, one human life is sacrificed for the sake of saving five human lives. But our instinctive moral judgments are not identical; they are in fact inconsistent with each other, since we somehow feel that the two situations are very different. And when people's brains are observed using neurological instruments, it is found that the two moral dilemmas actually activate different regions of the brain.

Was the Human Brain Designed to Be Religious?

Our brains did not evolve primarily in order to reveal abstract truths about the universe, or to be logically consistent in their treatment of moral dilemmas. The multifarious abilities of our brains, no matter how poorly or well developed they are, are there simply so that we can orient ourselves in our environment, figure out what is near us, and thereby manage to survive. These are the mental capacities that were selected for and step-by-step refined by evolution. Way back in the dark, dim prehistory of proto-human beings, certain simple questions kept on coming back and were unavoidably central to the lives of all: What can I eat? How can I reproduce? How can I protect my offspring? Who is the leader of my pack? How can I survive during the night? How can I keep warm in this freezing weather?

Contemporary research in evolutionary psychology (a branch of psychology that tries to explain the behavior of people and other primates by using ideas about evolution) has discovered a number of basic cognitive processes in our brains that arose long ago purely because of their survival value. These primordial brain processes can also have a number of side effects that don't *directly* increase our chances of survival but that nonetheless exert a significant influence on the course of our everyday lives. Religion is one of those processes.[15]

There are many people who assert, somewhat imprecisely, that human beings have a *religious drive*, without being any more specific than that. Some people say that religion arose as a way of helping people deal with the fact that they are going to die; others claim that religion is a kind of social glue binding people to one another. These statements may well be valid explanations as to why many people today are religious, but there is a much more puzzling question if one thinks about things from an evolutionary angle. Why, if there is no god, is there religion? How did religion and religious thinking ever come to be at all, if they don't correspond in any way with the existence of a divine being?

The origin of religion is a riddle that has long intrigued evolutionary psychologists. It's not clear that religious belief in itself has any survival value. Rather, it may well be the case (as many evolutionary psychologists believe) that religion is a natural side effect of other cognitive capacities that evolved over millions of years because they were crucial to our survival (and to the survival of earlier species out of which *Homo sapiens* evolved).[16] In other words, perhaps religion is like music—merely a side effect of other cognitive processes that have survival value. It's not clear that either music or religion has any survival value in itself.

We human beings are the outcome of a very long evolutionary story, a story of countless gradual changes and adaptations, all thanks to natural selection. It makes perfect sense that our nature should be defined by traits that favor our survival. Otherwise, those traits would long ago have been mercilessly selected out. One key trait that favors survival is the ability to *spot patterns and relationships*. All animals, of course, also do this, to varying degrees, but none of them comes anywhere close to humans. This capacity presumably made it easier for people to cooperate and thereby to survive better.

The problem is that we are actually a little *too* good at pattern spotting. Our brains are so good at seeing patterns and relationships that they do so even when there are no patterns or relationships to be found. They hallucinate things that aren't there. We see animals and faces in the clouds above us, and in random blobs of ink, as in the famous Rorschach test, developed at the beginning of the twentieth century for the detection of personality traits. It consists of a series of symmetric blobs of ink, and the subject is supposed to say what these pictures represent. Presumably the subject is actually projecting onto these meaningless shapes certain ideas that are floating about in their subconscious. (It is doubtful that the Rorschach test actually reveals key contents of the subconscious, but there is no doubt at all that it makes us "see" things that aren't really there.) When we look at splotches of ink, we can't help but see familiar shapes in them—perhaps the outline of Africa or of Australia, perhaps a horse's head, perhaps a flying bird, perhaps a dragon (etc., etc.)—even though none of these is really there. The brain always wants to see something meaningful; it abhors meaninglessness.

It is worthwhile to go a little more deeply into this question of seeing things that "aren't really there." As has so often been the case in this book, we run once again into that tricky word "really" and the elusive concept of *reality*. To put some perspective on the matter in this particular case, let's focus in on the familiar car game of trying to find all the letters of the alphabet, in alphabetical order, on signs, as the family car is driving down the highway. I spot an *a* in the word "pizza," then you spot a *b* in the word "beer," and someone else spots a *c* in the word "welcome," and so forth. We all agree that those letters are *really there*.

But what if the rules of the game were changed a little bit, so that the letters didn't have to be on human-created signs, but could be found in any roadside object, or in any visible entity at all? Would seeing a tall telephone pole then constitute a valid sighting of the capital letter *I*? Would seeing two crisscrossing contrail lines in the sky constitute a valid sighting of the letter *x*? Would seeing a car's wheel, or a roundish rock by the side of the road, constitute a valid sighting of the letter *o*? Would seeing a so-called "T" intersection of two roads constitute a valid sighting of the capital letter *T*? You can easily continue this game of letter spotting using your own imagination. What might constitute an *h*? What might constitute a *w*? And so forth and so on.

Now the question arises as to whether those letters of the alphabet are *really there*. Is a car's wheel *really* an instance of the letter *o*? Is a "T intersection" *really* an instance of the capital letter *T*? (If not, then why do we call it a "T"?) And keep in mind that many of the letters used on signs are extremely distorted (think of the arches making the *M* in "McDonald's" or the swirly *C*'s in "Coca-Cola"—and these are just the tip of the iceberg of distorted letters!), and yet we haven't the slightest hesitation in saying that those letters are *really there*. So why wouldn't a perfectly round car tire be a *superb* example of an *o*—a really-there *o*, so to speak? And how would we argue against somebody who claimed that a flat tire was an *o*? Or that a fencepost with three horizontal wires coming out of it was an *e*? Is there *really* an *e* there, or not? What does "really" really mean?

Let's now come back to the ink blots or the clouds. If one summer's day you see a jumping dog up in the clouds, and you point it out to me, and I see it too, then who's to say that it isn't "really there"? Of course no artist *drew* a jumping dog up there in the sky, but if you and I and our friends all agree that the cloud *looks like* a dog, then in some objective sense, there is nonetheless a picture of a dog up there in the sky, albeit unintentionally, and albeit fleetingly (since in a few minutes the cloud in question will surely have changed shape so much that the short-lived dog will, alas, no longer exist).

Or suppose the game were merely a competition to be the first person to see a square. Well, what constitutes a square? Is a rectangle whose perpendicular sides are *nearly* equally long square? How close in length do they have to be before we all agree that this *is* a square? Or if the goal were to spot a circle, does a full moon count as a circle? Or a nearly full moon?

Physicists will tell you that there is a good reason that the stars, planets, and moons are all spherical—a sphere is the natural shape that large objects assume because of the mutual gravitational attraction of their parts. Planets may start out as ungainly blobs, but throughout time, they *become* spherical. They really *are* spheres, and entirely because of natural law. And so, in some sense, we are perfectly correct if we see the sun's outline as constituting a genuine circle. It would be hard to find a better circle anywhere, in fact! So are we humans *hallucinating* circularity when we see the sun as a circle? Are we *projecting* the concept of circularity onto something where that concept doesn't apply at all?

And are we *hallucinating* a dog when not only we, but all the people around us, agree that the cloud in question looks like a dog? And what if I say that the stranger at the bus stop looks a lot like Chairman Mao, and you agree with me, and so do the other two friends who are standing with us? Are we all victims of a collective hallucination? Is it *false* that this person looks like Chairman Mao? Or are we validly perceiving a *fact* about this unknown person? Most likely, were I to go up to him and say, "Excuse me, sir, but you know, you look a lot like Chairman Mao!," he would look at me with annoyance and say, "Well, you're only about the fifty-billionth person to tell me that." At that point, I would feel pretty stupid, since obviously this is an *objective fact* about the person.

Altogether, then, what this discussion is meant to point out is that the statement that humans tend to spot patterns where there actually *aren't* any is not exactly true. Yes, we all spot patterns whose existence is sometimes dubious, but sometimes it must be admitted that there are hidden patterns that are "out there" even if no one intended them, and even if they aren't part of natural law. They are just interesting accidental facts, but that doesn't make them *unreal.*

Homo sapiens is by far the most talented animal at spotting causal relationships (that is, situations where one event is seen as a cause, and another event as an effect). But this great human ability, too, is sometimes carried too far. Psychological experiments have shown that humans often hallucinate cause-and-effect relationships where there really aren't any. We have an obsession with finding reasons for events, even when the events have no reason. If an evil person is killed by a stroke of lightning, religious people will see this as God's fitting punishment. And if a whole churchgoing family is wiped out in a fiery collision with a speeding truck whose driver was drunk, religious people will argue that this, too, happened for a reason, and will simply add, "God works in mysterious ways." In other words, this random-seeming event was actually part of God's plan; there was a cause and an effect, even if they are not apparent to us mere mortals.

Belgian psychologist Albert Michotte (1881–1965) studied cause-and-effect perception in a classic series of experiments that he undertook starting in 1946. He constructed a device that could project two moving circles onto a screen. Michotte could control the balls' sizes, colors, and movements very precisely. He could make either ball move to the left or to the right—and at any desired speed and starting moment. Using this device, Michotte was able to study how human subjects perceived the balls' movements. He first showed that people often described one ball's motions as causing the other ball's motions; he then demonstrated that very small changes in the balls' ways of moving gave rise to completely different interpretations of the causal relationships. We see causal relationships even where there are none.

In another experiment conducted by Itzhak Fried, Charles L. Wilson, Katherine A. MacDonald, and Eric J. Behnke and reported in *Nature*,[17] a subject puts on a helmet that can emit electromagnetic fields that stimulate certain parts of the brain, making the subject want to laugh. The moment the experimenter turns the helmet on, the subject starts to laugh out loud. But when the experimenter asks what the laughing is all about, the subject usually replies with clearly fabricated nonsense, such as, "Well, the woman who just walked into the room was wearing such a funny-looking outfit!" or "What do you expect? You were talking with such a weird accent!"

Our brains do not deal well with situations where something seems to happen for no reason at all. If someone wins the lottery, they will often say, "It was the hand of fate!," or perhaps, "The moment I bought the ticket, I just had this funny feeling that I was going to win . . .," even though their win was simply due to a random number.

Justin Barrett, a professor of psychology at the Fuller Seminary in Pasadena, California, has studied how our cognitive processes can give rise to religion. In his book *Born Believers: The Science of Children's Religious Belief*,[18] Barrett explains that even from the very earliest age, small children have a tendency to believe that *someone*, rather than *something*, is behind all events that happen. When a little girl bangs her head against the table, she may well hit the table back and angrily say to it, "You're so dumb!" She sees the event as having been intentionally caused by an *agent*—a conscious being of some sort—not simply as the result of a mechanical, physical process.[19] We humans are prone to anthropomorphize inanimate things in our environment and to identify agents around us. Barrett suggests that we have a special cognitive ability that he calls our "Hypersensitive Agent Detection Device."[20]

Sunset is approaching, and I go out and sit down on my porch to watch it. All at once, I catch a slight quivering movement out of the corner of my eye. Is it a just a bush blowing in the breeze, or is it a burglar? Well, I would do better to interpret it as an agent, rather than just an inanimate object, even if I'm wrong. The point is, it's better for one's survival and well-being in this world to mistake a quivering bush for a burglar than to mistake a burglar for a quivering bush. Certainly this general principle was very important on the savanna hundreds of thousands of years ago. Better to see a trembling bush and mistakenly take it for a tiger than to see a tiger and think it's merely a trembling bush. Our capacity to attribute agency to mere objects in our environment clearly has survival value.

This very human tendency of ours to see agents and intentions can be thought of as a built-in "better safe than sorry" mechanism, brought to us care of natural selection. This is why psychologist Barrett labels our in-built agent detector "hypersensitive." Like the little girl who banged her head on the "dumb" table, we all very easily mistake inanimate things for animate ones that have their own purposes, and not the reverse (which is lucky for us!). This kind of evolutionary useful cognitive hypersensitivity may lead us to fear the dark, or to believe in ghosts and aliens in UFOs, and so forth.

Our Very Human Theory of Mind

We need our "agent detector" to be able to figure out whether there is an agent, not just an object, in our vicinity. But to do this, we need to be able to detect *goal-drivenness*. Any agent (as opposed to a mere object) has one or more goals; it *wants* something, so to speak. We humans (and certain more highly developed animals) have what has been called a *theory of mind*. This means that we are constantly perceiving and "reading" agents (mostly other people) in our vicinity. What does this one want? What does that one believe? What does that other one think about me? Our theory of mind is the most crucial requirement for successful social interaction and for our ability to cooperate with one another. It thus has enormous survival value.

Of course we don't all do this in the same way, or with equal levels of skill. Some people, such as those who are autistic or have Asperger's syndrome, have deficits in their theory of mind. A classic experiment is often used on four-year-old children (the age at which a theory of mind standardly starts to take shape), when one wants to determine if they have autistic tendencies. The experiment involves two children, whom we'll call Carl and Lisa, and runs as follows.

Suppose we want to determine whether Carl has autistic tendencies. We show him a matchbox and ask him what he thinks is inside it.

"Matchsticks," he says. Then we open it up and show him that in fact it contains glass beads, not matchsticks. Then Carl's friend Lisa comes into the room. We say to Carl, "If we ask Lisa what she thinks is in this box, what do you think she will answer?" If Carl has Asperger's syndrome or autism, he will probably answer, "Glass beads," betraying his very limited understanding of other minds.

If one has only a rudimentary theory of mind, one is unable to imagine how another person could possibly have different beliefs from oneself. It's especially hard to attribute *false* beliefs to someone else (such as that the matchbox is filled with matches, when one has just learned that it isn't the case).

The ability to understand what other people believe and what they want is, just like our agent detector, a crucial tool that helps us survive and plan our interactions with our environment. But it's precisely these two cognitive faculties of ours that, wonderful though they are, can give rise to the unfortunate side effect of magical thinking.

Magical Thinking

In the oldest religions, rooted in nature, people often imagined spirits or magical powers that inhabited stones, trees, rivers, mountains, clouds, and so on. The environment as a whole was seen as alive, and nature was seen as possessing a soul. This is rather easy to understand, given that it was happening in a prescientific time when no rational explanations existed for such natural phenomena as thunder and lightning.

Let's imagine we are back in prehistoric times, long before any scientific knowledge existed. I live in a hut out on the savanna, in a simple hunting society. A stroke of lightning hits my hut, which rapidly goes up in flames. All at once, my hypersensitive agent detector snaps feverishly into action: "Who did this terrible thing to me? Some great invisible power high in the sky chose to make my hut burn down! The unseen monster must be far stronger than I am, and it seems very angry. Once I've rebuilt my hut, I'll have to be clever to protect it from being struck again. To stay in the good graces of this scary being, I'll offer it some delicious fruit, maybe even a goat . . ."

Out of nowhere I've conjured up an invisible, terrifying, monster-like celestial power, far stronger and angrier than any human could ever be, an agent to which, or to whom, I must make offerings, demonstrate my devotion, and ask for help and support in my life.

Well, what do you know—I have discovered, or perhaps invented, God!

Back before there was any scientific understanding of our world, this kind of thinking was plausible and sensible. It made good sense to posit powerful invisible agents in the sky—at that time; however, this sort of "reasoning" started becoming suspect once we recognized that there were many good reasons not to trust it.

The just-quoted monologue about the "agent" who caused the hut to burn down, were it to issue forth from the mouth of a university student today, would strike us as extraordinarily far off base, since educated people today have a much clearer understanding of what it is that causes damaging strokes of lightning.

Could it be, however, that even today, people's tendency to believe in God is a cousin to this irrational kind of thinking? Could it be that it is a residual side effect of overactive agent detection? That is exactly the claim of American cognitive scientist Paul Bloom (1963–) in a noted article he wrote under the title "Is God an Accident?"[21]

If God really existed, then it would not seem at all strange for religions to exist. But the existence of religions can also be explained as a natural result of cognitive and evolutionary processes in our brains—even if no god exists at all. In short, a widespread belief in God has natural, scientific roots that we can easily understand, even if no god exists to believe in.

~

INTERLUDE: ON HEAVEN AND PARADISE

What does the notion of heaven mean for you?

Heaven, or paradise, is a central concept both in traditional Christianity and in Islam. Tales of a place or an existence beyond life crop up in religious belief systems the world over, and there is good evidence for the idea that our fairly recent ancestors and even our very ancient predecessors had similar notions. We know, for example, that the Neanderthals, who, until roughly thirty thousand years ago lived contemporaneously with our own species, *Homo sapiens*, buried their dead in accordance with rituals that are very similar to our own funereal rituals today.

Many believers speak of their hopes of going to heaven or paradise, but their notions of what heaven is like are extremely varied. Some types of Christians, such as Mormons and Jehovah's Witnesses, paint a vivid but surprisingly dull picture of the place to which they so deeply aspire to go. At its most boring, their notion of heaven doesn't sound all that different from an Ikea catalog or a zoo.

Could it be that the belief in some kind of paradise beyond earthly life is humanity's most dangerous idea? People give up their lives in "holy wars" throughout the world to be martyrs and thus to wind up in paradise. Pol Pot, leader of the Khmer Rouge in Cambodia, tried to build a "paradise" on earth and declared that the reckoning of dates would henceforth start from his life. The upheaval he created cost millions of human lives. The current leader of North Korea, Kim Jong Un, is trying to do something similar today. And the two dozen or so Muslim men who hijacked the airliners that crashed into the World Trade Center towers on September 11, 2001, had all been assured that each of them would wind up in heaven with "seventy-two virgins" for their sexual delectation. In short, the goal of reaching paradise, or creating paradise on earth, has caused untold suffering.

What might the word "heaven" mean for those of us who do not believe in life after death?

Well, there is indeed a heaven—namely, what we see above our heads when we look into the sky. Sometimes it's cloudy up there; sometimes it's clear. Sometimes it's bright; sometimes it's dark. Perhaps peering up at the night sky inspired humans to begin to think abstractly, to ask questions that went beyond their daily concerns of food, warmth, and survival.

Perhaps one day long ago, the stars' mysterious silent nighttime twinkling set in motion a thought process that launched our collective curiosity about reality, going beyond what we see and hear directly. Perhaps that moment was the birth of human inquisitiveness, of the human desire to discover, of the human impulse to investigate, and thus of science itself.

If so, should we thank the stars above for our ability to muse about existential matters? We'll never know the answer, but it is a fact that the starry heaven, minus gods, angels, and eternal life, can be a source of profound inspiration for human beings. This is as clear as the Milky Way on a cold winter's night.

CHAPTER SIX

A NATURAL WORLD

Concerning Naturalism, Agnosticism, and Atheism

Even gods die when no one believes in them any longer.—Jean-Paul Sartre

The question of God's existence has bedeviled humanity for a very long time. If one considers the world as a natural rather than a supernatural phenomenon, governed by the laws of nature rather than by supernatural forces, it follows that one is not religious; rather, one has what could be called a *naturalistic* outlook on reality.

One consequence of naturalism is that one thinks of God, or gods, or any other kind of supernatural beings, as extremely unlikely. One believes that it is far more likely that such things are simply ideas or faiths springing out of the fertile soil of human imagination. In short, a person who has a naturalistic point of view about reality is automatically an atheist.

Atheism, however, is merely one part of the large picture of naturalism. In a naturalistic world, not only divine beings are implausible, but so are supernatural forces, occult phenomena, and New Age claims. In compensation, though, in place of such imaginary "magical" stuff, when one ponders the natural world's enormous subtlety and beauty, one feels a powerful sense of awe and fascination, without any need to resort to supernatural interpretations or explanations.

Naturalism and Materialism

The concept of naturalism is actually rather complex, and it is a broader concept than that of materialism. A strictly materialistic stance implies that all that exists is matter and energy, following the laws of nature. (Thanks to Einstein, we know that matter and energy are two views of the same thing, so that rather than speaking of matter *and* energy, we might say "matter/energy" or just "various forms of energy.") One can of course also wonder whether the laws of nature actually exist, or whether they are just mental descriptions of reality—crutches for us humans to use to reason about the world.

By "naturalism," I also mean the idea that everything in the universe can in principle be explained using the methodology of the sciences. Naturalism rejects the idea that there are phenomena or beings that lie beyond the range of scientific investigation (such as the gods of most religions).

But a naturalistic outlook on the world can also include more than a purely materialistic stance. Certain philosophers believe, for example, that mathematical objects (such as prime numbers) exist in the natural world, without being material. Other philosophers believe that prime numbers are solely a construction of human thought.[1] Some people believe a naturalistic stance includes absolute moral values, while others feel that moral values have no objective existence.

In both Swedish and English, the word "materialism" has two contrasting meanings, which unfortunately are often confused with each other. In common parlance, "materialism" means the desire to own and flaunt many material belongings—luxury cars, expensive artworks, fancy houses, and so on. (This could also be called "consumerism.") This, however, has nothing to do with the philosophical stance called "materialism," which, as stated previously, means believing that nothing exists except matter/energy, following the laws of nature.

A materialist in this philosophical sense might well choose to go off to a remote grotto and meditate like a monk, never giving a moment's thought to the acquisition of physical possessions. And contrariwise, someone who believes that the world is permeated by some kind of mystical spirit, be it God or some other force, can still be madly obsessed with the quest for high status and material success in life. One needs merely to think of the so-called "theology of success" associated with certain revival movements in the United States. (The notion variously known as "theology of success," "positive confession," and "theology of prosperity" is a brand of religious belief that posits that the Christian faith is a pathway to success in all areas of life.) The best-known Swedish church following this theological tradition is Livets Ord ("The Word of Life"), a megachurch headquartered in the ancient university town of Uppsala. Ironically, though, the founder of Livets Ord, Ulf Ekman, eventually converted to Catholicism, leaving his followers to their own devices.

Atheists as Provocateurs

Many people today, especially in the United States, think of atheism as a troubling provocation. This is because belief in some sort of a god is deeply rooted in Western history and deeply ingrained in Western culture. Although few people in today's Sweden have any interest in the church, in religion, or in God, it's still common for an atheistic outlook to be looked upon with suspicion by Swedes. The negative way atheism is viewed throughout the world has significant repercussions.

It is quite common for people, in order to avoid negative connotations, to declare that they have "some kind of faith," without thinking carefully what they mean by this. Of course, it is fine for anyone to make such a claim, but if one is concerned with philosophical questions about life and the nature of the universe, one really ought to reflect carefully about whether God exists or not, and about what one truly believes in, and then have the courage of one's convictions.

There are all sorts of common objections to atheism, and if one doesn't believe in God, one inevitably winds up hearing them frustratingly often. "Are you an atheist simply because you can't see or understand God? But surely you believe in *wind*, even though you can't see it, don't you? And don't you believe in love, as well, even though you can't explain it? So why don't you believe in God?"

How should one react to such arguments? Consider the following exchange:

> "Say, do you know that there's an invisible ghost in my house?"

> "What?! An invisible ghost? Why on earth should I believe in any such thing?"

> "Because I told you! Why would you doubt in the ghost's existence merely because you can't see it or understand it? That's no reason to doubt its reality!"

This argument for believing in a ghost is analogous to the argument for believing in God, and it's obvious that such flimsy "reasoning" cannot be taken seriously.

Atheism in American Life

The word "atheism" comes from two Greek roots—the prefix "a-," meaning "not" or "without," and the word "theism," meaning "belief in god" ("theos" means "god"). Thus "atheist" means simply "non-god-believer." It has no trace of aggressiveness or evil.

In the United States, however, the word "atheism" tends to be heard as a deep *hostility* toward religion. Instead of sounding like it means "a lack of belief in a divine being," which is quite gentle, it comes across, to many religious Americans (and perhaps even nonreligious ones), as meaning *anti-theism*—in other words, as an active struggle, run by the devil, aiming at the absolute and total destruction of God and of all religion (especially Christianity).

And thus, atheists in the United States are often perceived as *attacking* God, when in fact all they are doing is exercising the fundamental American principle known as "freedom of religion" (an unfortunate phrase, since it is unclear whether it includes the freedom *not* to be religious). Freedom of religion was one of the most central tenets underlying the foundation of the United States, when Europeans from many countries desperately crossed the Atlantic to flee the oppression of alien religious beliefs crammed down their throats.

Some eighty years after the American Revolution, waged in large part in the name of religious freedom, the Reverend M. R. Watkinson begged the secretary of the treasury to print God's name on coins, saying, "This would relieve us from the ignominy of heathenism and would place us openly under the Divine protection we have personally claimed." Since the secretary liked the sentiment, he ordered the U.S. Mint to find a suitable phrase. "In God We Trust" was chosen, echoing a line from "The Star-Spangled Banner," and today that overtly religious slogan is printed on all pennies and nickels, as well as bills of all denominations. And then, in 1956, the U.S. Congress passed a resolution declaring "In God We Trust" to be the official national motto, replacing the Latin phrase "E pluribus unum" ("Out of many, one"). How much had changed in only eighty years!

American atheists do not approve of the motto "In God We Trust," as it clearly presumes that some sort of divine being exists, and that all Americans (or at least all red-blooded ones) believe in "Him." Such a presumption, no matter how noble it may sound to some, clashes violently with the spirit in which the United States was founded. And the fact that the slogan is nondenominational (not Catholic, not Baptist, not Jewish, not Buddhist, and so forth) does not mean that it is not religious.

In the early 1950s, the Cold War between the United States and the officially atheistic Soviet Union had erupted, and Senator Joseph McCarthy was at the height of his notoriously ruthless crusade against "godless communists." At that time, all American schoolchildren were required to recite, each morning, a Pledge of Allegiance to the American flag, with their hands held on their hearts. In the pledge there was no allusion to religion, but even so, as McCarthyism flourished, the American flag was becoming ever more an object of worship, and patriotism was becoming ever more conflated with religiosity.

In those years, various patriotic/religious organizations, such as the Knights of Columbus and the National Association of Evangelicals, were growing increasingly active and influential. When, in 1953, the latter organization issued its "Statement of Seven Divine Freedoms" (a decree that basically declared that the United States was founded on biblical principles), the newly elected (and newly baptized) President Dwight David Eisenhower was the first to sign it.

These same organizations then started agitating for a variant of the Pledge of Allegiance to be adopted and recited in schools. The new version added just two words—"under God"—in a strategic spot, so that it ran as follows: "I pledge allegiance to the flag of the United States of America, and to the republic for which it stands—one nation, under God, indivisible, with liberty and justice for all." The reference to God was added in order to make impressionable children think that the United States (in contrast to the Soviet Union) was a nation protected by God.

A bill was introduced in Congress to adopt the revised Pledge of Allegiance; it was passed into law in June 1954. On the day that President Eisenhower signed the law that added this "innocent" phrase, he stated:

> From this day forward, the millions of our school children will daily proclaim in every city and town, every village and rural school house, the dedication of our nation and our people to the Almighty. In this way we are reaffirming the transcendence of religious faith in America's heritage and future; in this way we shall constantly strengthen those spiritual weapons which forever will be our country's most powerful resource, in peace or in war.[2]

It is quite shocking that the highest officeholder of a nation that was founded on the sacrosanct principle of religious freedom could proudly declare that all Americans are forever dedicated to "the Almighty," but that's what happened.

For obvious reasons, American atheists did not approve of the noble-sounding language "one nation under God." On numerous occasions they launched lawsuits to get the phrase overturned, but the public perception of such "militant atheists" was that they were all dangerous anti-American subversives. Needless to say, the two key words have never been deleted from the Pledge.

In one of the most recent rulings (May 2014) on such a case, the highest court of the Commonwealth of Massachusetts ruled that the Pledge of Allegiance does not discriminate against people who do not believe in a god. The court went so far as to declare that having children recite the words "one nation under God" on a daily basis constitutes merely a *patriotic* practice, rather than a religious one. Once again, it is simply astonishing how blind a group of people can be toward something that is staring them in the face; it is very hard to fathom why such sanctimonious people are unable to call a spade a spade.

Atheists as Beings with No Morality

The world over, atheists are often presumed to be feelingless or cold, and they are also seen as believing that life has no meaning; it is even taken for granted that they are incapable of appreciating life's spiritual values.

For example, many religious Americans believe that atheists by definition lack morality. In 2012, a Gallup poll[3] in the United States asked what sorts of people would be unacceptable as president. Of those who replied, 4 percent said that they would not vote for a black person to be president, 5 percent would not vote for a woman, 6 percent would not vote for a Jew, 30 percent would not vote for a homosexual, 40 percent would not vote for a Muslim, and 43 percent would not vote for an atheist. As this vividly shows, there are significant prejudices in the United States against many different groups, but the strongest of all are those against atheists.

Agnostic or Atheist?

As the notions of *atheism* and *agnosticism* are often confused with each other, we will explore the contrast here.

The word "agnostic" comes from the Greek word "*agnostos*," which means "unknown, unrecognizable, hidden." Agnostic thinking, taken in this sense, played a major role in philosophy in ancient Greece; however, the expression was first used in English in the nineteenth century by Charles Darwin's colleague Thomas H. Huxley.

Agnostics say that there is no reason to have any belief at all about whether God exists or not; they see the question as simply unanswerable. Of course, it makes perfect sense to be agnostic about all sorts of matters other than God's existence, such as whether life exists on other planets, how the universe was born, how life started on earth, whether all people have exactly the same inner experience when they see the color red, or whether we today know all there is to know about some specific phenomenon.

One should distinguish between two types of agnosticism: strict (or permanent) agnosticism and empirical (or temporary) agnosticism.

Strict agnosticism is the attitude that one not only lacks an understanding of some phenomenon but also that one considers it to be forever impenetrable, which is to say, that it is in principle impossible to take any position on it. There are some scientific questions toward which we must adopt a stance of strict agnosticism. Consider, for example, the theory of electromagnetic waves. If we ask ourselves, "Do we know everything today that could ever possibly be known about electromagnetic waves?" we would have to answer in a strictly agnostic manner. Although we know a great deal about electromagnetic waves, we can never, even in principle, know whether we know *everything* there is to be known about them. And the same holds for most everything around us. Any time we learn something new about some scientific phenomenon, we are forced to recognize, as a result, that there was something about it that we didn't fully understand previously. But now that we know this new thing, aren't we done? Of course, we can never be sure of that. We just have to remain strictly agnostic about our state of knowledge.

By contrast, *empirical agnosticism* means that one lacks an understanding of some phenomenon, but the lack is presumed to be temporary. All we can say is that *up until now* we haven't understood the matter in question, and that's because our empirical experiences and knowledge aren't yet sufficiently deep to allow us to do so. A sample question toward which it would be wise to adopt an empirically agnostic attitude is the following: "Is there life on other planets?" We have no direct evidence of life on other planets, but we don't have any evidence of the opposite, either. (This discussion, of course, takes for granted that there is a clear consensus about the definition of the concept "life." Exactly how such a definition should run is far from self-evident. But for the purposes of this discussion, we merely assume that there is such an agreed-upon definition—not necessarily that it is the correct one.) Of course, we can still make educated guesses about the matter, based on probabilistic reasoning: "The universe is unimaginably huge, with untold billions of stars and planets. It would thus be very unlikely that life exists only on our own planet." But we can also adopt an empirically agnostic stance on the matter: "I have no opinion on the matter because I have no facts on which to base such an opinion."

It's obvious that this amounts to agnosticism of the empirical rather than the strict sort. At any moment, scientific researchers might discover life, or clear signs of life, on another planet. If so, we would have good reason to change our minds, and to drop our temporary agnosticism. (The European Southern Observatory [https://www.eso.org/public/] recently decided to build the world's largest optical telescope on a mountain peak in northern Chile, at an altitude of roughly three thousand meters. This remarkable device—a reflecting telescope whose main mirror will be thirty-nine meters in diameter—is expected to be ready by the year 2022. It will be called the "European Extremely Large Telescope," or "E-ELT," and its mirror will collect more than fifteen times as much light as today's largest telescope can. This should allow direct observation of planets circling other stars, as well as studies of their atmospheres. One of the first purposes of the telescope will be to search for earth-like planets in our galaxy. Stay tuned!)

As far as the existence of God is concerned, agnostics say that people shouldn't form any conclusion about the question. If asked, "Does God exist?" a strict agnostic would reply: "I have no idea, nor will I ever be able to form any opinion on the matter." By contrast, an empirical agnostic would reply: "I have no idea at present, but maybe in the future I'll be able to form an opinion on the matter."

Many Agnostics Are Actually Atheists

When someone says, "I don't believe in God," it most often means "I believe God doesn't exist"; however, some people say instead, "I don't believe God exists." Note the subtle distinction. Someone who asserts "I don't believe that God exists" can also assert "I don't believe that God *doesn't* exist" without self-contradiction. Such a person simply holds no belief in either direction about the matter. This is an agnostic stance. By contrast, someone who asserts "I believe God doesn't exist" is revealing an opinion on the matter. This is an atheistic stance.

It turns out that many people who call themselves agnostics are actually atheists. Their basic attitude is that, although they find it improbable that

God exists, they are still open to the possibility. But this, in truth, coincides with the atheistic stance. After all, no one can know with absolute certainty that no imaginable type of god exists. But as long as one *believes* that no such thing exists, then one is an atheist.

It is quite common to think that agnosticism constitutes a kind of "soft" intermediate position, located somewhere between theism and atheism; however, agnosticism is not a position at all. In fact, it's the *absence* of a position. Nonetheless, many people seem to feel more at ease calling themselves agnostics than calling themselves atheists. This is probably a consequence of the negative connotations that the word "atheism" has taken on in cultures that are deeply steeped in religion. For this reason, it is a misconception to think that an agnostic outlook is somehow less "hard" or less dogmatic than an atheistic one.

It's rare for people to take an agnostic stance when it comes to other types of supernatural beings, such as elves or unicorns; in fact, most people are very secure in their skepticism about such beings, and about old-time gods as Thor, Odin, and Zeus. Most people are thus *atheistic* about such gods rather than *agnostic* about them.

Clearly, if one has a scientific outlook, one should be prepared to change one's mind whenever new facts come to light that suggest that strange entities of some sort exist. This holds not just for atoms, photons, quarks, and so forth but also for elves, unicorns, ghosts, trolls, the Loch Ness monster, and so on. But having such scientific open-mindedness does not constitute any reason to declare oneself agnostic (i.e., opinionless) about the existence of elves, trolls, and other fairy tale sorts of beings. In such cases, atheism seems like the more reasonable stance. At least Ockham's razor would suggest this, since it's the *simplest* reasonable position.

Sometimes it is claimed that God exists beyond the range of science, not subject to empirical testability. If in fact God's existence is not a scientific or experimentally testable claim, then we should all abstain from drawing any conclusions about the matter. Well, there's no doubt that one can formulate visions of God that fit in with this vision. For example, a deistic theory positing that God first created the Big Bang and then withdrew totally from the world is clearly beyond testability.

In that case, however, the question of God's existence becomes uninteresting. If God plays absolutely no role in our universe, then why should we worry our heads about whether God exists? Today's monotheistic religions throughout the world, however, do not see their diverse gods in this manner.

For each of them, God is like a person—someone who *can* intervene in the world. God listens to prayers. God performs miracles. And if one has this vision of a personal, prayer-listening, miracle-performing, life-intervening God, then the matter of God's existence immediately becomes an empirically testable, and thus scientific, question.

It's not too hard to imagine experiments whose results would help either to support or to undermine the hypothesis that there exists a God. Consider, for example, scientifically investigating the statistical correlation between prayers and actual outcomes. Experiments of this sort have been carried out, but no such correlation has ever been detected. Of course, this lack of correlation does not constitute a watertight proof that God is a fiction dreamed up by human desires and is unrelated to reality, but it adds credibility to such a viewpoint.

What Does an Atheist Actually Believe?

Let's take a closer look at what being an atheist really means. Many people have the impression that atheists claim to understand more deeply than others, and to have more knowledge than others. To be sure, this may be true of some atheists; after all, arrogance arises in all sectors of society. But there is nothing inherent in atheism that makes it more arrogant than theism (the belief that God exists).

Another common misunderstanding about atheism is that it rigidly and categorically claims to have *proven* that God doesn't exist. Well, God's nonexistence is in a strict sense unprovable. But the nonexistence of *any* entity, such as unicorns or faster-than-light particles, is also unprovable, and for the same reasons.

It will be easier to understand the atheistic standpoint if we take a concrete example (and we'll borrow a vivid one from Bertrand Russell). If someone asserts that there is a teapot orbiting the moon, it would seem reasonable to be highly skeptical of the claim. Furthermore, it would seem unreasonable to adopt an agnostic stance toward such a claim. Would you be likely to say, "Hmm . . . maybe there is a teapot orbiting the moon, and maybe not—how could I possibly know?" Such wishy-washiness would sound rather silly.

Sane people (such as you and me) are inclined to take an "atheistic" attitude about the existence of such a teapot. We actively believe that there is *not* a teapot orbiting the moon, even if we cannot *prove* this to be the case. For better or for worse, we can't fly off into space and go scouring the vacuum for moon-orbiting teapots—and even if we could, we might easily miss the teapot, if it's up there. And yet, although we can't do this, we still believe—indeed, we *strongly* believe—that no such teapot is up there at all! After all, how would it have gotten up there? Why on earth would it be there? Might some astronaut have carried a teapot all the way to the moon? Even if so, how would the teapot then have gotten from the moon's surface into an orbit around the moon?

The claim that there's a teapot orbiting the moon simply has no plausible underpinning. But compare this now with the assertion that there's a teapot on my kitchen counter. I believe the latter claim in part because I *see* one there. This gives me a strong reason for believing that there *is* a teapot in that location. But no analogous fact lends support to the claim that there's a teapot orbiting the moon. I thus draw the conclusion that this is a very unlikely possibility. The burden of proof always lies with the person who asserts that something actually exists.

An atheist believes there is no God in much the same way as a normal person believes there is no teapot orbiting the moon. The atheist cannot *prove* the truth of this belief but feels justified in sticking to it until someone presents good evidence to the contrary. Sensible atheists are not dogmatic; they are ready to reconsider matters if new evidence turns up favoring the existence of God.

To my mind, having an *agnostic* viewpoint about God's existence—that is, being unwilling to speculate, even tentatively, about whether God might exist or not—is a wishy-washy position to adopt in this day and age. After all, we all have opinions about the reality or irreality of all sorts of phenomena in our universe:

1. We all believe that the earth is round (even though few of us have actually checked it out empirically).
2. We all believe that there is no teapot orbiting the moon.
3. We all believe that a stone will fall to the ground whenever we release it, even though we cannot *prove* this will happen the next time we try it out.

Given that such beliefs, none of which we can prove in an absolute sense, seem to lie beyond reasonable doubt for virtually all thinking humans, then why should we be so timid—as agnostics would urge us to be—about the question of God's existence? Upon consideration, the agnostic stance seems more like an irrational insistence on absolute, total certainty before letting oneself believe anything.

On the other hand, the atheistic stance is not a very radical one. For instance, atheists agree completely with believers in standard religions that the vast majority of gods that various peoples have believed in, down through the ages, do not exist. The world's major religions are based on many different sorts of divine beings. Hinduism is said to have hundreds of gods, if not millions, all of them being incarnations or avatars of one single God. (In Hinduism, an *avatar* is the appearance on earth of a god such as Vishnu, in the form of a person or an animal. The word has acquired a second meaning in the world of computer games—namely, the character one adopts when playing an online game.) Most Christians firmly reject all the gods that historically have been claimed to exist, except for one—namely, the Christian one. Well, an atheistic stance is very similar to the Christian stance; it simply involves adding one more god to the list of gods whose existence you deny.

Why do so many languages have a specific label for people who deny God's existence? Is such a word necessary? A naturalistic outlook on the universe leads naturally to an atheistic stance, but it also leads to the rejection of many other supernatural beings. In a naturalistic world, *no* kind of supernatural being exists—so why don't we have special words for people who reject other kinds of beings?

Greek mythology spoke of sirens—birdlike feminine beings that bewitched sailors with their enchanting songs, luring them to their island, Anthemoessa. For thousands of years, people have told tales of sirens, both orally and in writing. Despite all these tales, most of us have no belief whatsoever in the existence of sirens; we think of them as purely imaginary mythical creatures. Yet there is no special label attached to us siren-doubting folks, such as "asirenist," parallel in its etymology to the word "atheist."

The reason no such word exists, while the word "atheist" does exist, is that since time immemorial, human societies and cultures have been deeply intertwined with belief in some sort of god or gods. A belief in God today has precedents going back thousands of years. To believe in God, in other words, is historically normal—it is the "default" stance of humanity. It is *disbelief* in God that is historically exceptional. Moreover, down through the centuries it has been very dangerous to be an atheist, and even today being an atheist is extremely dangerous in many parts of the world. The idea that sirens live (or used to live) off the coast of Greece is not the default belief of anyone today, whether they are Swedish, Greek, or American. For this reason, there is no need in any language for a special word denoting those who are committed to the idea that sirens don't exist.

Thirteen Prejudices about Atheism

There are many prejudices and misimpressions about atheism, and in this section we'll take a look at a good number (13 is a very good number!) of them. I've already touched on some, but now let's go into the topic more systematically. Although some of the following prejudices overlap a bit, I feel that each one deserves its own separate comments.

1. Atheists Are a Special Brand of People

Atheists are not in fact a special brand of people who are all alike in some way. Atheists can believe in anything whatsoever (except for God) and have any imaginable value system. To imagine that all atheists are alike is similar to imagining that all people who don't play baseball are alike. Not playing baseball is not a clue as to someone's inner nature. In order to get to know a particular non-baseball player, you have to interact with that person over time and gradually come to understand their likes and dislikes.

What are all non-Catholics like? What are all non-feminists like? What are all non-gay people like? What are their values? Such questions sound ridiculous. To know what any non-Catholic or non-feminist or non-gay person is like, you have to go deeper than the surface. The same holds for atheists. An atheist can be a secular humanist, in which case we can guess at least a little bit about their values and attitudes. But an atheist can also be completely different from that.

2. Atheism is a Philosophy of Life

A frequent misimpression about atheism is that it is a complete philosophy of life. This is not so. To be an atheist is simply to hold a certain belief about one particular factual matter. A person can be an atheist and also believe in astrology or homeopathy, in acupuncture or flying saucers, in ghosts or unicorns. All bets are off.

A number of schoolbooks that are used in Sweden to teach about religion list atheism alongside the great world religions, such as Christianity, Judaism, Islam, and Hinduism. This classification, though well intended, is based on a misunderstanding. It is like having a textbook for a course on sports that has chapters about basketball, tennis, soccer, golf, and so on, and that then features a chapter about people who don't engage in any sport that involves a ball. What purpose can such a chapter serve if it doesn't describe what such people *do* engage in instead? Perhaps they engage in horse racing, or in diving competitions, or in archery. These activities can and should be handled as their own categories.

3. Atheists Hate God or Are Angry at God

The Anglican archbishop in Sydney, Peter Jensen, made the following statement about atheists:

> As we can see by the sheer passion and virulence of atheists, they seem to hate the Christian God. . . . Atheism is every bit of a religious commitment as Christianity itself. It represents the latest version of the human assault on God, born out of resentment that we do not in fact rule the world and that God calls on us to submit our lives to Him.[4]

The notion that atheists are angry at God is a common prejudice. But it should be obvious that one cannot be angry at someone or something that one believes doesn't exist.

On the other hand, atheists can certainly be extremely upset at the oppression and the heinous acts that take place throughout the world in the name of God. The god that is described in the Old Testament is certainly not a very pleasant figure, but atheism is not about one's relationship to God, since for an atheist this relationship doesn't exist at all.

4. Atheists Are Extremists, Are Arrogant, Are Militant, Are Intolerant . . .

Being atheistic does not imply being an extremist or an arrogant person. Yes, there are atheists who have these qualities, but so are there religious people who have these qualities. Having any such personal trait is not a consequence of theism or atheism. The notion that atheism is arrogant may spring from the religious notion that one should never dare to compare oneself to God.

Certain of atheism's critics speak of "militant atheists." But "militant" is a word that is usually applied to those who are violent or who advocate violence. One can read in newspapers about militant Islamists who carry out terrorist attacks, or militant Christians in the United States who murder doctors for carrying out abortions; however, as far as I am aware, the only weapons wielded by "militant atheists" are their pens and their right to free speech.

Perhaps "militant" is meant only in a figurative manner, and the complaint is merely that atheists argue fervently for their stance. But in that case, we would also have to complain about militant *theists* of many stripes, including fundamentalist Christians, ultraorthodox Jews, and so forth.

Calling atheists "intolerant" is a bit less harsh than calling them "militant," but it's still a strong term. It's clear that in this world of ours there are plenty of things that one should be intolerant of, such as the genital mutilation of young girls, the chopping-off of thieves' limbs, the "honor killing" of family members, the forcing of rape victims to bear the rapist's child, and the stoning to death of adulterers, but probably most atheists are neither more nor less intolerant of these things than other people are. If atheists are distressed by such gruesome acts carried out in the name of religion, it probably comes more from their concern about human rights than from the fact that they don't believe in God.

5. Atheism Is a "Harder" Attitude Than Theism

Atheism is neither "harder" nor "softer" than theism, nor is it more rigid or more flexible than theism. Each is the other's mirror image. A theist believes that God exists; an atheist believes that God doesn't exist. Neither stance can be more rigid or more flexible than the other, since they are symmetrical positions; however, atheism is by nature a humble stance: it's the idea that since there is no convincing reason to believe in any sort of god, one should not draw the conclusion that God exists. An atheist believes that God doesn't exist, until there is a powerful reason to believe otherwise.

6. Atheism Is Just Another Religion

It doesn't make sense to suggest that atheism is a religion. Such a characterization is no more accurate than saying "not collecting stamps is a hobby" or "not playing baseball is a sport." To do that would stretch the meaning of the word "hobby" or "sport" way beyond its normal bounds. Similarly, calling atheism a religion stretches the meaning of "religion" way beyond its normal bounds.

If, on the other hand, we take the word "religion" in a philosophical sense (i.e., something that one believes in but cannot prove), then the sentence is true in a trivial sense. All it means is this: "Atheism is an idea that some people are convinced of." Well, obviously atheism is an idea that some people believe in, just as is any other claim about reality.

Every belief about the world amounts to a faith, in an uninteresting sense of the word. Atheism expresses a faith, just as saying "I think the sun will rise tomorrow" or "I believe the Loch Ness monster doesn't exist" or "I expect the next election will radically change the makeup of Congress" expresses a faith. But none of these faiths has anything to do with religion.

Statements of faith can have very different levels of reasonableness. One faith might be extremely implausible, while another is essentially self-evident. Compare these two statements: "I believe that there are trolls in these woods" and "I doubt that there are trolls in these woods." Both express comprehensible visions of the world, and in that sense they both express faiths, but they are faiths of very different levels of plausibility. So if someone tells you, "Yes, but you're just expressing a *faith*!," don't let yourself be fooled by this facile label.

7. Atheists Claim That Religion Is the Root of All Evil

Atheists are perfectly aware that the world is full of all sorts of cruelty and suffering that have nothing to do with religion. Natural catastrophes—"acts of God," as they are often called in insurance policies in the United States—can hit people very hard without any kind of human cruelty being involved. Atheists know that human cruelty manifests itself in many forms, including some varieties that come from religion and some that have nothing to do with religion. And it goes without saying that religion can be a positive force in some people's lives. Atheists realize that a personal belief in God can give one hope and can offer consolation. There is no question that religions sometimes inspire people to do good things, and an atheist will certainly accept this fact.

One should also distinguish between a religion as a *personal philosophy of life* and a religion as a *social institution*. A moral criticism that can be leveled at religions as institutions is that they very often try to indoctrinate outsiders into their belief systems, and then to subject them to its laws and rites. This is a behavior that can legitimately be criticized—by atheists and theists alike.

8. Atheists Totally Lack Morality

Some people believe that atheists cannot have any kind of moral standards and do not believe in the goodness of human beings. In many countries, including the United States, an atheist is often perceived as a person who lacks moral principles. This is why so few Americans are open to the idea of an atheistic candidate for president.

The idea that all atheists lack morality is way off target, however, for the simple reason that morality is not correlated with belief in any kind of god or other supernatural being. One's morality can and should be judged solely by *how one treats other people and animals*. If you observe a wide spectrum of people, you will find that atheists, just like religious people, can behave morally or immorally. Neither type of believer has a corner on the market of moral behavior.

One strategy for painting atheism in a somber light, common in religious circles, is to assert that, since atheists reject the divine origin of moral principles, they are totally open to, and willing to engage in, any type of behavior whatsoever. Such reasoning is fallacious. Obviously atheists do not think that moral principles come from on high, since they don't believe that there is any being "on high" for them to have come from. But this doesn't imply that atheists reject the notion of moral principles, or that they view any arbitrary kind of behavior as acceptable. The conclusion simply does not follow from the premises.

9. Atheists Believe Only in What They Can See and Measure

Some people try to portray atheism as a narrow and pedantic attitude that dismisses all ideas except those that can be demonstrated in a scientific laboratory. This is a misimpression, to put it mildly. Like anyone else, an atheist believes in all sorts of things that cannot be measured or seen in a lab (or anywhere else). For instance, many atheists believe deeply in human rights, equality, and love. And some atheists may believe that life exists on other planets. Secular humanists (who of course are also atheists) have a clearer way of defining what they believe in: they believe in anything for which there is a convincing argument, regardless of whether the argument involves measurement or not.

10. Atheism Is an Ideology

Atheism is often falsely accused of being an ideology. Not so. Atheism is no more and no less than the thesis that gods don't exist. This is purely a matter of truth or falsity, and has nothing to do with any ideology. One cannot derive either good values or bad values from atheism alone. For this reason, one cannot draw any conclusions about where atheists stand on other important issues, such as stem-cell research, homosexuality, abortion, gun control, sustainability, and so on. It goes without saying that merely denying the existence of a supernatural being does not condemn a thinking person to accepting a whole set of predetermined opinions and values about all other questions in life. In short, atheism is not an ideology.

On the other hand, religions, which certainly do embrace moral values, are ideologies *par excellence*, since they prescribe sets of moral attitudes and ideas about how the world and society should be and about how people should act.

11. Communism Is Atheistic and Odious, and Therefore Atheism Is Odious

The evil deeds of communist governments are often attributed to atheism, but this mental link is a result of illogical thinking. Communism is an ideology, very much like a religion in numerous respects, in that it includes cult figures and the worship of leaders. (Just think of North Korea today.) It is true that some communist leaders such as Joseph Stalin were both atheists and mass murderers, but there is no cause-and-effect connection. Stalin also had a mustache, but that doesn't mean we should conclude that all mustachioed men are mass murderers.

As is well known, there are plenty of mass murderers who believe in God—and ardently so. The atrocious acts committed in the past few years by ISIS in Iraq and Syria are just one example among many. Obviously, we cannot conclude from such examples that all people who believe in God are mass murderers.

The logical error that leads to such conclusions is that of confusing a statistical correlation of two phenomena with a cause-and-effect relationship. For example, there is a strong statistical correlation between days when there is an unusually high amount of ice-cream eating and days when unusually many drownings take place. But this correlation does not reveal a cause-and-effect connection. The reason people drown is not that they eat ice cream. The problem is that a certain key variable—namely, hot weather—has been left out of the picture. Whenever it's hot, people tend to eat more ice cream, and people also tend to go swimming more. Therefore, when it's hot, the risk of drowning goes up. Much the same kind of confused reasoning falsely pins the blame for the evils of communism (an effect) on atheism (a cause). Though there may be a statistical correlation, it doesn't follow that there is a cause-and-effect link.

This type of "reasoning" recalls the attention-grabbing headlines in tabloid newspapers that blare such things as "Stuck All Alone in an Elevator, Got Pregnant!" (The actual situation in this case was that a woman was trapped in an elevator for so long that she wasn't able to take her morning-after pill in time.)

12. Nazism Was an Atheistic Ideology

Not just communism but also Nazism is often trotted out as an example of the horrors that atheism leads to. But as we saw with communism, this is the result of a logical *non sequitur*. And the fact of the matter is, Nazism was anything but atheistic! Although Hitler himself had a complex and blurry relationship with the Catholic Church, many Catholic priests were Nazis, or willingly collaborated with the Nazis. In *Mein Kampf*, Hitler describes his admiration for the Catholic Church:

> Here the Catholic Church presents an instructive example. Clerical celibacy forces the Church to recruit its priests not from their own ranks but progressively from the masses of the people. Yet there are not many who recognize the significance of celibacy in this relation. But therein lies the cause of the inexhaustible vigor which characterizes that ancient institution. For by thus unceasingly recruiting the ecclesiastical dignitaries from the lower classes of the people, the Church is enabled not only to maintain the contact of instinctive understanding with the masses of the population, but also to assure itself of always being able to draw upon that fund of energy which is present in this form only among the popular masses. Hence the surprising youthfulness of that gigantic organism, its mental flexibility, and its iron will power.[5]

It is clear that Hitler believed himself to have a divine mandate to safeguard the purity of the Aryan race. In *Mein Kampf*, he writes about the Jews:

> Systematically these black parasites of the nation defile our inexperienced young blond girls and thereby destroy something which can no longer be replaced in this world. Both, yes, both Christian denominations look on indifferently at this desecration and destruction of a noble and unique living creature, given to the earth by God's grace. The significance of this for the future of the earth does not lie in whether the Protestants defeat the Catholics or the Catholics the Protestants, but in whether the Aryan man is preserved for the earth or dies out. Nevertheless, the two denominations do not fight today against the destroyer of this man, but strive mutually to annihilate one another. The folkish-minded man, in particular, has the

sacred duty, each in his own denomination, of making people stop just talking superficially of God's will, and actually fulfill God's will, and not let God's word be desecrated. For God's will gave men their form, their essence and their abilities. Anyone who destroys His work is declaring war on the Lord's creation, the divine will.[6]

Here we see that Hitler believed that the Protestants and the Catholics should stop fighting each other and should instead fight the Jews, *on behalf of God*. He was convinced that he was acting in line with God's will.

Unfortunately, many priests obeyed Hitler's injunction. Prominent Protestant bishop Martin Sasse (1890–1942) actively encouraged the burning of synagogues in Germany. He considered it only proper and fitting that Adolf Hitler should punish the Jews for having killed Jesus. On the other hand, certain priests demonstrated great courage and opposed Hitler, a prime example being Dietrich Bonhoeffer (1906–1945), who vehemently spoke out for years against the Nazi regime and was part of the German resistance movement. On April 9, 1945, just weeks before VE-Day, Bonhoeffer was hanged by the Nazis.

Religion was actually central to Nazism, which to a large extent turned its own ideology into a new kind of religion. In each German soldier's life-jacket was found the inscription "Gott mit uns" ("God is with us")—a slogan that is disturbingly reminiscent of "In God We Trust" (the national motto of the United States).

On the other hand, Hitler was highly critical of some aspects of Christianity. Whether he himself was a Christian is hard to pin down, but there is considerable evidence suggesting that he had a religious outlook on the world.

Last but not least, the Nazi ideology included many superstitious, pseudoscientific, and occult elements. Everyone knows that the Nazi "morality" was beneath contempt, and it is therefore deeply dishonest to smear atheism by trying to link it with Nazism.

13. Atheists Are Cold and Emotionless

There are those who believe that atheists are less able than religious people to appreciate art, music, and other aesthetic creations. This is another bizarre *non sequitur*. There is nothing in the naturalistic stance toward the universe that would make an atheist less receptive to the beautiful and the sublime in art, music, or nature. In fact, sometimes it is quite the reverse. Someone who sees the world as natural, and who strives to fathom the inner workings of nature, can more deeply appreciate the amazing phenomena that surround us.

The teeming world of life on the African savanna does not lose any of its fascination as a result of our understanding the conditions of nature. A bumblebee is no less astonishing a creature when we understand how it flies. Music does not lose its beauty or power when we understand how different notes and harmonies are related to each other in the terms of music theory. The mysteries of physics are incredible, but they are natural. The development of quantum physics in the last hundred years gave us even more incredible mysteries to ponder. We still are completely unable to explain some of them, but that doesn't mean that the phenomena transcend nature. The universe, just as it is, is sufficiently wonderful and mysterious, without any need for magic or supernatural beings.

Atheism's Rebirth and Disappearance

The concept of atheism is actually not terribly interesting. It just means the lack of belief in a god, or in gods. And yet, down through the ages, the word has become so emotionally charged and so fraught with evil connotations that, even today, many people have a profoundly negative knee-jerk reaction to any mention of it. It is quite fascinating to see how much evocative power a word can be imbued with, thanks to history. These days, however, it may be the case that the word is slowly losing its stigma and is slowly becoming neutral sounding, as well it should.

When this happens, that would be the proper time to abandon the words "atheism" and "atheist" because they have a *raison d'être* only in a culture that is thoroughly steeped in religion. When the day comes around that we truly have a secular society, we will no longer need to focus our mental and linguistic energy on things that we don't believe exist. When that day comes, it will be better to focus on things in which we genuinely believe.

INTERLUDE: ON SUPERSTITION AND HOCUS-POCUS

What is superstition? Superstition is believing that there are lucky and unlucky numbers, that walking under a ladder or seeing a black cat or breaking a mirror brings bad luck, that disasters are more likely to happen on Friday the 13th, that carrying a magical charm such as a rabbit's foot will bring the bearer good luck, that knocking on wood or crossing one's fingers will ward off bad luck, that you can hurt someone far away by sticking pins into a doll, that buildings should be constructed so as to optimize the flow of "soft" and "hard" energy, that placing razor blades underneath a pyramid-shaped object will sharpen them overnight, that your secret birthday wish will come true if you blow out all the candles on your cake with just one puff, that the message inside a randomly chosen fortune cookie will reveal the fate of the cookie-opener—and so forth.

What is shared by all these ideas is that there is no good reason to believe any of them, and yet they all give the believer a sense of security, and an illusory feeling of control of the chaotic flux of events that we see around us.

Some people say that a horseshoe hung above one's door will bring good luck (provided, of course, that it is oriented in the right direction!). There's an old story about a journalist who noticed a horseshoe hanging above the door of the cottage belonging to the great Danish physicist Niels Bohr. The journalist's curiosity was immediately aroused. "Why," asked the journalist, "did you hang a horseshoe up there? Does that mean that you believe, at least a little bit, in this old superstition?" As the story goes, Bohr smiled at the journalist's question and replied, "Oh—that? No, of course I don't believe in silly superstitions of any sort, but you know, some people say that these things work even if you don't believe in them!"

Bertrand Russell had a very clear notion of what superstition is. He said, "Superstition is other people's religion." By this he meant that no religious belief system is considered by its adherents to be in the least superstitious, yet other people's religious beliefs are often cursorily dismissed as mere superstitions, even though the rival set of ideas may be just as reasonable, or just as unreasonable, as one's own.

In the early 1950s, an unknown children's magazine editor named Martin Gardner came out with a collection of essays called *Fads and Fallacies in the Name of Science* (1952). Astoundingly well written, insightful, profound, shocking, and amusing, this book was devoted to the art of debunking of nonsense. In it, with the highest artistry and skill, Gardner tore apart some two dozen highly diverse types of pseudoscience and superstition. A few years later, in the opening paragraph of the preface to this classic work's second edition (1957), he poker-facedly writes:

> The first edition of this book prompted many curious letters from irate readers. The most violent letters came from Reichians, furious because the book considered orgonomy alongside such (to them) outlandish cults as

dianetics. Dianeticians, of course, felt the same about orgonomy. I heard from homeopaths who were insulted to find themselves in company with such frauds as osteopathy and chiropractic, and one chiropractor in Kentucky "pitied" me because I had turned my spine on God's greatest gift to suffering humanity. Several admirers of Dr. Bates favored me with letters so badly typed that I suspect the writers were in urgent need of strong spectacles. Oddly enough, most of these correspondents objected to one chapter only, thinking all the others excellent.[7]

How telling is that last sentence—how deeply human—and how unforgettable!

Within the traditional Christian faiths, especially those that send missionaries to remote lands, the religious beliefs of indigenous peoples have always been seen as superstitions. For instance, the belief that trees or stones have spirits, or the performance of rain dances and various forms of animal and human sacrifice to placate the gods of nature, are considered to be superstitions or magical thinking, and to constitute a dangerous flirtation with evil spirits and demons.

But Christianity's branch of Catholicism involves a number of beliefs that would be hard not to see as superstitious. Consider, for example, the notion of transubstantiation, which was established in the Fourth Lateran Council, in 1215. This doctrine asserts that, "in a way surpassing understanding," an ordinary piece of bread and some ordinary wine literally turn into the body and blood of Christ during the Eucharist portion of a mass, although their appearance, taste, and odor remain completely unaffected. In a church council in the thirteenth century, it was determined that "Christ's body and blood are actually found in the sacrament, in the form of bread and wine, and then the bread is transubstantiated into the body and the blood into wine through the power of God." In the sixteenth century, Martin Luther's Reformation movement branded this type of thinking as magical and superstitious—and rejected it.

Another example taken from traditional religious practices is the cult of saints, whereby parts of the bodies of historical figures are preserved for centuries and are worshipped, as they are believed to be imbued with magical powers.

For nonreligious people, such practices can only be called "hocus-pocus."

The words "hocus-pocus filiocus" are a magical incantation that goes back at least as far as the seventeenth century. All indications are that this phrase has its origins in Christian rites. To be specific, the Catholic ritual of transubstantiation includes the Latin phrases "Hoc est enim corpus meum" ("This is my body") and "filioque" ("and of the son"). The mystical-sounding phrase "hocus-pocus filiocus" is a shortened and garbled version ("hoc est . . . corpus . . . filioque") of the full Catholic incantation, collectively invented by the minds of myriad pious mass-goers who loved the ritual but spoke little or no Latin. The garbled phrase automatically took on all the delicious flavors of magic, incomprehensibility, and miraculousness that were attached to the original phrase by the priests who conducted masses.

What is called "superstitious" changes over time. In Sweden, for instance, goblins and elves were once generally believed to exist; today, however, any Swede who believes in such things is looked upon as a nutcase.

In a hundred years' time, how will people look upon today's common beliefs in ghosts and spirits, or the dogma that Jesus turned water into wine and was resurrected from the dead? Or the dogma that the entire world was created in but six days only a few thousand years ago? Or the idea that one's zodiacal sign tells secrets about one's character and fate, or that the lines on one's palms reveal one's destiny? Or that aliens in flying saucers have visited our planet and had contact with some humans?

Or, as believers in ESP maintain, that certain gifted people can communicate mind-to-mind at thousands of miles of distance? Or, as believers in telekinesis maintain, that some people can bend keys and spoons by the sheer power of thought? Or, as believers in the so-called "Singularity" maintain, that, in the next fifteen to twenty years, we will all be able to upload our brain's microcircuitry into computer memories and thereby live forever as software creatures?

And who knows—might today's stock-market prediction techniques and miracle weight-losing methods eventually come to be seen as nothing but flim-flam? What will people think, in a hundred years, about today's theories of supersymmetric particles, and of dark matter and dark energy? What about string theory and the theory of "branes"? What about the Big Bang and its complementary companion, the Big Crunch? What about multiverses and wormholes and time travel? Well, we'll just have to wait to see. And if the Singularitarians are right, many of us will be alive not just a hundred years from now, but even a thousand, and a million, and a billion years from now . . .

CHAPTER SEVEN

BEING GOOD WITHOUT NEEDING GOD

Concerning Goodness, Evil, and Morality

> Is God willing to prevent evil, but not able? Then He is not omnipotent.
> Is He able, but not willing? Then He is malevolent.
> Is He both able and willing? Then whence cometh evil?
> Is He neither able nor willing? Then why call Him God?
>
> —Epicurus

In a naturalistic world, there will inevitably be both good and bad. People who accept the idea of a natural world without supernatural forces will not brood about how evil's existence is compatible with God's grace. For them, the existence of evil is rooted in natural causes, and the battle against evil in its many forms must simply be waged without stop. But for people who are religious, a serious problem arises: How can both good and evil exist at one and the same time?

The Problem of Evil

God died in Auschwitz.

Only through such a metaphor could one attempt to convey a sense for how unfathomable were the atrocities carried out against Jews and others in Nazi concentration camps. Evil of that magnitude seems totally irreconcilable with belief in an all-powerful and benevolent God. Consider the following horrific event, which was witnessed by a friend of mine who somehow survived imprisonment in the camp at Auschwitz:

> In our camp there was a woman who had a very young child. She tried desperately to keep her child alive, but in vain. When the child died, she was distraught and tried to escape. She was quickly captured and we prisoners were all forced to stand in a large circle around her. In the middle,

standing near the captured woman, there were two SS men. We were made to watch how they handled runaways. First they chained her tightly to a tree and whipped her. Then they pulled her arms, her hands, and her shoulders out of their sockets. They continued torturing her so brutally that in the end she collapsed and died before our eyes.

How is it possible to believe in a god who would permit such things to take place in our world?

If there is a god, an omniscient and omnipotent god, then why is there so much evil and suffering in the world? This mystery is known as the problem of *theodicy*, a Greek word meaning "God's justice." Theologians and philosophers have grappled with it since time immemorial; indeed, it is one of the strongest arguments against the existence of the god who is described in the holy books of the three Abrahamic religions: Judaism, Christianity, and Islam.

Omnipotent—Yes or No?

Religious people, in trying to justify their belief in God, have dreamed up numerous ways of solving the problem of theodicy. One solution is simply to accept the idea that God is not all-powerful. In creating humanity, God endowed each person with free will; God thereby limited his own powers. Each person had the freedom to make morally right or wrong decisions. Thanks to their free will, people could carry out evil deeds, and God would be helpless to prevent them.

But one can easily imagine a world where people had free will and God was also all powerful. In such a world, God could intervene in human affairs, when needed, to prevent evil deeds stemming from people's freedom. For instance, God could intervene to keep people from killing each other. People could use their free will as long as they were choosing between morally good alternatives. But if someone were about to choose to do a morally bad act, then an all-powerful God could jump in and nip the evil deed in the bud before it was carried out.

God might also have created the universe in such a way that whatever people chose was always for the good. In principle, it's conceivable for a world to exist in which people's free choices would always bring about good. Such a world would, in fact, be more perfect than a world in which choices sometimes brought about good and other times evil. If an omnipotent God existed, capable of producing the best possible world, then why would he not have created a world of this sort?

This type of attempted solution to the theodicy problem only solves the problem of *moral* evil, which is to say, the problem of evil acts that are intentionally committed by people, such as murder, rape, treason, and war. But it doesn't even touch on the problem of *natural* evil, which refers to bad things that happen that are not the result of human intentions, such as earthquakes, landslides, fires, tidal waves, hurricanes, volcanic eruptions, epidemics, stampedes, collapsing buildings, sinking ships, airplane crashes, and so forth. There is a great deal of suffering of this sort in the world—absolutely devastating, but happening for no good reason. If someone is struck by lightning, or is buried in an avalanche, or comes down with a fatal brain tumor, or falls from a rooftop while helping a neighbor repair a leak, their death is certainly tragic, but it has nothing to do with anyone's evil intentions.

An all-powerful god should be able to prevent such suffering, and so should a god whose powers are more limited because of the fact of having granted people free will. The upshot of all this is that the theodicy problem is not resolved by bringing in free will as part of the solution.

The Lord Works in Mysterious Ways

A different theological attempt to explain why there is evil in the world runs roughly as follows: We humans cannot understand God's reasons for allowing evil to exist. What we interpret as evil and suffering actually serves a far higher purpose, and when it is viewed from a sufficiently long-term perspective, it turns out to be for the good, even though we, right now, may not be able to see how that could possibly be the case. In a familiar slogan, "The Lord works in mysterious ways."

Unfortunately, such a "nugget of wisdom" not only is untenable but also immoral and repellent. To begin with, it would be deeply offensive to recite this phrase to someone whose child has just been murdered by a school bully or killed in a car accident. Where lies the long-term good for the person whose life has been brutally ended? If someone dies because of another person's cruelty or because of a natural disaster, the argument is absurd. To be sure, going through terrible ordeals can occasionally strengthen people, but people who die in an earthquake or a terrorist attack can hardly be said to have been made stronger.

Certain advocates of this approach to the theodicy problem might suggest that the sacrifice of one person's life somewhere serves the purpose of creating a good effect elsewhere. This seems to be quite cynical. Are we perhaps rewarded for our earthly suffering after we have died and have gone to heaven? This does not solve the problem of theodicy, since great suffering still exists on earth.

A benevolent and all-powerful God would set up a world in which we did not suffer either before or after death.

Some Kinds of Evil Aren't Real, despite Appearances

Let's take a look at one further way of trying to explain away the coexistence of evil and God. This will involve extending the first approach to solving the theodicy problem, and the basic idea goes along the following lines.

Moral evil is a genuine part of this world, and it is a result of human beings' free will; however, *natural* evil simply does not exist. It is not an actual thing; it is, rather, the *absence* of a thing—namely, the absence of good. Much as black is not a true color, but merely the absence of light and color, so what looks to us like natural evil is merely the absence of good.

This approach to the theodicy problem is often called the "privation response," from the Latin *privatio boni*, meaning "taking away of good." The privation response is a kind of play on words—a desperate attempt to save God from the treacherous theodicy problem.

Suppose that the privation response were correct, meaning that what we humans perceive as natural evil is merely the absence of good. Fine, but how does this kind of semantic nitpicking matter to those whose lives have been shattered by an earthquake, or to someone whose family was swallowed up by a tsunami? Their suffering is overwhelmingly real, whether we describe it as natural evil or as absent goodness. If God truly existed and were all-powerful and good, he would be able to create a world that contained *greater* good—or if you prefer, a greater absence of suffering. Such a god could also intervene wherever an absence of goodness caused suffering and could release humans (and animals) from terrible agony.

Despite all the flaws in religious attempts to explain away the theodicy problem, such attempts are often cited approvingly in so-called Free Churches, both in Sweden and elsewhere. In an event commemorating the liberation of Auschwitz, I once took part in a debate with a leading Swedish theologian who was associated with a Swedish organization of Free

Churches. The view this man advocated was that what the Nazis did to prisoners in Auschwitz might be explained by the fact that once God had given people free will, He could no longer intervene. (I use the masculine pronoun, since it reflects the nearly universal image that religious people seem to have of God. And in this particular context, I even used a capital *H*, just to make clear that I wasn't referring to the theologian.)

Another view that this theologian proposed was that there may be a higher moral goal that only God understands, and that the genocide conducted by the Nazis in Auschwitz and in their other death camps might simply be the price that had to be paid to reach that higher goal.

To my mind, this type of view is not only logically fallacious but also deeply cynical. If there truly were an omnipotent and good god who was watching what was going on in Auschwitz, would he have been able to prevent it? Or would preventing it have required that God override the free will of the prisoners or of the guards? The answer is actually very simple (at least if you are an all-powerful god). All it would take is a fairly minor miracle—for instance, the gates of the camp might all have suddenly sprung open, allowing the prisoners to escape. A different miracle that would have turned the trick equally well would have been to make all the prison guards suddenly come down with the stomach flu, keeping them from carrying out their bestial executions. Neither of these interventions would have infringed on human free will, and to make either of them happen would have been child's play for any supernatural being that was able to work true miracles.

What about the idea that God had a higher moral goal in mind, which would justify the mass murders in Auschwitz, but which would be incomprehensible to our feeble human minds? Well, even if this were the case, it could never justify the death camps. A god who was all powerful could certainly reach any kind of higher moral goal that might exist without needing to inflict inconceivable suffering on millions of innocent people.

Cancer researcher Georg Klein, who escaped from Nazism in Hungary, drew what I consider to be the most sensible lesson from the theodicy problem: "God is so merciless and malevolent that the only excuse for His behavior is that He doesn't exist."

Is God Actually Good?

What makes people think God is good, anyway? Many if not all religious people believe that God is 100 percent good; it's just that his powers, although great, are not so great as to be able to prevent all the evil that people, using their free will in ways that God disapproves of, wreak on other people in this world. Very few religious people ever express doubts about whether God is good.

In fact, though, why couldn't the truth be exactly the opposite? Imagine instead that God is 100 percent evil; it's just that his powers, although great, are not so great as to be able to prevent all the kindness that people, using their free will in ways that God disapproves of, lavish on other people in this world. The combination of human free will and a totally evil God could easily give rise to just the world that we see around us today. Even though this version of God is constantly doing everything he can so as to produce as much suffering as possible, there is still a great deal of goodness and pleasure in the world, and it's all thanks to the existence of human free will.

These two descriptions of the world are mirror images, and each of them is just as likely as the other. But the latter image is less appealing. A far more likely conclusion, as the theodicy problem so clearly shows, is that an all-powerful and good God is a chimera.

British philosopher Derek Parfit[1] told me, in a conversation we had in October 2014, that up until the age of eight he was religious, even to the point of wanting to become a monk; however, as a result of thinking hard about the theodicy problem, he slowly gave up his belief in a good and omnipotent God, even though this was what he had been taught in school. He felt deeply misled, and he lost respect for the authorities who had taught him such things. At age nine, he became an atheist, and in 2014, he told me that, in his opinion, the world would be a better place if fewer people believed in God.

Morality Has No Need for a God

Let's try a thought experiment. Imagine asking someone who believes in God, "What motivates you to act morally? Is it because morality was dictated to us lowly humans by God on high?" If the person's answer is "Yes, my morality is derived from on high," then it would seem that they would have to agree with the following: "If God were to ask you to injure or murder another person, you would willingly carry out this request from on high, wouldn't you? Or suppose that one day you were suddenly to lose your faith in God. Then I should be terrified to be anywhere around you, shouldn't I, since your morals would have vanished into thin air?"

If instead the person's answer is, "No, my morality doesn't derive from on high," then it would seem that they would act morally whether or not God existed. The reason for their moral behavior would not be tied to their belief in the existence or nonexistence of God.

We could reformulate the question in the following way. Is an act morally right because God advocates it, or conversely, does God advocate it because it is morally right? This issue was already carefully explored by Plato (427–360 BCE) in his dialogue *Euthyphro*. In it, Socrates, who is just about to be brought to trial for the crime of atheism, debates with Euthyphro, who acts as if he knew exactly what piety and justice are. Euthyphro is in fact so sure of himself that he is about to bring his own father to court because an act of the latter resulted in a death.

> *Euthyphro: Yes, I should say that what all the gods love is pious and holy, and the opposite which they all hate, impious.*
>
> *Socrates: Ought we to enquire into the truth of this, Euthyphro, or simply to accept the mere statement on our own authority and that of others? What do you say?*
>
> *Euthyphro: We should enquire; and I believe that the statement will stand the test of enquiry.*
>
> *Socrates: We shall know better, my good friend, in a little while. The point which I should first wish to understand is whether the pious or holy is beloved by the gods because it is holy, or holy because it is beloved of the gods.[2]*

Socrates asks this rhetorical question so that he can later reveal that there is only one reasonable answer to it:

> *Socrates: And what do you say of piety, Euthyphro: is not piety, according to your definition, loved by all the gods?*
>
> *Euthyphro: Yes.*
>
> *Socrates: Because it is pious or holy, or for some other reason?*
>
> *Euthyphro: No, that is the reason.*
>
> *Socrates: It is loved because it is holy, not holy because it is loved?*
>
> *Euthyphro: Yes.*
>
> *Socrates: And that which is dear to the gods is loved by them, and is in a state to be loved of them because it is loved of them?*

Euthyphro: Certainly.

Socrates: Then that which is dear to the gods, Euthyphro, is not holy, nor is that which is holy loved of God, as you affirm; but they are two different things.

Euthyphro: How do you mean, Socrates?

Socrates: I mean to say that the holy has been acknowledged by us to be loved of God because it is holy, not to be holy because it is loved.[3]

Socrates thus shows Euthyphro that an act's being good is not merely because the gods want it to happen; indeed, quite to the contrary, the gods want it to happen because it is good. It follows that what defines the good is independent of the gods.

British philosopher Simon Blackburn (1944–), in his book *Ethics: A Very Short Introduction*[4] writes the following about *Euthyphro*:

The point is that God, or the gods, are not to be thought of arbitrary. They have to be regarded as selecting the right things to allow and to forbid. They have to latch on to what is holy or just, exactly as we do. It is not given that they do this simply because they are powerful, or created everything, or have horrendous punishments and delicious rewards in their gifts. That doesn't make them good. Furthermore, to obey their commandments just because of their power would be servile and self-interested. Suppose, for instance, I am minded to do something bad, such as to betray someone's trust. It isn't good enough if I think: "Well, let me see, the gains are such-and-such, but now I have to factor in the chance of God hitting me hard if I do it. On the other hand, God is forgiving and there is a good chance I can fob him off by confession, or by a deathbed repentance later . . ." These are not the thoughts of a good character. The good character is supposed to think: "It would be a betrayal, so I won't do it." That's the end of the story. To go in for a religious cost-benefit analysis is, in a phrase made famous by the contemporary moral philosopher Bernard Williams, to have "one thought too many."

The detour through an external god, then, seems worse than irrelevant. It seems to distort the very idea of a standard of conduct.

If an act is considered to be good merely because God has so labeled it, it would mean that the act, when carried out, had no intrinsic moral value. This would imply that God can dream up any random moral principles whatsoever, and these principles will henceforth eternally define what is called "good" simply because God laid them out and dictated them from on high. In that case, if God were evil, couldn't he invent a set of moral standards that encouraged evil acts? Such a scenario, in which what is deemed "good" or "moral" is merely a function of God's arbitrary whims, runs deeply against our ingrained intuitions. Don't we all believe that torturing innocent people should *always* be considered morally wrong and evil, even if some god on high were to command us to carry it out?

Religious people would surely object that God would never command people to carry out cruel acts. They would say that God considers torture to be wrong, and that he therefore commands us not to torture others. They would also say that God considers happiness to be good, and that he therefore urges us to make others happy.

This sounds very nice, but unfortunately it contradicts the idea that moral principles stem from God on high. In fact, it presumes that there are *objective* moral standards that God himself has to respect. And thus even a deeply religious view winds up recognizing that morality is independent of God. A deeply religious view of morality is that God wishes us to act in certain ways *precisely because those ways are morally correct*; there is no need for an added-on divine seal of approval.

For those who don't believe in any god, this reasoning is unproblematic. But for believers in God, it constitutes a very important dilemma because it shows that morality does not depend on God or come from on high. Morality exists in this world, whether God exists or not.

The Fight between Zeus and Shiva

Let's explore this theme a little further. Suppose it were the case that God dictates what is morally good. Then what would it mean to claim that God is *good*? After all, if goodness consists merely in obeying God's orders, then stating that God's will is good is tautologous—it's nothing but a circular statement. It merely asserts that God's will agrees with God's will, which is a vacuous statement. To be able to assert *meaningfully* that God is good, one needs an *independent* sense of what goodness consists in.

To illustrate how absurd it would be if moral goodness were determined merely by a divine edict, let us imagine a competition between two gods: Zeus and Shiva.[5] The winner's prize will be eternal omnipotence. Shiva hopes to win this prize in order to vastly improve the lot of human beings. Specifically, he wants to bring about eternal peace, happiness, and justice in the world. Zeus, too, hopes to win, but he intends to use his omnipotence in a totally different fashion. He looks forward, first of all, to brutally murdering the vast majority of humanity, and then enslaving all the remaining humans, putting them to work in a giant concentration camp where, under unbearable conditions, they will work themselves to death.

As fate would have it, Zeus wins the competition. One would naturally be inclined to think that the most unthinkable of events has occurred: an evil and fearsome deity has become all powerful and is now going to throw the world into an endless era of inconceivable evil and suffering.

But although this is how one might first react, it misses the key point. The first way that Zeus uses his newly won omnipotence is to change the definition of moral principles—that is, to change the definition of what is considered "good." He turns the violent massacre of innocent people into an act of great moral worth and goodness, and he makes the inflicting of random suffering on innocent people epitomize fairness and justice. He makes being a slave in chains become the most virtuous life to which one can aspire. And last but not least, he converts himself into a morally perfect being—not by altering his own behavior, but by altering the very nature of morality. The world of torture, slavery, and mass murder that he has created thus becomes, by definition, morally laudable and highly desirable. .

Well, there is just one problem with this story: it sounds totally crazy. If God is the sole possible source of morality—that is, if God is the embodiment of goodness—then in a universe with *no* god, there would by definition be neither good nor evil. And this means that there couldn't be anything negative or unpleasant about a godless universe, since in it, neither morally good things nor morally bad things would exist. It is thus a contradiction to state, "Without God, there would be no morality in the universe; how immoral such a universe would be!" If moral truths or standards don't exist, then there can't be anything *immoral* either.

If God is the sole source of morality, then a godless universe should more aptly be called an *amoral* universe, in the sense that in such a universe there would be a lack of morality. But "amoral" doesn't mean either "moral" or "immoral." A fish is neither moral nor immoral; a fish is simply amoral.

All of this is of course merely philosophical game-playing, though, since we all know perfectly well that both happiness and suffering do exist in this world. We are left with only one conclusion to draw—namely, that a god cannot be the source of moral principles. To be sure, one can easily imagine that a god could help people *understand* and *follow* moral principles without being the source of those principles.

The key question, then, is actually the following: Could people ever find and understand moral principles *without any help from any sort of god*? As we will see, there is much evidence that they can. But first let us look at how things would be if we all actually acted on the basis of what we believe is God's will.

Eve Is Always Right

Do all people deep down share the same moral values? Just take a look around you and you will find the answer, and it is clearly "no." Every day, many people die for the simple reason that they don't happen to have the "right" views of God's moral truths. Every day, for "moral" reasons, women are oppressed, religious or ethnic minorities are discriminated against, atheists are persecuted, homosexuals are murdered or thrown in jail, people spread and contract AIDS because they have been denied access to "unholy" birth control methods. The list is dismally long of abhorrent ways in which people treat each other—in God's name.

But let us turn to the question of what it would truly mean to follow God's moral code. Let's assume, for simplicity, that there really exists a god who, from on high, has handed us a set of moral rules. Is it moral, then, to follow them? Or might it be that someone who follows these divinely given rules is acting amorally?

Suppose that I have a good friend called Eve. Eve is a bit unusual, in that she claims that she is always right in everything she does. She sincerely believes that all her opinions, views, and values are inevitably correct, and she doesn't hesitate to tell everyone she encounters that this is so. But that's not all; I myself fully accept the idea that Eve's opinions, views, and values are always right on the mark. In a word, I believe in Eve.

One day, another friend of mine, of course named Adam, asks me for my opinion about a number of questions about morality. For instance, he asks me what my view is of homosexuals' rights and of women's rights. When I reply that I advocate equality for all people, and therefore I believe that there should be no discrimination against women or homosexuals, Adam asks me why I hold this belief. I tell him I hold these opinions because my friend Eve holds them, and because Eve is always right about everything. Adam then says, a little suspiciously, "But how do you *know* your friend Eve is always right?" I candidly reply, "Because I asked her very specifically, and *she personally told me* that she's always right. I remember the conversation like it was only yesterday. I've never forgotten it. It's burned into my memory. Eve is always right. I believe in Eve."

Despite the sincerity of my explanation, Adam feels that this is an inadequate response. Even if all the consequences of my moral beliefs turn out to be good, I should still have a more objective and more plausible reason for holding these beliefs than just that some friend of mine spoon-fed them to me. Adam suggests to me that I should be able to ground my views in something more solid and universal, not just in circular reasoning. I might, for example, compare Eve's dictates with moral principles that seem to flow from basic ideas about human worth and human decency. He says that only if I myself first think about and select certain basic moral principles (and they may well wind up coinciding with Eve's moral principles) can I reasonably consider myself to be a moral person.

I agree with the hypothetical Adam. I feel that a moral agent should build up a set of morals through an ongoing series of *moral checks and tests*. If I choose to act in a certain way simply because my friend Eve told me to, I am acting *amorally*. Can this way of running my life even be called a *choice*? If I am unwilling or unable to test the rules of conduct that govern my behavior, then I am not an autonomous creature, and I have no genuine morality at all. In a word, I am amoral.

Slavishly following another person's moral dictates without engaging in one's own reflections and ruminations is a type of intellectual surrender. One is obsequiously bowing to someone else's presumed superior wisdom, almost as if one were merely a programmed robot. This is a mindless way to act.

And by exactly the same token, it is mindless to constantly quote God's rules like a robot and cite supposedly sacred texts as the basis for all of one's moral views. Even if someone believes that God has spoken directly to them and has told them what is morally correct, it still behooves that person to carefully consider whether what God dictated to them from on high agrees with what they instinctively feel is the right way to treat other people. Genuine morality is not robotic.

A Secular View of Evil and Morality

What, then, is a *secular* way of thinking about the existence of evil in our world? First of all, if no god exists, then the knotty problem of theodicy does not arise. People are simply responsible for their own behavior. Only *human beings* can create a set of standards for their own moral behavior and ways of behaving that will improve the lives of others.

Moreover, careful reasoning and the constant growth of scientific knowledge can contribute to the flourishing of goodness in the world. With the help of medical research, we can cure more and more diseases. Using new forms of technology and precise engineering, we can design and construct buildings that are more able to withstand earthquakes and other natural disasters. The "natural evil" that is inevitable in our world can at least be reduced through science and technology.

But what about *moral* evil—the myriad sufferings caused by human beings? It seems unlikely that we can totally obliterate that. People will always be subject to fits of anger, and unkind acts will always be part of human life. Nonetheless, through the constant development of democracy, equality, education, and respect for human rights, we can step by step *reduce* the amount of moral evil in our world. In today's increasingly globalized climate, numerous cooperative organizations have sprung up to combat oppression, starvation, and poverty. International politics today involves fewer decisions made unilaterally and more decisions made through mutual agreements. All this is clear progress, but of course all these good things could go considerably faster, and we should be doing still more.

If and when humanity comes to see *itself* as responsible for the world that it is shaping, if and when people realize that there is no supernatural being to turn to for guidance, then secular morality will be able to help the world move in positive directions.

"The Problem Is Merely That Humans Misinterpret Religion"

In discussions of religious oppression and other evil acts perpetrated in the name of religion, I have often been told, "All that bad stuff is not the fault of religion itself—it's just that certain people misinterpret the message of religion."

This is a remarkable argument—not because it is false, but because it is trivially true. Religion is, in fact, nothing but human opinions. What else could it be? If one is a fundamentalist, one can of course claim that the holy scriptures were dictated by God and that they consist of absolute truths—and in that case, the argument might be valid. But quite surprisingly, many liberal and moderate believers in God bring this argument forth to defend their religion against accusations that it is in any way responsible for bad things.

Another common defense of religion's intrinsic goodness is to make a list of good people who were religious. Such people as Martin Luther King Jr. (1929–1968), Mother Teresa (1910–1997), and Desmond Tutu (1931–) often show up in such lists. Martin Luther King Jr., the American minister and civil rights leader, advocated nonviolence and fought segregation for his entire life. For this, he was awarded the Nobel Peace Prize in 1964. Also for this, he was assassinated in 1968. Mother Teresa, actually named Ajnezë Gonxha Bojaxhiu, was a Roman Catholic nun, born in what was at the time Turkish Albania, now Macedonia. She lived most of her life in India and garnered the Nobel Peace Prize in 1979. Desmond Tutu is a South African priest and church leader who became famous for his struggle against apartheid. He received the Nobel Peace Prize in 1984.

Many Christian people say, "Desmond Tutu was good because he was deeply religious." Well, Tutu is an admirable human being, and he deserves without question to be hailed for his long-term fight against apartheid in South Africa. But Tutu was not the only Christian force acting in the long battles over apartheid. Consider the Dutch Reformed Church—the dominant branch of Protestantism in South Africa—which strongly *defended* the country's apartheid laws, even using arguments taken from the Bible. This was the church that insisted that South Africa should forbid mixed-race marriages. This was the church that saw apartheid as *sacred*, basing its views on the Christian Bible. Christians conveniently don't ever mention this when claiming that religiosity brings about morality. It's just swept under the rug.

As these contrasting examples show, very often, when religious people do *magnanimous* acts, plenty of credit is dished out to religion, but when religious people carry out *horrid* acts, it's claimed not to be religion's fault but merely the unfortunate result of human misinterpretations.[6] It seems hard to deny that there is a double standard going on—a kind of selective history-writing—when religious people, in talking about how Christianity impacted apartheid, cite only Desmond Tutu, the favorable side of the coin, and neglect the other side, which was just as evil as Tutu was good.

One really has to make up one's mind. Either religion is a driving force for both good and evil, or it is a driving force for neither.

The Natural Roots of Morality

If there is no god who created the principles of moral behavior, is morality then completely subjective and up for grabs? In a godless world, why would it be immoral to make another human suffer, or to torment a guinea pig to death? *Can there be any moral principles at all if they are not handed down to us from on high?*

A starting point for thinking in a secular way about morality is the following simple trio of axioms:

1. Happiness is preferable to unhappiness.
2. Freedom is preferable to captivity.
3. Not suffering is preferable to suffering.

From this primitive-seeming starting point, we can derive many further and more specific moral norms. Such norms have been clearly articulated in numerous turning points of human history, such as the French Revolution, marked by its memorable watchwords *liberté, égalité, fraternité,* and the earlier American Revolution, marked by its famous phrase "the right to life, liberty, and the pursuit of happiness."

Research in evolutionary psychology has clearly shown that other animals, such as chimpanzees, have the ability to act morally.[7] And thus morality is not an exclusively human trait.

But how did morality come out of nowhere?

Research has demonstrated that the origins of moral behavior lie in evolution. If living beings are guided by moral principles and act unselfishly or altruistically toward other members of their group, the chances for the group's survival go up.

A rudimentary form of morality can be found in certain other animal species, especially apes, but even vampire bats exhibit a limited sort of morality. (I am referring here to the animal species, not to the vampires featured in horror stories, which of course are not particularly moral creatures.) If they behave altruistically in certain types of situations, their chances for long-term survival will be enhanced. Vampire bats hunt for food during the night, sucking blood from cows and other mammals. Because they have a high metabolic rate, they have to seek food every single night, but not all of them find food every night. Successful vampire bats therefore share their food with their less successful conspecifics, and the latter, on subsequent nights, will return the favor.

It also happens that certain kinds of animals warn one another, using various sounds, when they spy a predator approaching their flock. They do this even though, by making the warning sounds, they will draw the predator's attention to themselves and thus will increase their own chances of being caught and eaten. It seems quite reasonable to assume that such animals are not religious, and from this it would follow that altruistic behavior, such as warning others at one's own peril, has its origin in biology, rather than in sacred texts handed from Providence.

The human race, like all other animal species, must continually struggle to survive, and cooperation has shown itself to be a very effective strategy in this struggle. It seems likely that humans, in an early stage of their development, realized that they could undertake and accomplish large projects if they banded together and worked peacefully, side by side. They could also protect one another from threats if they cooperated. Unselfish members of a tribe lived longer than selfish ones.

Thanks to natural selection, then, the tendency to cooperate became incorporated in the human strategy for survival—that is, in the human genome. One of the main functions of genes is to propagate our set of traits to the next generation. Although genes themselves always act "selfishly," we humans who carry selfish genes can, thanks to that very property, act in self-sacrificial ways in certain situations. For instance, we will lay down our lives in order to save our children, because we are "programmed" to propagate our own genes, and our children share our genes. From a strictly evolutionary perspective, the survival of an individual's genes is more important than the survival of the individual.

Researchers who investigate the origins of moral principles often speak of *reciprocal altruism*. An example of this is the following idea: "If my group succeeds in bringing down a mammoth, then we will have more meat than we need. But we don't often have such good luck. If we share our food this time with the neighboring group, then we'll increase our chances of getting a little food from them the next time they bring down a mammoth." The general principle can be formulated as follows: "If I help you now, then hopefully you'll help me on another occasion." This type of altruism arises often in the animal realm, and most of all among our nearest relations, the chimpanzees.

A skeptic could argue that if altruism is built into our genes, then it is almost as if altruism had been handed down from "on high" (where natural selection plays the role of Moses's tablets). If that's the case, then when we are kind and generous to others, we aren't following our free will; we are actually just behaving mechanically, because we were *programmed* (by evolution) to behave in that type of kind and generous manner. Instead of the Bible or the Koran being our holy scripture, it's our DNA. In which case, maybe we humans aren't so noble after all . . .

Empathy and Identification

But there is another and very different way that altruism, generosity, and compassion arise in the human world, and that phenomenon comes directly from our intelligence. Douglas Hofstadter, in a chapter called "How We Live in Each Other" in his 2007 book *I Am a Strange Loop*, describes a threshold that is reached when a brain (or mind) becomes representationally rich enough to be able to mirror *other* brains (or minds) in a deep way. The minds of most adult humans (aside from sociopaths) have crossed that threshold, which gives them a property that Hofstadter calls "representational universality," and about it he writes the following:

> Once beyond the magic threshold, universal beings seem inevitably to become ravenously thirsty for tastes of the interiority of other universal beings. This is why we have movies, soap operas, television news, blogs, webcams, gossip columnists, *People* magazine, and *The Weekly World News*, among others. People yearn to get inside other people's heads, to "see out" from inside other crania, to gobble up other people's experiences.
>
> Although I have been depicting it somewhat cynically, representational universality and the nearly insatiable hunger that it creates for vicarious experiences is but a stone's throw away from empathy, which I see as the most admirable quality of humanity. To "be" someone else in a profound way is not merely to see the world intellectually as they see it and to feel rooted in the places and times that molded them as they grew up; it goes much further than that. It is to adopt their values, to take on their desires, to live their hopes, to feel their yearnings, to share their dreams, to shudder at their dreads, to participate in their life, to merge with their soul.[8]

Examples of this kind of spontaneous empathy—of reaching out to others who are in tough plights—are rife in the lives of nearly all humans. Such acts are all rooted in the ability—in fact, the irresistible inclination—to map oneself onto another sentient being. Some fairly random but typical examples include not swatting a bee that one finds buzzing about in one's house, but carefully catching it and escorting it outdoors; picking up a lost dog, reading its nametag, and phoning the number on it; running after someone who has dropped their wallet and making sure that they get it back (rather than selfishly keeping its contents for oneself); or more mundanely, carefully cleaning up a repulsively soiled toilet that one finds when using an airplane lavatory, or simply leaving a big tip for a never-seen maid when one checks out of a hotel.

In none of these cases is there the slightest hope that the kind act will be reciprocated later; the desire to perform the act comes simply from *identifying* with the unlucky bee; or with the lost dog, or with its frantic owner; or with the unsuspecting wallet-dropper, who might well be panicked upon finding that it was missing at some later point; or with the anonymous future lavatory user, who one imagines would be viscerally repelled by the filthiness; or with the unseen maid, who will appreciate having a little extra spending money.

All these types of vulnerability or fear or shock or neediness in someone else are negative experiences that one has been through oneself at earlier points in one's life, and one thinks to oneself, "I certainly wouldn't want *that* to happen to *me*!" This is exactly what identification with another creature means. Or to put it another way, all these types of caring or joy or relief that one can choose to give to another being are positive experiences that one has been the recipient of oneself earlier in life, and so one thinks to oneself, "How much I'd appreciate it if *that* were to happen to *me*!"

Once again, this projecting of oneself into another's life (or vice versa) is the meaning of identification, and identification with others is the deep facet of our human nature that gives rise to empathy. It has nothing to do with having been genetically programmed to be altruistic—and even less to do with "sacred" moral codes that have been handed down from on high.

All of the foregoing shows clearly that people (as well as animals) can be good, compassionate, and moral, without any need for a divine being by whom they have been instructed to act in certain ways.

~

INTERLUDE: ON CRIME, PUNISHMENT, AND RESPONSIBILITY

When is one responsible for one's actions? The tempting answer is: always. But are things actually that simple?

The idea of moral responsibility presumes that we have free will or, more precisely, that we have the possibility to choose what actions we will take. If everything that we did were predetermined, then we could never make choices, and as a result, we could not be held responsible for our behavior. Let us thus presume that humans can make choices. As long as we are not under threat of being shot or tortured, whenever we make a choice of how to act, we are morally responsible for what happens. But are there exceptions to this rule? I once read of a legal case that made me think long and hard about this, leading me to wonder if things are actually as simple as one's first instinctive reaction would suggest.

A mild-mannered man—a carpenter by trade—was sitting on a bench in a park with his fiancée. He had never committed any crime and had never exhibited any tendencies toward violence. And yet suddenly, without any warning, he began to speak in a disjointed fashion, stood up, went behind his fiancée's back, pulled out a knife that was in a pocket of his work clothes, and stabbed her.

The woman somehow managed to escape and notified the police. Her fiancé was captured and underwent a series of psychiatric and neurological tests, in the course of which a highly aggressive brain tumor was detected. The tumor turned out to be operable, and after it had been successfully removed, the man regained his gentle personality and showed no further violent tendencies. According to the doctors, it was almost certain that the tumor had caused his aberrant behavior. And so, should the man be imprisoned for his tumor-induced crime?

Things can get considerably more complicated than that. Not long ago, a gene called "mao-A" was discovered, which is responsible for the production of a protein that regulates the rate of uptake of the chemical serotonin in the front part of the brain. A tiny percentage of people suffer from Brunner's syndrome, which is a certain type of mutation in the mao-A gene that results in a vast overproduction of serotonin and related chemicals. As a result, such people lack the ability to control their impulses, and they act destructively, aggressively, and violently toward people or animals in their vicinity. They also frequently injure themselves. Such people are extremely dangerous.

To what extent should individuals suffering from Brunner's syndrome be held morally responsible for their actions? Since they cannot be called mentally ill, they can be brought to court and tried. And yet, thanks to today's scientific understanding of the brain, one can definitively say that their ability to make choices has been greatly reduced, if not totally destroyed, by the genetic mutation that afflicts them. How should society deal with such people? How should they be judged if they commit crimes?

Today's rapid progress in biology and neurology is presenting us with an ever-increasing number of such ethical dilemmas, and in the future, we will find ourselves facing all sorts of thorny questions about moral responsibility. In the light of an inevitable flood of unpredictable discoveries about the brain, and thus about the essence of human nature, many of our traditional ideas about moral responsibility, often rooted in ancient religious texts and teachings, will start to look simplistic and incomplete. Many of them will go completely by the boards. As a society, we will have to learn to think in profoundly new ways about what it means to say that a person is morally responsible for what they do. Only when one is freed from the shackles of old dogmas can one hope to tackle this type of challenge honestly and seriously.

CHAPTER EIGHT

NEW AGE BELIEFS AND
THE CRISIS OF REASON

Concerning People's Beliefs in Strange Ideas

Every error is due to extraneous factors (such as emotion and education); reason itself does not err.—Kurt Gödel, November 29, 1972

We now shift our focus from traditional religions to what is often called "the new spirituality" or, more commonly, New Age beliefs. In certain respects, there is no sharp boundary between the older and newer styles of belief, but overall one can say that, compared to traditional religions, New Age ideas amount to a less systematic ideology and have fewer theoretical bases to rest on. In fact, the set of New Age ideas could be likened to a *smörgåsbord* (literally, a "sandwich table" in Swedish), where belief-hungry folks are free to pick and choose those items that they find most appetizing and ignore all the rest. The New Age culture is individualistic and, as we'll see, quite often self-centered. Moreover, it is also characterized by a cynically commercialistic aspect, for it turns out that there is a great deal of money to be made by exploiting the belief-hungry.

Unreasonable Beliefs

A recent poll in the United States showed that roughly 32 percent of the population believes in ghosts, 31 percent in telepathy, 25 percent in astrology, 21 percent in communication with the dead, and 20 percent in reincarnation.[1] None of these notions enjoys any support coming from direct knowledge of the world, or from experience with the world's workings. Rather, such ideas are almost always based on uncorroborated hearsay and rumors about events and happenings of various sorts. Nonetheless, many people believe wholeheartedly in such things.

Pop culture is filled with myths that many people unreflectingly accept. For instance, it is often said that people go crazy when there's a full moon, that people who suffer from schizophrenia have "split personalities," that cancer can be cured by a positive attitude, or that infants will become smarter if Mozart is frequently played to them as background music.[2] Although pop psychology abounds with mistaken or speculative ideas, self-help books incorporating such ideas are commercially very successful.

But shouldn't we all just believe in whatever we wish to believe in? Of course we can do this. Every person has the indisputable right to believe in any idea at all. But the key question is this: Is there any utility in striving to believe in ideas that are true rather than in ones that are untrue? I personally believe that there is, and that it makes sense to use well-developed strategies to attain this goal.

If one wishes to believe in what is plausible rather than what is implausible, then one must carefully consider the nature of the difference between plausible and implausible beliefs. Slightly oversimplified, the difference is this: Implausible beliefs are those that have no support in, or that run counter to, contemporary knowledge. Typical examples of implausible beliefs include the following:

1. beliefs that are contradicted by most experts in the area;
2. beliefs that are logically impossible or extremely improbable; and
3. beliefs that are based solely on anecdotes and hearsay rather than on scientific findings.

Positive Thinking and Pseudoscience

In New Age circles, one often encounters a distinct scorn for critical thinking. Instead, New Agers speak of the importance of positive thinking (as if critical thinking and having a positive outlook on life could not coexist): "Don't be so negative! Don't go around spreading negative energy! Think positive instead!"

Now who wouldn't want to have positive thoughts? It is clearly important and worthwhile to try to be upbeat and to try to think creatively and along new pathways. But that is not the same thing as being uncritical. Refusing to formulate one's own judgment about whether a claim is believable is hardly positive thinking. In fact, it is actually adopting a negative stance about one's own powers of thought.

The expression "critical thinking" is also often interpreted erroneously. Some people think that it means being negative, whiny, and contrary. But that is far from accurate. In this connection, it is helpful to think about what a film critic does—namely, writing reviews of movies. Sometimes a critic will say that a movie is wonderful, adding glowing comments about the actors and other aspects of it. Other times a critic will write that a film was poorly directed or that the story it tells stretches one's beliefs a bit too far. Some films will get high grades, others low grades. But to be a film critic certainly doesn't mean that one is constantly critical. If that were the case, then movie reviews would be of very little use to anyone.

Critical thinking means no more and no less than that one thinks about claims one hears and tries to judge them oneself. It is based on using one's ability to reason carefully and cautiously. If a claim is checked out and passes all of one's critical tests, then it should be accepted.

Advocates of the ideas of the "new spirituality" keep their distance from the scientific way of testing ideas for truth. They don't think that their claims of miracles or other strange phenomena need to be verified in a laboratory. And when you think about it, taking such an attitude is perhaps not so weird, since if the phenomena in question were actually carefully investigated, they would be revealed as groundless.

Sometimes the New Age culture uses what sounds like scientific language. Various claimed phenomena are described using supposedly advanced theories of "energies," "meridians," "zones," and "chakras" in the human body (a chakra, in Hinduism and Buddhism, is an alleged "center of vital energy"). In certain New Age movements, there are even instruments that supposedly measure "energy fields" or other forces unknown to science.

When someone makes a claim for which there is no scientific evidence but uses scientific-sounding jargon to describe it, they are engaging in pseudoscience. The language being used sounds on the surface like that of a scientific article, but behind the words there is no substance. Often practitioners of pseudoscience will borrow various terms from physics, particularly from quantum physics, to make their ideas sound very sophisticated. Unfortunately this cheap technique misleads many people since most people's scientific knowledge is quite scanty.

Nikken as a Case Study

A well-known example of a business that exploits pseudoscience is the Japanese firm Nikken (https://www.nikken.com), launched in 1975. This company specializes in selling allegedly scientific products of various sorts, and it has had considerable success around the world, including in Sweden. Nikken's sales strategy works through pyramidal schemes, where each franchise dealer recruits new sellers "below them," thus building up a hierarchical network of their own sellers.[3]

Nikken organizes sales meetings in which newly recruited sellers are given pep talks by more experienced sellers, who attempt to demonstrate the great success of Nikken products. The feverish atmosphere that pervades such meetings is often reminiscent of the revival meetings of fundamentalist religious sects. I myself attended such a meeting, at which we attendees were all told that if we started to work with Nikken products, we would soon be not only much happier but also much richer; however, as soon as I raised some skeptical questions, I was rapidly escorted out of the meeting.

One of Nikken's best-selling products is KenkoInsoles, a type of magnetic insert for shoes. These inserts are claimed to possess a healing force and to emanate various sorts of energy. Let's take a look at how Nikken exploits scientific-sounding language to give the impression that its products have a serious scientific basis. The following text is taken from a brochure advertising a number of Nikken products.

> Magnetic energy is an intrinsic part of the body's natural processes and part of the natural environment for all living things.
>
> An ancient principle and significant component of Far Eastern cultures and traditional Chinese medicine (TCM—the foundation of healthcare in China and Japan for thousands of years) is that of flows of energies in the body. This has led, for example, to the West recognising the value of alternative practises.
>
> In addition, the magnetic field of the earth has changed substantially over human history, even reversing many times over the planet's history. Modern civilisation has also played its part, with the growth of electromagnetic fields and radiation, since the end of the 19th century, all of which inter-react with the body's natural energy fields. Approximately 3% of the population in Sweden are officially recognised as suffering functional impairment due to electromagnetic fields in the environment; this is a growing problem.
>
> The use of magnetic technology was developed by Nikken to counterbalance this effect, and make our surroundings more like the natural magnetic field that has protected human beings for millions of years. Nikken's magnetic products help to restore the conditions in which human beings are meant to live and thrive.

The first application of Nikken's Magnetic Technology was the design of the Nikken KenkoInsoles®, the company's first product when launched in 1975. Since then, Nikken's research teams have continued the process of research and found that the more complex the fields, the more effective they are on the body. As a consequence, we now offer four major developments:

Field Gradient Technology: Nikken products such as the Power-Patch®, Kenkoseat®, and KenkoDream® Quilts use bipolar magnets, either singly or in a multiple-magnet array. The arrays are precisely spaced to produce a pattern of magnetic flow (flux lines) between the magnets, creating a complex field gradient of magnetic energy utilising multiple polarities and flux.

EQL Magnetic Technology: Equilateral Magnetic Technology uses a patented design that increases the magnetic surface activity by maximising the number of flux lines possible on a flat surface. It multiplies the number of positive and negative poles, thus creating more lines of force crossing this surface. The practical result of this approach is that the magnetic field is perfectly uniform over the entire area.

Dynamic Magnetic Technology: A dynamic magnetic field is more complex than a static field, and the Magboy® and MagCreator® achieve this effect with two rotating magnetic balls. This produces a deeply penetrating 3-dimensional magnetic field—and a soothing and relaxing massage roller as well!

Magnetic Tension Technologies: Two recent product developments include Nikken's patented Magnetic Tension Technologies, which start to bring some of the 3-dimensional complexity of the Dynamic Technology into a static form. RAM® (Radial-Axis Magnetism) Technology Modules feature small hi-tech magnetic spheres in a stepped alignment that produce a series of overlapping magnetic fields in tension, while the latest Dyna-Flux® Technology has been incorporated into the new PowerChipTM.

Is everything crystal-clear so far? Well, if it isn't, don't worry. . . . No professor of physics can make head or tail of any of this doubletalk, either.

In the Nikken sales meetings, sellers learn, through a simple procedure, how they can convince would-be buyers of the fantastic effect of the magnetic inserts:

1. Have the subject (the person to be persuaded to buy the product) stand on the floor in their stocking feet. Ask them to place their hands behind their back.

2. Place yourself behind the subject and gently pull their hands down and back further. The subject will quickly lose their balance and will have to take a step backward to keep from falling.

3. Repeat this procedure but this time with the subject wearing Nikken magnetic inserts. This time around, amazingly enough, the subject will feel stronger and will not need to take a step backward!

The seller will now explain with pleasure that the magnetic inserts provide both energy and force, allowing the person to maintain their balance. This seems to be a highly effective and reliable demonstration, and the magnetic inserts seem to have done their job very well.

There are a few aspects of this experiment, however, that give pause for thought:

1. The idea that these magnetic inserts can create or provide some kind of unknown energy or force that opposes gravity and allows a person to stay standing when thrown off balance runs completely against our knowledge of the laws of nature. There is no known force of this sort. This doesn't mean that we have proven that the whole claim is pure poppycock. Indeed, there just might be some kind of totally unknown new phenomenon and some new kind of force, even if this is extremely unlikely. In any case, thinking along these lines gives us further reason to reflect critically on the test as described.

2. The experiment was carried out in such a way that in the second trial, the subject was prepared for what was going to happen, and thus, whether consciously or unconsciously, adjusted their body in such a way as to stay standing.

3. We cannot be sure whether the second time around, the seller—either consciously or unconsciously—pulled the subject's hands a little less vigorously than the first time around.

These kinds of subtle factors can seriously influence the experiment. For this reason, we cannot be sure whether trickery is taking place, or whether some sort of placebo effect is involved. Of course nothing so far stated proves definitively that the experiment is fraudulent; however, if we are critical thinkers, we are certainly led to be on our guard and to insist on making more careful tests before accepting the claim that these inserts actually work.

Is there any way to determine the truth definitively in such a situation? Yes, there is. We can do a double-blind test. First, we make a second pair of inserts that look exactly like the Nikken inserts, but that have no magnets in them. Next, we bring in a tester who does not work for the Nikken organization. Then we have a third person (again not involved with Nikken) hand out the pairs of inserts in a randomized fashion, so that neither the subject nor the tester knows whether the pair being currently used is magnetic. Then we carry out the experiment several times in a row, always randomly choosing which inserts to use.

When the test is carried out in this fashion, we soon discover that there is no longer any effect. The subject is equally likely to lose their balance in either case. Whatever cause might have produced an effect before this double-blind test was carried out is irrelevant. Was it a placebo effect in the subject's mind? Was it shameless fakery on the part of the seller? Or could it have been wishful thinking in the seller's mind? We don't know, and it doesn't matter. All we know for sure is that the effect does not arise from any kind of special forces produced by the magnetic inserts, and that's what we were trying to ascertain.

This case study demonstrates a very important idea. Taking a skeptical attitude toward the world does not mean that one should be able to account scientifically for every phenomenon that is claimed to take place. A more modest stance is simply to check carefully whether or not the claimed phenomenon actually exists.

There is no reason for us to analyze carefully the pseudoscientific mumbo-jumbo that supposedly explains the wonderful powers of the magnetic inserts if in fact those powers completely vanish when a double-blind experiment is carried out. Such fancy-sounding phrases as "Field Gradient Technology," "DynaFlux Technology," and "Radial-Axis Magnetism Technology Modules" don't really matter if people are equally likely to lose their balance with or without the inserts labeled by those terms.

Selling magnetic inserts is just one of innumerable unscrupulous techniques for making a quick buck off of gullible people. Mastering the art of clear thinking helps one not to fall into such traps.

Homeopathy

Another example of a commercially successful business based on quack medicine is the "alternative medicine" known as homeopathy. It is based on a theory proposed by German doctor Samuel Hahnemann (1755–1843). At that time, it was popular to use the bark of plants such as Cinchona to combat high fevers. Hahnemann noted that Cinchona bark produced symptoms in healthy people that were similar to those in sick patients whom the product was supposed to cure. This observation suggested to him a notion that he expressed in Latin as "Simila similibus curantur" ("Cure an ill by something resembling it"), and from that came the disease-fighting approach that he called "homeopathy" ("similar disease").

According to the homeopathic theory, various symptoms and illnesses can be cured by using substances that cause these problems, but in a very diluted form. Hahnemann believed that excessive doses of a medicine could do more harm than good. This idea in itself was an intelligent insight; however, the way he then proceeded went in a completely different direction. He claimed that the more diluted a medication was, the more effective it would be. He coined the term "potentization" for increasing the dilution (and thus, in his theory, the effectiveness) of a given medication. The process of dilution was also supposed to be carried out in a particular manner, following certain rituals, to work properly.

Homeopathic medications are thus diluted over and over again, by a factor of many millions—in fact, in the end, not a single molecule of the original curative substance remains in the preparation. Homeopaths explain that the water that remains after full potentization retains a "memory" of the curative substance and thus that it inherits magical powers from it.

This kind of medication is popular in such countries as France, England, and Germany. In Sweden, homeopathic medicines can be found in many health-food stores, as well as in other outlets for alternative medicine.

The only kind of effect that a homeopathic medicine can have is as a placebo. This has been carefully researched and demonstrated. Double-blind experiments have been carried out both in Europe and in the United States. Defenders of homeopathy often say, "Maybe science hasn't yet been able to show that it works, but that doesn't mean it doesn't work. There are lots of things we don't understand." But this is fallacious logic. Science has indeed been able, thanks to very careful research, to show that homeopathy does not work. (The central tenet of homeopathy—that the more diluted a substance is, the stronger is its effect—leads to some quite absurd consequences. For instance, if you are supposed to take a homeopathic treatment every day, then you should be extremely careful not to miss any day, because if you do miss a day or two, you might die of an overdose! This kind of viewpoint is clearly nonsensical.)

Exploiting the Cachet of Antiquity and that of Science

Many theories build on ancient superstitious myths about how the world works and how people work. The advocates of such theories eagerly point out how their ideas are grounded in venerable traditions, as if this heritage in some way or other makes them more legitimate or more trustworthy.

Unfortunately, the fact that teachings and methods are old is not an automatic guarantee that they are good. For a very long time, people tried to cure sicknesses by bloodletting. The blood of sick people was sucked out by leeches in the hopes of rendering them healthy. The widespread use of this method was a consequence of erroneous theories of the role of blood in diseases. The fact that people believed in bloodletting for a long time doesn't mean, however, that it constitutes a valid method of curing diseases. The people of the long-gone past can be excused for believing as they did, because they had no access to a systematic body of medical and biological knowledge as we do today. We should not criticize them for that lack. But anyone who today would advocate bloodletting as a general cure for illnesses is doing so despite vast amounts of contrary evidence. (Leeches are still used for medical purposes in certain very rare situations, but for completely different reasons than in the old days.)

Today, the New Age culture tries hard to make hay from the high esteem in which science is held in society. Scientific concepts such as energy, vibrations having certain frequencies, and even quantum mechanics are glibly exploited by people who don't have the faintest idea what they are talking about. We are awash in treatises written by "prophets" who try to buttress pseudoscientific ideas using complex and abstract concepts (or at least words!) borrowed from genuine science.

Methods and procedures that build on the genuine knowledge that we have today are to be preferred. Of course this doesn't mean that today's medicine never makes mistakes, but merely that the methods that are used today are at least better than those that are built on ignorance and superstition.

Sometimes pseudoscientists exploit purely linguistic trickery to mislead the public. Thus in numerous books of New Age inspiration we can read about the nutritive needs of cells and the importance of eating well. Cells require energy (energy in its technical sense, in terms of physics and biology) to be able to carry out their work. So far so good. But in the same books we will read about the dangers of associating with people who "take energy away" from you. If you hang around such people too much, you will "lose energy." We are told that we have to protect ourselves from "energy leakage, not just from our own bodies but also from someone else who is trying to steal our energy"—in fact, such a person is called an "energy vampire." Here, the word "energy" is being used in a distorted fashion—as a metaphor for a psychological and emotional characteristic.

A similar example of linguistic confusion occurs when we say, "After I exercised at the gym today I had so much energy!" From the point of view of physics and biology, one cannot gain energy from exercising; indeed, quite the contrary—one uses up energy while exercising. Afterward, of course, one can feel more fit, quicker, livelier, happier, and more "filled with energy."

To use the notion of energy in this kind of metaphorical sense is not wrong in itself; however, if one inserts an equals sign between these two notions of energy, one scientific and the other metaphorical, then one has left the realm of science and verged over into that of pseudoscience. Seldom if ever, in New Age writings, is a distinction drawn between different meanings of the same term. The conclusions involving such notions are therefore erroneous and unscientific. Unfortunately this type of slippery linguistic usage is very typical of the New Age culture. Concepts that sound scientific but that in reality are just sloppy metaphors are systematically exploited, and such usages mislead many people, tricking them into believing that something serious lies behind what is being asserted.

Quantum Physics and the New Age

It's popular among New Agers to allude to research in quantum physics. The basic idea is that research into such fundamental domains justifies such trendy notions as "collective consciousness" and the idea that we can leave our bodies via "astral projection."

These days, books that purport to find deep links between quantum physics and New Age ideas are a dime a dozen. In them we read about the implications of quantum physics for leadership techniques, for clear thinking, for good health, for the understanding of consciousness, and on and on. Writers such as Danah Zohar and Deepak Chopra have promoted such views, and with enormous commercial success. But the actual content of the books they write is unfortunately not science at all; it is a hodgepodge of pseudoscientific claptrap and trivial truisms.

Quantum physics deals with the most basic constituents of matter, and in it one encounters phenomena that genuinely defy normal understanding. Moreover, quantum mechanics has forced physicists to rethink certain key assumptions about the innermost nature of matter. But this does not mean that from quantum physics one can validly draw simplistic conclusions about other scientific areas, such as biology, psychology, or research into the nature of consciousness.

Quantum mechanics does not tell us that God most likely exists, or that when one meditates sufficiently deeply, one can levitate off of the ground and hover in the air. To claim that the laws governing infinitesimally small particles apply also to human beings is to commit a serious error. It makes no sense to make such vast leaps. What does make sense is to test new theories that spring out of observations and experiments. People can always dream up theories based on anything that springs to their minds, but if a theory doesn't find confirmation in experiment, it must remain in the realm of sheer speculation.

Category Errors and the Ladder of Sciences

In a crude approximation, one can think of the various scientific disciplines as describing different hierarchical levels of reality. Thus at the highest level, we are formed by our culture, our society, our economy, our aesthetics, and our laws. One level down lies the psychological level, where we are formed by and influenced by thoughts and emotions. Below that comes the level of biological processes, which in turn are made up of chemical processes that obey the laws of molecular interactions. And then molecules are made up of atoms, which are described by the laws of physics. The processes inside atoms are described by quantum mechanics, which is a branch of physics. Overall, then, our perspective on the world is a multitiered structure, a hierarchy, which can schematically be represented in the following fashion:

Society
Culture, economy, law, art, literature, aesthetics . . .
⇓
Psychology

Decision making, perception, dreams, emotions . . .
⇓
Biology
Cellular processes, proteins, synapses,
metabolic processes, evolution . . .
⇓
Chemistry
Molecules, chemical reactions, chemical bonds . . .
⇓
Physics
Gravity, electricity, magnetism, quanta,
fission, fusion, atomic energy . . .

Of course, what happens on any particular level depends on what happens on lower levels. In that sense, the different levels are linked to each other. But there are phenomena on each level that have their own special qualities, and so we cannot automatically assume that concepts and phenomena that belong to one level also belong to other levels. The fact that we humans have thoughts and experiences does not imply, for instance, that such microscopic entities as electrons, carbon atoms, and DNA molecules, or such macroscopic entities as universities, companies, and nations, also have thoughts and experiences. To draw such a conclusion would be to make a "category error." This concept can be further explained with the following example.

Suppose that we have in safekeeping a very old parchment document dating back to the early eighteenth century. On the parchment is written, in dark black ink, our country's Constitution. Indeed, this precious document is the foundation of our entire legal system. Suppose further that early in the twentieth century, researchers were fascinated by the way the ink had remained readable after such a long time. After analyzing it using the best techniques available to them, the researchers unanimously concluded that the reason for the ink's durability was the presence of a certain chemical in it.

Now fast-forward to the present. A hundred years more have passed, and our knowledge of chemistry has greatly developed. Researchers once again analyze the ink in the old parchment. This time, however, they conclude that the ink doesn't have any of the chemicals in it that earlier researchers had thought. The secret of the ink's continuing darkness, legibility, and durability turns out to be a totally different chemical constituent than people had thought a hundred years ago. And so we have to completely overhaul our explanation of the ink's enduring blackness. So far, no problem . . .

New knowledge forced us to revise our story about the ink on the parchment. But does that mean that we also have to revise our interpretation of the legal meaning of the text written using that ink? Should we now feel obliged to totally overhaul our legal system? No, obviously not. To jump to such a conclusion would be to make a grievous category error. The chemical composition of the document's ink has no bearing whatsoever on the document's legal content, which resides on a higher, more abstract level.

The notion of a category error can perhaps be conveyed more simply and more vividly as follows: A poem does not become less beautiful for being written in ink of a different color.

When is it justifiable to take a high-level perspective, in the sense of focusing on the whole rather than on its basic constituents? Isn't everything tied together? If we think of science as being organized into an unambiguous, hierarchical set of levels, doesn't that make science seem unduly rigid?

To be sure, considering entities as wholes is very important. We see this clearly in medical research, which constantly reminds us that mind and body exert influences on each other and are intimately coupled together—and are, in fact, one and the same thing. Scientific research should always try to take a broad perspective and should, above all, be daring and creative.

New theories are being formulated all the time, and among them some will surely be quite radical. But to become valid parts of science, they must pass muster empirically. No scientific conclusion is valid without being confirmed by numerous laboratory tests. The idea of category errors has to do with the kinds of conclusions that can and cannot legitimately be drawn from new scientific findings. It says nothing about what is true and what is not true. It could perfectly well happen that at just the same time as some chemistry researchers make new discoveries that give us a much better explanation of the properties of the ink used in an old parchment, legal scholars might also discover that there are reasons compelling us to drastically revise our understanding of the legal meanings of the text on the parchment—but if that should happen, the two events would be unrelated. The legal finding would be totally independent of the chemical finding. The two findings, concerning two very distant levels of the parchment, would have nothing to do with one another.

Similar comments apply to quantum physics and ordinary life. New Age people enthusiastically talk about quantum physics and its "holistic" perspective. Well, in quantum physics there are certainly many remarkable phenomena that we still do not understand, but this does not mean that these mysterious aspects of quantum physics will necessarily explain mysterious phenomena in other areas of life. For instance, the quantum phenomenon known as "entanglement," whereby the states of two particles that are widely separated in space are intimately linked together, is undeniably extremely remarkable, but it does not lead to the conclusion that we are all just tiny parts of one great collective consciousness, or that thought control is possible, or that the cosmos has a long-term plan for humanity as a whole. If we fall for such ideas, we are simply making elementary category errors.

None of this belies the fact that the quantum world has given us beautiful new metaphors that we can use in talking about all sorts of everyday phenomena. But when they are applied to our human level of existence, these are just metaphors and nothing else. Unfortunately, quantum physics and other areas of science are abused by New Agers who like to paint the world in more mystical and thrilling terms (and some of them also like to make a quick buck off of naïve and gullible folks). For example, it's sometimes claimed that there is no reality other than what our senses tell us, and that our consciousness governs what happens in the world, or that all phenomena all across the universe are somehow mystically tied together. All of this is stated using pseudoscientific gibberish that to the untrained ear sounds like it is expressing profundities of quantum physics.

Pseudoscientists often borrow scientific concepts without being in the least troubled by doing so in a sloppy and slapdash manner. Although it is merely window dressing, their seeming use of science makes their viewpoint come across to scientific novices as fresh and exciting, without the writers having to seriously back up any of the claims being made.

A Case Study: Astrology

One should never underestimate the powerful lure of pseudoscience. The ancient doctrine of astrology is an excellent case in point. Many people take astrology with a big grain of salt, reading horoscopes in the newspaper merely for entertainment, but for other people it is a very serious matter. Astrologers earn their living by giving advice and supposedly telling people what they really are, how they feel, and what fate will befall them. All this, of course, is courtesy of the distant stars.

Horoscopes are an ancient manner of fortunetelling based on the positions of heavenly bodies at the time of someone's birth. Advocates of astrology wish to call the act of divination by horoscopes a science, but it is a far better example of a pseudoscience.

A poll conducted by Swedish magazine *Vetenskap & Allmänhet* (Science and the Public) found that roughly one Swede in four considers astrology to be a science. This figure is roughly the same as in other Western nations, and in Africa and certain parts of Asia, it is even higher.

In India, where belief in astrology is nearly universal, astrological "data" can dictate marriages, affect travel plans, and determine business decisions. On the *Skeptical Inquirer* website, one finds the following rather amazing news item:

> The lobby for Indian astrology had its crowning glory when, in February 2001, the University Grants Commission (UGC) decided to provide funds for BSc and MSc courses in astrology at Indian universities. Its circular stated: "There is urgent need to rejuvenate the science of Vedic Astrology in India . . . and to provide opportunities to get this important science exported to the world." Within nine months of the UGC's announcement, forty-five of India's 200 universities had applied for the UGC grants of 1.5 million rupees (about \$30,000) to establish departments of astrology.[4]

Believers in astrology sometimes defend their views by saying, "The moon influences the tides, so why shouldn't it affect human beings as well? After all, our bodies are mostly made of water! And why shouldn't the planets and the stars also be able to influence our lives?" All sorts of reasons can be adduced for the silliness of this argument, among which here are four:

1. Astrology is not primarily about the moon, but about the arrangement of the planets and stars in the sky, in so-called "signs of the zodiac." The planets and stars are enormously much farther away from us than is the moon, and none of them has the slightest influence on the tides.
2. No doubt our bodies have much water inside them, but the influence of the moon on the oceans is because each ocean contains a huge volume of water. The moon doesn't have any noticeable effect on such minuscule quantities of water as are found in our bodies.

3. Even if the moon affected the water in our bodies in the same way as it does tides—namely, by pulling us a little bit with its gravitation—this would have nothing whatsoever to do with determining our personalities or the events that will befall us in the course of our lives. In any case, the earth's gravity has a far greater effect on us than does the moon's.

4. If the moon affects the water in our bodies, and if that effect somehow determines our personality, then why should it be that this crucial effect takes place only at the exact moment of our birth and at no later time during our life?

To cite how the moon brings about the tides as an argument for the validity of astrology is completely off base. But astrologers can always try to counter scientific objections by saying, "We don't know all the laws of the universe. Science can't explain everything. So why couldn't things be as we say they are?"

This is an argument that in principle makes sense. Astrology might conceivably say something true about reality, even though it runs totally against the grain of today's understanding of the nature of the universe and despite the fact that we couldn't even begin to explain why it was true. If astrology had even a grain of demonstrable truth, however, we would have to profoundly revise our views of the nature of the universe. The chance of astrology being true is as close to zero as one can imagine ("a snowball's chance in hell," as an old phrase goes), but in principle it could happen. Nonetheless, it makes sense to dismiss astrology, as we shall see in the explanation that follows.

Astrology has been the subject of comprehensive scientific studies in the past fifty years. A typical piece of research might have the following form:

1. Determine what astrology considers to be a definite link between, on the one hand, a specific sign of the zodiac and planetary alignment with, on the other hand, a specific personality trait or bodily feature.

2. Collect data concerning a large number of people's horoscopes and their personality traits and bodily features.

3. Statistically evaluate whether these traits and features occur more frequently among people with the "right" horoscopes than would be expected if the distribution were completely random.

Many such studies have been carried out, and they all show that the astrological predictions are no better than random guesses.[5] What is interesting about these kinds of studies is that they do not involve in any way the mechanisms through which astrology might work, and they are thus independent of whether we understand astrology.

There are also studies that show that people eagerly take astrological claims as true and as confirming their own prior opinions. For instance, when five hundred people were given an identical horoscope-based personality description, and each of them was told that this description was uniquely their own, the majority of them claimed to recognize themselves in the description. (Such an experiment was once carried out in France, and in it, all the subjects were given a horoscope drawn up on the basis of the birth data belonging to an infamous French murderer.) This kind of effect gives us all the more reason to dismiss astrology.

Two British social psychologists, Martin Bauer and John Durant, looked into the question of what kinds of people are drawn toward a belief in astrology. As might be expected, they found that there was a clear statistical correlation between skepticism and the subjects' level of scientific knowledge. The more scientifically knowledgeable a person was, the less likely it was that the person would believe in astrology.[6]

To sum up, astrology is a pseudoscience that doesn't merely lack experimental support, but that has in fact been decisively refuted by a large number of experiments. The "scientific" theory that astrology has the power to tell people about their personalities and their fates is thereby falsified.

Second Case Study: The Myers-Briggs Test

Since time immemorial, we human beings have needed to simplify and schematize our lives by trying to put each other into neat little categorical boxes. Our feeling of control is greatly enhanced when we believe that we are able to stick other people into unambiguous, sharply defined pigeonholes. Perhaps having the sense that we know exactly who we are dealing with serves to calm us and to reduce our anxiety. Astrology has lent a helping hand to this activity for thousands of years.

In the nineteenth century, people tried to satisfy this psychological need with new methods. The biology of race became popular, as did phrenology (the doctrine that claims that a person's personality can be read from examining the shape of their skull). Today we know that these notions not only are pseudoscientific and groundless but also are morally harmful. Nonetheless, the psychological need to pigeonhole others has not gone away.

One of today's most popular personality tests, both in Sweden and in other countries, is the Myers-Briggs Type Indicator (MBTI). In companies, the MBTI is often used to characterize the psychology of the members of a work group, the goal being to make the group function more smoothly. This test was designed in the 1940s in the United States by two amateur psychologists, Isabel Briggs Myers and Katharine Cook Briggs, who were inspired by theories that had been proposed by Swiss psychologist Carl Gustav Jung. According to the MBTI, there are sixteen distinct personality types (four more than in astrology, which has twelve, one for each sign of the zodiac).

The MBTI involves four scales or dimensions, each of which is characterized by two opposing extremes, as follows:

Extrovert (E)—Introvert (I)
Sensing (S)—Intuition (N)
Thinking (T)—Feeling (F)
Judging (J)—Perceiving (P)

This means that for each type there are roughly 450 million people on earth today (assuming the sixteen types are equally distributed).

If you choose one of the two opposites for each of the above four dimensions, this will of course yield sixteen (= $2 \times 2 \times 2 \times 2$) distinct personality types (see figure 8.1).

The basic idea of the Jung-inspired MBTI involves contrasting types of behavioral traits, such as "realistic/imaginative," "rational/empathetic," and "thinking/feeling." Such contrasts, though they sound appealing, are not genuine opposites. For instance, to do well, a scientific researcher needs to be both imaginative and realistic. Even though modern psychological research has long since rejected Carl Jung's theories, that seems to have had little or no effect on the commercial success of the MBTI.

But even if these theories are mostly just New Age silliness, couldn't the Myers-Briggs test still reveal something about an individual's personality?

Given that the MBTI is one of the most widespread personality tests in the world, it has been thoroughly investigated by independent researchers. The upshot is that the test lacks reliability, in the sense that the results are

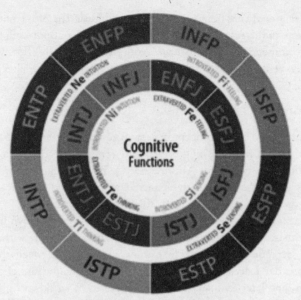

Figure 8.1. Distinct Personality Types

not repeatable. The simplest way to show this is simply to give the same test to a given individual on two different occasions, with a sufficiently long interval between them, and of course without telling the individual in advance what is being looked for. Even if the interval between the two tests is as small as five weeks, as many as half of the subjects wind up on the second test with a different personality type than on the first go-round. And yet the theory behind the test is that personality traits are innate and do not change during the course of a person's life.[7]

The MBTI test also lacks validity. Advocates of the test claim that certain MBTI personality types crop up much more often in certain professions, such as that of teacher. Comprehensive studies have shown, however, that the distribution of MBTI profiles in the teaching profession is not significantly different than in a randomly chosen subset of the population.[8]

In sum, the MBTI test belongs to the same pseudoscientific class as astrology. And yet Swedish corporations, and other corporations all around the world, are perfectly happy to fork over large sums of money to MBTI consultants.

It is conceivable that the test acts a bit like a placebo, in the sense that subjecting the members of a work group to the MBTI might give them the sense that they are being observed, and this could lead them to be more careful about how they treat each other. That would be nice! But to tell the truth, I would recommend astrology instead. Not only is it equally effective (which is to say, not at all, aside from the placebo effect), but it is far cheaper.[9]

Synchronicity

A commonly tossed-about concept in New Age circles is synchronicity (from the Greek prefix *syn-*, meaning "co-," and *chronos*, meaning "time," thus essentially a Greek version of "simultaneity"). It's not unusual for someone who is not otherwise particularly attracted to New Age ideas to have the sense that there may be something to this notion.

So what is synchronicity, then? The basic idea is that of two intimately related events taking place at exactly or nearly the same time, without there being any clear or understandable causal connection between them. For example, one day I decided to look up an old schoolmate from elementary-school days, and later that very day, she happened to sit down right next to me on a bus. Now one part of me thought, "Oh, what a nice random coincidence!" But another part of me thought, "This is just too improbable! It can't just be random! There must be some hidden meaning in this event!" That's what is meant by "synchronicity."

Like MBTI, this theory harks back to Carl Gustav Jung, and it has been eagerly seized upon by certain "life coaches" and in self-help courses of various sorts. The notion of synchronicity is, however, based on misunderstandings of probability. Let's take a simple example: Suppose that at the very moment I'm thinking about a dear friend she calls me on the phone. "How amazing! Synchronicity!" But clearly, the key question is this: "How many times have I thought about her when she didn't call me?" The fact that I happened to be thinking about her on one of the innumerable occasions when she has called me is not all that surprising.

The same thing holds for my running into the old schoolmate on the bus. It might seem highly improbable, but the question is this: "How many other improbable coincidences might have happened to me that day but simply didn't?" When seen in this light, the meeting in the bus wasn't such a remarkable event. As a matter of fact, very unlikely events are happening all around us all the time, yet we don't even notice the vast majority of them (and we shouldn't). Consider, for example, learning that the latest winning lottery ticket's combination was (7, 2, 5, 21, 13, 25, 37) and exclaiming, "Oh, how incredibly improbable!" You would be right, in a sense, since for that particular combination to win is extremely improbable, but of course it's extremely improbable for any preselected combination to win the lottery. The fact is, in an enormous lottery, something that is enormously improbable will happen, without fail. And in such cases, something highly improbable is not just probable but even inevitable!

We can try out a thought experiment that shows why the idea of synchronicity is implausible. Let's define an "extremely unlikely event" as being one whose chance of occurring is one in a million. That is, its probability is 1 in 1,000,000, or otherwise put, 0.0001 percent. Surely we all agree that an event with this low a chance of happening is a very improbable event. In our thought experiment, let us also assume that our brains can register roughly one conscious moment per second. That is, our brains register roughly two million separate moments per month (the time when we are sleeping is not counted). An "extremely unlikely event" should thus happen roughly twice per month, without any help from "synchronicity."

Why, then, is it so easy and so tempting to believe in synchronicity? Perhaps because it puts our own ego in the center of the picture. You and your life seem to become important to the universe itself: "This can't be random; the universe must have intended for me to have this experience just now. The universe's vast magical powers have colluded—and all just for me!" As this shows, the New Age culture is narcissistic, and belief in synchronicity is a particularly clear revelation of this fact.

Case Study: Hug and Bug

Let's take a look at a typical case of what some might call synchronicity. I have a friend who has confided in me that she surprisingly often sees cars whose license plates feature the letters "HUG" in that order. She sees such licenses so often that she's convinced that it can't be mere chance. Out of all the thousands and thousands of three-letter combinations that exist, why should "HUG" show up so often? She says there must be some hidden meaning in it, some occult reason that explains why "HUG" is pursuing her everywhere. Is it a secret message of love? After all, love and hugs are tightly linked.

Well, many cognitive errors are blurred together here. For starters, consider all the license plates that my friend sees day after day and doesn't notice consciously. Most of them, of course, go by without being read or thought about at all, although a few will be fleetingly read and immediately forgotten—in one ear and out the other. The thing is, if you start obsessing about "HUG" license plates, you are almost sure to notice all the "HUGs" that come within your field of view. This is a straightforward consequence of the nature of human perception. (If you choose to buy a red car because red cars seem very rare to you, the first thing you'll notice, once you've bought it, is that all of a sudden everybody seems to have a red car. How annoying!) We tend not to see things in which we have no interest. And so my naïve "HUG"-obsessed friend will pay no attention when "BUG" licenses, "HAG" licenses, "HUB" licenses, and "HWG" licenses by the droves go sailing by her eyes.

The question as to whether "HUG" is overrepresented in the set of currently valid license plates is purely a matter of fact, and should not be hard to check, by contacting the appropriate official government agency. Of course, it's always possible that that three-letter combination actually occurs on licenses more often than some others, for some obscure reason. But if that's the case, there would be a mundane, boring, and perfectly logical reason for all the "HUG" sightings—and that's the last thing that a synchronicity believer wants to hear about! Such a person would far sooner believe that the universe (or some cosmic force, or Fate with a capital F, or karma, or dharma, or whatever) wants them to come across license plates that say "HUG" at an outrageously unlikely rate. But if "HUG" actually is more frequent than other three-letter combinations, it's not a proof of synchronicity but simply a sign of sloppiness on the part of the government agency that manufactures and distributes license plates.

The notion that the universe "desires" to send a love letter to little old me, by arranging things so that, as I wander hither and thither, I will bump into a mysteriously high density of "HUG" cars is, to put it mildly, self-centered in the extreme. If "the universe" really took it into its head to send little old me a message, surely it could do so in easier ways than by making four-wheeled vehicles with certain license plates materialize out of the blue, right before my eyes, at just the right moment. Why couldn't the universe just send me a text message on my cell phone, instead? Of course that wouldn't seem so mysterious and "cosmic," but it would be a lot more efficient.

Each of us should remain humble before the subtle challenge of judging probabilities, and each of us should be wary of reading nonexistent, mystical-seeming links into the scheme of things, especially when belief in such links would lead us to taking our very own self for the universe's be-all and end-all.

New Age Prejudices

Reincarnation, past-life therapy, crystal healing, astrology, and numerous other notions falling under the rubric of the "New Spirituality" are today more the rule than the exception in many people's attitude toward their lives and their work. Those who have the courage to declare that their worldview doesn't include supernatural phenomena, supernatural beings, or mysticism are often branded "boring," "limited," and "shallow" by people who consider their own stance to be more open minded, more inclusive, and less arrogant. But sad to say, such people are actually ignorant and filled with prejudices.

In previous chapters we have touched on some of these prejudices, but in the next few pages, I will explicitly exhibit a few specific ones and show how they can be countered. One such prejudice about nonbelievers might be expressed as follows:

You believe only in things that science can explain. Your dogmatic viewpoint rejects any phenomenon that science cannot explain. In your rigid, closed mind, there is no room left for the unexplained or the unknown.

A retort to this might go as follows: Science does not need to be able to explain why a claim is true. What science asks for is simply evidence for the claim, which is to say, some reason for believing in it. Obviously, the more a claim runs against known laws of nature or against standard views, the stronger the evidence presented for it must be, since standard scientific beliefs have great masses of evidence supporting them already.

Suppose I see an ad in a magazine that describes a marvelous new natural cure for snoring. Should I believe it? Well, in itself, the claim doesn't run against any prior law of nature that I know, or any set of experiences I've had. So perhaps I should believe it! On the other hand, I know that there are many companies that want to make a quick buck using striking claims that, unfortunately, are utterly ungrounded. We've already seen a number of such examples. It would therefore be naïve for me to believe in this claim merely because I saw it in a magazine. But what would it take for me to start thinking it might be true?

It would be ideal to have a simple test that could be carried out by an impartial and trustworthy team, such as giving a natural medication to one group of heavy snorers and sugar pills to a second group of snorers, without telling either group what kind of treatment they have been given. If members of the first group do in fact stop snoring, while members of the second group keep on snoring away, then there are at least some logical grounds for me to imagine that the natural medication might actually work.

In short, I don't have to know why the snoring cure works in order to believe that it works; I just need evidence. But as soon as it's known to work, then science's next task is to try to find out why it works.

Here is another typical New Age prejudice about rational people that one needs to know how to combat:

> *You are narrow and limited. You are so determined to reason everything*
> *out that you aren't like the rest of us. You are cold, closed minded,*
> *and hyper-analytic. You are next to feelingless, and you deeply*
> *mistrust all your gut feelings.*

Well, of course there is no connection at all between being rational and being cold, narrow minded, and emotionless. In fact, people who have clearly thought-out reasons for their beliefs are much less likely to fall for superficial trickery or con schemes. That's a safer way to live than to rely solely on one's gut feelings.

It's important to be aware of the distinction between what is plausible and what one wishes were the case. Take mind reading, for instance. I don't believe in mind reading, although I sometimes wish it were real. If I were convinced that certain individuals could read other people's thoughts, I would find that fantastic, wonderful, fascinating, and thrilling (although of course very worrisome as well). But wishing that it could happen doesn't make me a believer in mind reading. It is not intellectually honest to believe in something simply because one wishes it were so. If one follows that pathway, then one can believe anything at all.

> *You seem to think that one should make one's decisions through reason alone. You*
> *don't understand that people are in fact totally governed by feelings.*
> *In thinking that people make decisions and form opinions solely*
> *through rational processes, you are fooling yourself.*

Of course human decisions are not arrived at solely via reasoning. Many careful psychological experiments have demonstrated that decision making is far more driven by emotions than most people suspect. Feelings and desires play an incalculable role in all human decisions. But it is crucial to find a happy medium between pure feeling and pure intellect. It's precisely because we are so strongly driven by emotions that we need to cultivate our ability to reason and help it to develop systematically.

A doctor, thanks to rational thinking and intensive training, can become convinced that a certain medical treatment may cure a patient. But above and beyond this intellectual realization, the doctor also wants the patient to recover through the treatment and hopes it will happen. Indeed, a central factor in many people's choice to become doctors is not just their attraction to rationality but also their powerful desire to bring help and relief to people who are suffering.

A politician can, thanks to rational thinking, see that a certain action can help bring about greater social equality. But this fact alone is not sufficient motivation for the politician to decide to carry out this action; it also requires a strong desire, on the politician's part, to achieve greater equality. Without feelings and desires, no decisions would be made and no actions would be carried out. The optimal way of reacting to complex political questions is to combine one's feelings with one's reasons—to use both head and heart in a judicious mixture. After all, as Blaise Pascal poetically declared, "Le cœur a ses raisons que la raison ne connaît point" ("The heart has its reasons that reason knoweth not"). But when feeling takes over completely from reason, then decision making can become problematic.

There are, of course, some sorts of decisions that should be made fully by feeling and not by logic. Thus, for instance, one would surely decide on the basis of purely emotional factors which painting to put up in the living room or what kind of music to listen to. And the choice of one's life partner is mostly an emotional one, although reason may play a subordinate role in this decision. (Concerning mate choice, there are many parts of the world where pressures lying completely out of one's own control, such as the preferences of one's parents, may control one's decision.)

On the other hand, there are many types of decisions where feelings should definitely not be given free rein. This applies most clearly to decisions that involve the fates of other people. How would society look upon a doctor who decided how to treat patients on the basis of nothing but fleeting whims, rather than on the basis of extensive technical training and long years of experience? Or imagine if laws concerning the rights of homosexuals in society were 100 percent determined by the legislators' feelings.

You always believe that you know best! But no one can ever
have better reasons for their beliefs than anyone else does.

This is a common opinion, but it is completely wrong. One person can have far better reasons for their beliefs than another person does for theirs. Good reasons have to do with publicly available facts. If your viewpoint is in agreement with the information that is available to you, whereas my viewpoint was just plucked at random out of the blue, then your ideas are more grounded than mine are. People who state that crystal healing and crystal therapy do not work are not more dogmatic than those who state that crystal healing and crystal therapy do work. Although the former is the more grounded and thus the more reasonable attitude, such an attitude seems to trouble many people.

You who are so skeptical are actually frightened of the unknown,
frightened of losing your footing and plummeting into an abyss.

This is hardly true. Someone whose worldview is grounded in factual judgments is an open-minded and curious person. Curiosity, the pursuit of knowledge, and scientific research are in fact the lodestars of reason. New discoveries are constantly altering our picture of the universe. But people who will jump on board any old belief bandwagon, without requiring careful justification for their leap, are most likely frightened of taking responsibility for their opinions.

The New Age movement has hijacked certain notions because they have positive connotations. For instance, the phrase "I am a seeker" is one that New Age–infatuated people often utter with pride. It suggests, however, that we who are not New Agers are not sufficiently open, curious, deep, or spiritual. Instead, we are superficial, limited people who aren't interested in the world around us; we are people who may have sought the truth for a while, but who have given up. And yet the truth is often precisely the reverse. Perhaps we non–New Agers should say, "We love to learn new things; we delight in exchanges with people who know things of which we know nothing. It's positive and healthy to be convinced that one should always test one's viewpoints about various topics. We are curious and eager for new knowledge. And that's what makes us seekers!"

We who are not New Agers are relentless seekers of facts, of knowledge, of insight, of understanding, and of experience. We seek not to be taken in by quacks or charlatans. We do not seek security in dogmas or irrational sets of beliefs. We do not seek answers where none can be found.

Those who espouse New Age culture are happy to think of themselves as deep, while thinking of us as merely superficial. But why should the New Age culture be the reference point for defining what counts as deep and what counts as shallow? Many New Agers are seekers in a quite different sense of the term; it is not truth about reality that they are seeking. What they are seeking is a type of faith that will give them peace of mind, as opposed to a faith that is consistent with the way the world truly is.

～

INTERLUDE: ON THE MYSTIC LAND OF SHANGRI-LA

What is the most beautiful land on earth?

Shangri-La is a fictitious paradise that was dreamed up by British writer James Hilton in his 1933 novel *Lost Horizon*. It was described as lying in a deep valley somewhere in the Tibetan Himalayas. The myth of Shangri-La has been taken over by popular culture, and it has turned into the vision of a beautiful, paradisial land where people never age and who live in permanent harmony with nature and with magic. Tibet itself has often been portrayed as being the true Shangri-La, a land whose traditions, going back many thousands of years, are filled with ancient wisdom and magic.

I spent one month of summer 1999 living out of a tent in the Tibetan wilderness. I had traveled to Tibet to "smuggle" a few computers into a little country school high up in the Tibetan plateau. I was eager to see for myself this myth-enshrouded land before it was too late.

Tibet is the most beautiful and most exotic place I have ever visited in my life. The light, the mountains, the vast expanses, the flora and fauna, and the Tibetan monasteries gave me a wonderful experience, far beyond the ordinary. It is easy to understand why, in the Western world, there is still a romantic image of Tibet as the New Age's promised land.

During my trip, I sometimes met other foreigners who had come to Tibet to realize themselves and to develop their spiritual side. I remember particularly well an elderly lady I met. She wanted to help me with the dead batteries in my camera, since there was no electricity in the region of the Tibetan plateau where we were camping (we were at about seventeen thousand feet above sea level). This lady was convinced that she had magical powers, and moreover, since we were in Tibet, they should be even stronger than usual.

She sat by the side of a spectacular mountain lake, holding my batteries in her hands to recharge them with her spiritual energy. For some reason, though, my camera was resistant to her spiritual energy. The camera refused to work despite her attempt to charge its batteries, but luckily, my memories of the landscape, the light, and the natural beauty are still just as strong today as that day, despite my having failed to capture them in photos.

Tibet was occupied by China in 1950, and in 1959, the country's religious and political leader, the Dalai Lama, fled the country and went to Dharamsala in northern India. The Dalai Lama is no longer the country's political leader; in 2011, he handed over that duty to Lobsang Sangay, who today is the prime minister in the exiled Tibetan government. The Dalai Lama remains, however, the country's religious leader.

Tibet has an unmatched physical beauty, but in terms of human living conditions, things are far less pristine. Tibet is an extremely poor country; in fact, its level of development is comparable to that of Sweden in the seventeenth century. The Chinese oppression of Tibetans is severe and brutal.

In connection with my trip, I came to know a Tibetan man who was working for Tibetan refugees in exile. Many years later he became the finance minister in the exiled government under the leadership of the Dalai Lama. Several years ago he invited my wife and me to Dharamsala to meet the exiled Tibetan government. As part of that visit, he set up a meeting for us with Gyalwang Karmapa (the so-called "Karmapa Lama"), a Tibetan Lama who was living in exile in India.

Gyalwang Karmapa is the leader of the Karma Kagyu school, one of the most important schools of Tibetan Buddhism. Each new Karmapa Lama is believed to be the reincarnation of a predecessor. Gyalwang Karmapa is thus considered to be the seventeenth Karmapa Lama and, thus, "three reincarnations older" than the fourteenth Dalai Lama.

In many ways, this was a fascinating encounter. A man who grows up surrounded by constantly repeated statements that he is divine cannot avoid being influenced by this viewpoint. The question is whether the effect is positive or negative.

Since I did not believe that he was divine in any way except in the eye of believers, our meeting was soon taken up by political questions.

The impression that a Lama of this great level of "dignity" should virtually glow with spiritual wisdom is very strong, even in our part of the world. Some weeks before the trip took place, the following dialogue played itself out over a lunch with some friends.

"You'll have to be sure to ask some deep questions about spirituality and the mystical quest and the inner voyage when you get the chance to meet this reincarnated god, this holy man!"

Well, I couldn't let this comment go by without answering in a slightly harsh tone: "No, I don't plan to ask superficial and self-absorbed questions about self-realization and the inner voyage. On the other hand, I will certainly ask him some touchy questions about the political situation involving China and Tibet, and whether there has been any progress in terms of China respecting human rights in Tibet."

This kind of romanticization, in New Age circles here in the West—especially when it involves a very remote country that is suffering from oppression and underdevelopment—can sometimes be shocking.

To this day, I retain the memory of Tibet as a natural experience of the rarest sort, at the very edges of mysticism.

Part II

THE PATHWAY TO A NEW ENLIGHTENMENT

CHAPTER NINE

WHEN RELIGION RUNS OFF THE RAILS

Concerning Fanaticism, Extremism, and Christian-Style Talibanism

Think not that I am come to send peace on earth: I came not to send peace, but a sword. For I am come to set a man at variance against his father, and the daughter against her mother, and the daughter-in-law against her mother-in-law. And a man's foes shall be they of his own household.— Matthew 10:34–36 (King James Bible)

In this book's first half, I have tried to describe the basic requirements that are needed to be able to think clearly. My thesis is that to think clearly, we humans need some understanding of the nature of reality, of the nature of truth, and of what it means to believe in something. We also need to understand the way that science works: its procedures and its methods. I have pointed out many ways in which our brains are fantasy prone and unreliable, and I have argued for basing one's beliefs on a reality grounded in nature, a reality in which goodness comes not from heaven above but from within oneself.

And yet many people, even many nonreligious people, are convinced that religion and morality necessarily depend on each other. This simply is not the case. Religion and morality can freely exist, each one without the other. To state it in very blunt terms, good people can be believers or nonbelievers; bad people likewise can be believers or nonbelievers. Or, to turn it around, pious people can be moral or immoral; atheists likewise can be moral or immoral. Recognizing these simple facts is a key precondition for paving the way toward a new enlightenment.

Before charting out the coming pathway, we need to familiarize ourselves with the most extreme sorts of religious beliefs that are found around the world today. Once we dare to face the problems squarely, we will realize how acute the need for a new enlightenment truly is.

Does God Hate Women?

Even today, religion is often seen as the fountainhead of all "moral principles," yet these principles consistently degrade human dignity all around the world.

The universal oppression of women in the name of religion is our era's most pervasive violation of human rights. And yet the widespread silence about this fact is proportional to its extensiveness. This is a truly incomprehensible situation: after all, even as infants we all learn to protest against oppression. Many of us still remember the South African apartheid regime in the twentieth century; many nations protested against South Africa's oppression of black people, rooted in the division of humanity according to skin color. How, then, is it possible that we today accept a similar type of apartheid in the name of religion—namely, gender-based apartheid and sexual oppression, which in some ways are even more severe than the apartheid in South Africa?

Most religions have, at their core, a story of creation that winds up portraying woman as inferior to man. From this it follows that men and women should be separated and should be treated in different ways—and always to the advantage of men. This leads to males having the full weight of religion backing up their "right" to control society and its laws, which implies that women's interests usually come in second place.

This, of course, hits women hard. On a daily basis, we see evidence of how various religious dogmas, in many different ways, trample and squelch the self-evident right of women to be respected as human beings and citizens. In much of the world, women have no rights at all. This is particularly true in countries where religion permeates society and legislation, or where religion has recently taken a foothold. Typical of the anti-female discrimination in such lands are the outlawing of abortion, the legalization of forced marriage, the condoning of rape and sexual assault, and strict limitations on the mobility of women. Additionally, women are presumed to have less mental power than men, and they are given less legal control over their bodies, their lives, their education, their economic self-sufficiency, their choice of whom to love, and their children.

The Catholic Church's Opposition to Abortion

A good example of all this is one's attitude toward abortion. In many parts of the world, the right to abortion is unheard of or exceedingly limited.

This type of abortion control through law leads to terrible tragedies. Thus in 2007, in the Catholic country of Nicaragua, a law was passed that totally prohibited abortion. As a result, Nicaragua today shares the honor, together with Chile and El Salvador, of having the strictest laws in the world concerning abortion. Doctors who carry out abortions are condemned to jail. This means not only that women are denied abortions but also that they don't dare seek medical help for complications involved in pregnancy. In these countries, women are terrified of being accused of trying to induce their own abortions, because this, too, is punishable by law.

In El Salvador, a woman who had a life-threatening complication during her pregnancy was found guilty of having tried to induce her own abortion, and for this crime, she was sent to prison for thirty years.[1]

The organization Human Rights Watch, in one of its reports, published an interview with the mother of a young woman:

> Angela M.'s 22-year-old daughter is another case in point. Her pregnancy-related hemorrhaging was left untreated for days at a public hospital in Managua, despite the obligation, even under Nicaraguan law and guidelines, to treat such life-threatening emergencies. In November 2006, only days after the blanket ban on abortion was implemented, Angela M. told Human Rights Watch of the pronounced lack of attention: "She was bleeding. That's why I took her to the emergency room but the doctors said that she didn't have anything. Then she felt worse [with fever and hemorrhaging] and on Tuesday they admitted her. They put her on an IV and her blood pressure was low. She said. 'Mami, they are not treating me.' They didn't treat at all, nothing."
>
> From comments made by the doctors at the time, Angela M. believes her daughter was left untreated because doctors were reluctant to treat a pregnancy-related emergency for fear that they might be accused of providing therapeutic abortion. Angela M.'s daughter was finally transferred to another public hospital in Managua, but too late: "She died of cardiac arrest. She was all purple, unrecognizable. It was like it wasn't my daughter at all."[2]

Sexual Apartheid in Judaism

In Jerusalem in 2008, a young woman was assaulted by several ultraorthodox young Jewish men. They broke into her apartment and demanded that she immediately move out of it, claiming she had been observed in the company of an unmarried man. They made her shut her eyes and told her they would stab her and blind her with tear gas if she opened them. Then they forced her to reveal the name of the man she had met. The group that made the forced entry into her apartment belonged to the ultraorthodox "Morality Police."[3]

In certain ultraorthodox districts in Jerusalem, women are required to sit at the very back of buses, totally apart from men, and they are also told that they cannot walk on the same side of the street as men.[4] Writer Jackie Jakubowski wrote, in a chronicle about the rising influence of ultraorthodox rabbis in Israel:

> The orthodox rabbinacy has become Israel's own little papacy, which considers itself to have a monopoly on Judaism, and which claims the right to influence both domestic politics (women's status and their civil rights) and foreign policy (settlements, occupation, peace negotiations)—basing their ideas on statements by certain ultraorthodox rabbis, belonging to a completely different era.[5]

In short, whenever religion's influence on its surroundings increases, women invariably wind up being the biggest losers.

Sharia and the Cairo Declaration

In traditional Islam and its political branch known as Islamicism, there is a clear separation between men and women. Islamicism maintains the oppression of women through so-called "Sharia laws." Roughly translated, "Sharia" means "the divine laws," and it consists of the Koran and the *sunna* (traditions) of the prophet Muhammad, which are found in the *haditherna*—texts that describe Muhammad's life and way of living. The Koran and the *sunna* contain Islam's precepts, both general and detailed. The Koran is claimed to be the word of God, as revealed to Muhammad. The *sunna* act as a complement to the Koran, clarifying, interpreting, and making more precise its overarching ideas.

The jurisprudential part of Sharia is called *fiqh*, which grew from two sets of primary sources called *usul* ("roots") and were intended to be used in the settlement of social disputes.[6]

There are many different schools of interpretation of Sharia, and the interpretations differ greatly.

The Arabic word *sunna* refers to traditions that were formed and based on what Muhammad said, did, and approved of during his lifetime. Sharia laws can be classified under three main rubrics:

1. *Ibadat*: Norms regulating worship.
2. *Muamlat*: Civil and legal obligations.
3. *Uqubat*: Punishments.

Sharia law applies to different degrees in different Islamic countries. In Saudi Arabia, for instance, Sharia constitutes the most basic set of laws, while large parts of Iranian criminal law and family law build on Sharia; and in Egypt, Sharia law is merely seen as a source of inspiration for actual legislation.

In certain Islamic countries, Sharia is applied mainly to family law, while in others, such as Iran, it applies to crimes and sentencing. Although its interpretations vary greatly, Sharia acts as a powerful normative force all across the Islamic world today. This implies that Sharia controls the day-to-day lives of hundreds of millions of people in Islamic lands.

These days there is an ongoing debate among intellectuals in the Muslim world about how Sharia should be interpreted and applied. Some claim that Sharia laws were valid only in the era in which they were created, and that today they should be adapted to modern understandings of human rights. Others maintain that the laws are "God given" and are thus eternal, and cannot and should not be reinterpreted in a modern context.

The most debated of Sharia laws are those that concern the punishments of whipping, mutilation, and stoning. One of Europe's leading debaters on this topic is Islamic professor Tariq Ramadan at Oxford University, whose research area is Islam and its interpretations in modern days. In 2005, he urged the Islamic world to suspend all such punishments until Muslim scholars could come to an agreement about how to interpret Sharia in a modern context.

In Western lands, there is general advocacy of freedom of speech in debates, which includes not only the right to criticize ideologies and actions but also the duty to maintain respect for individuals. In other words, individuals have the right to control their lives, their bodies, and their consciences, as well as the right to express their opinions, but they have to accept criticisms of their opinions.

According to conservative interpretations of Sharia law, by contrast, the laws of Islamic society cannot be questioned—they are inviolable. According to these interpretations, anyone who commits blasphemy or who lives un-righteously can be punished, which can mean a loss of their social status, or of parts of their body, or possibly even of their entire life. Respect for Islam and for the group is thus more important than respect for individuals.

In a meeting of the Organisation of Islamic Cooperation (OIC) (https://www.oic-oci.org) in Cairo in 1990, some fifty-seven Islamic member countries ratified the Cairo Declaration of Human Rights in Islam.[7] This declaration was to some extent modeled on the United Nations' Universal Declaration of Human Rights, and there are long passages where the two documents resemble each other greatly, but in certain key areas, the Cairo Declaration diverges radically from the United Nations declaration. The Cairo Declaration states that Sharia law is the "only source of reference."

While the Hanafi School of Islamic Law (one of the four Sunni schools of *fiqh*) explicitly bars the death penalty for blasphemy committed by non-Muslims, it states that any Muslim who is convicted of blasphemy should be executed.[8]

In particular, human rights are essentially made subordinate to Sharia laws. For example, in article 22 of the Cairo Declaration, we read the following:

a. Everyone shall have the right to express his opinion freely in such manner as would not be contrary to the principles of the Shari'a.
b. Everyone shall have the right to advocate what is right, and propagate what is good, and warn against what is wrong and evil according to the norms of Islamic Shari'a.
c. Information is a vital necessity to society. It may not be exploited or misused in such a way as may violate sanctities and the dignity of Prophets, undermine moral and ethical values or disintegrate, corrupt or harm society or weaken its faith.[9]

Clauses (a) and (d) from Article 2 of the Cairo Declaration state:

a. Life is a God-given gift and the right to life is guaranteed to every human being. It is the duty of individuals, societies and state to protect this right from any violation, and it is prohibited to take away life except for a Shari'a prescribed reason.
d. Safety from bodily harm is a guaranteed right. It is the duty of the state to safeguard it, and it is prohibited to breach it without a Chore-prescribed reason.[10]

Adama Dieng, a Senegalese Muslim jurist and former secretary-general of the United Nations Commission on Human Rights declared in 1991 that, in the name of the defense of human rights, the Cairo Declaration of Human Rights in Islam introduced intolerable forms of discrimination against non-Muslims and women, while deliberately taking a restrictive approach concerning certain fundamental rights and freedoms. He also opined that the Cairo Declaration, under the cover of Sharia law, gave legitimacy to practices like corporal punishment while attacking the integrity and dignity of humans.[11]

The Cairo Declaration severely limits the right to free expression. Since one loses all one's rights as soon as one violates Sharia laws, the declaration's protection for individuals is highly misleading. Moreover, all individuals are urged to promote Sharia and to warn of anything that is not considered correct, which turns each individual into a guardian of morality, or a member of the moral police.

Sharia makes a distinction between men and women. In certain interpretations of Sharia, a woman doesn't have the right to make her own decisions. Instead, she must rely on male "guardians" or "protectors," meaning either her father, her husband, her brother, or the next-closest male relative.

Some observers have taken pains to point out that Sharia also contains laws *in favor* of women, such as the fact that the husband is required to provide for his wives (!) and pay for their upkeep. But this is hardly relevant; the system maintains sexual apartheid, regardless of who is "favored" in various manners. In practice, this means that a woman who has been rejected by her husband and her other relatives is completely without support.

A married woman must accede to her husband's sexual desires. Longstanding traditions define the steps the husband should take if his wife refuses to satisfy him. First, he should try to reason with her; if that does not change her mind, then he and she should start sleeping in separate rooms; and finally, if the problem persists, then he has the right to beat her. Incidentally, under Sharia, rape inside marriage is not considered to be a crime.

According to the standard interpretation of Sharia, and according to all four fundamental schools of Islamic law, a Muslim woman is not allowed to marry a non-Muslim man (in such a case, the marriage is invalid unless he converts to Islam), whereas a Muslim man is allowed to marry any woman at all, as long as she belongs to one of the three "Abrahamic faiths" (that is, Islam, Judaism, or Christianity). In certain Muslim countries, a man has a better chance of suing for divorce than his wife does; she can divorce her husband only if he is not paying for her upkeep, is mentally ill, is dependent on drugs, or is impotent.[12] But the husband can divorce his wife in her absence, which means that a woman can find herself divorced from her husband without any forewarning—all of a sudden out in the cold and without any kind of support system.

A consequence of some Hanafi interpretations of Sharia is that a woman is considered to have half the worth of a man. Brothers, since they are considered to be family providers, inherit twice as much as their sisters do. This same bias also turns up in certain legal situations, such as testifying about a crime. In Iran, for example, any man is considered to be a reliable witness, but it takes *two* women to count as one reliable witness.[13]

This attitude toward women can lead to absurd consequences—for instance, when it comes to murder trials. A man who murders another man will be sentenced to death. But if a man murders a woman and is sentenced to death, then before the judicial system will carry out the execution, the woman's family first has to pay the murderer's family half of the man's financial worth. As the man has merely killed "half a human," the woman's family is required to make up for the imminent financial loss to the murderer's family.[14]

In Iran, a man and a woman who have been sentenced to be stoned to death will not be punished identically. The man will be buried in the ground up to his waist, leaving his arms free, whereas the woman will be buried all the way up to her neck.[15]

Certain countries have a custom about how the stoning should be carried out: no stone can be larger than a thrower can hold in a single hand. This ensures that the victim will not die after the throwing of just one stone. Now a man who is only half buried obviously has a greater chance of protecting himself, using his arms, than does the woman. If by some chance an accused person survives being stoned, then that person is considered to be innocent—and of course it is men who survive much more often.

Religion and Tradition: A Study in the Oppression of Women

According to the Koran, any woman must cover herself, but to what extent is a matter of debate. Each woman must be veiled in order to hide herself from the view of men who are not her "protectors." The reason, of course, is that she should not arouse other men's sexual drives. It is thus the woman who must take the responsibility for not getting raped.

Traditional religious articles of clothing, such as yarmulkes for Jews and turbans for Sikhs, have to do with one's relationship to one's religion and one's community. But a veil, by contrast, is a symbol of a woman's relationship to a man. If she doesn't cover herself, then she is assumed to be trying to lure men to herself. This assumption is of course offensive to the woman concerned, but it is also an insult to men, who are presumed to be incapable of controlling their lust. When a young girl is made to don a veil, that act sexualizes her, which is to say, it implies that she is already considered to be a sexual being. A veil is thus more than just an article of clothing; it is a symbol of the entire Sharia system's view of women. (Of course, bikinis and many other types of revealing garments worn by women in the *Western* world also sexualize women—and thereby symbolize the West's collective view of women, which is no less sexist than the Sharia-based view. There is thus a strange sort of "symmetry" in this profound asymmetry.)

In the 1960s, Middle Eastern societies were freer than they are today. This fact could be seen in such things as the way women dressed. In those days, women wore skirts, used makeup, and attended universities. Today, however, the streets are filled with women whose faces are covered. Ever since the Iranian revolution in 1979, and similar Islamist seizures of power during the 1980s, Sharia has made even more inroads. In only two decades, the pendulum has swung over to the mandatory wearing of veils and, thus, to female submission to men.

Religious oppression of women runs like a red thread through the history of our Western societies. Women have been burned at the stake, pushed toward chastity, forced to remain silent in gatherings, denied the right to vote, and cast to the winds for having children out of wedlock. It has taken Western societies hundreds of years to transcend such views of women, but very slowly, the old traditions have given way to human rights, and laws have been created that form a more equal society.

It is not justifiable to defend anti-female traditions and aspects of culture simply because the doctrine of male supremacy stretches way, way back into the past. The mere fact that something is a long-standing tradition doesn't always mean it is defensible. Quite to the contrary, in fact: if something is a tradition, that is all the more reason to question it and give it even greater scrutiny.

The systematic subjection of women in rigidly religious societies is among the most highly charged of all political topics today. Many people speak of "religious freedom," but precious few speak of what that might actually mean. People should take the question seriously and define the limits of religious freedom; otherwise we will wind up returning to the shameful period of our history, analogous to South Africa's shameful apartheid. If we keep in mind how women are treated in the name of religion throughout the world, it's hard to avoid drawing the conclusion that God must hate women.

Blasphemy, Apostasy, and Freedom of Speech

According to a Gallup poll conducted in 2012,[16] roughly 13 percent of the world's population call themselves atheists, while 23 percent say they are nonreligious. Approximately 59 percent describe themselves as religious.

The proportion of atheists and nonreligious people seems to be growing throughout time. Since 2005, the number of people calling themselves atheists has gone up by three percentage points, and the percentage of religious people has gone down by nine points.

Despite these trends, criticizing religion today can be very dangerous. There are many countries in which it is actually *illegal* to question God's existence or to criticize various religious interpretations and expressions. In recent years, people have been arrested for the crime of blasphemy (insulting something holy) in Bangladesh, Egypt, Indonesia, Iran, Kuwait, Pakistan, and Turkey (to name just a few examples).

The illegality of blasphemy in certain lands' judicial systems, however, doesn't capture the broad range of risks associated with it. The sources of danger to those who commit blasphemy are of three different sorts: (1) actors at the state level, (2) non-state actors, and (3) the society at large.

In 2010, in Pakistan, Asia Bibi, a Christian woman, was sentenced to death for insulting Islam's prophet; this is a typical example of repercussions at the state level for blasphemy. Asia Bibi was held on death row for eight years, but in 2018, she was at last released by Pakistan's Supreme Court. This decision sparked furious protests by Islamist groups demanding her public execution. She is currently reported to be detained in an undisclosed but secure location inside Pakistan. Many Western countries have offered her asylum.

There are other very dangerous aspects of this case, coming notably from non-state and individual actors, which in some cases are even more dangerous than the sentence imposed by the state's legal system.

On January 4, 2011, Salman Taseer, at the time the governor of Punjab province, and one of the most influential politicians in Pakistan (he had been a minister of the Pakistan People's Party), was killed in an open marketplace by Mumtaz Qadri, one of his own bodyguards, who shot Taseer twenty-seven times with an AK-47 rifle. The *Guardian*[17] described the killing as "one of the most traumatic events in Pakistan's history." In its wake, the Pakistani government proclaimed a three-day nationwide period of mourning.

The reason behind Taseer's assassination was his outspoken criticism of Pakistan's blasphemy law and, in particular, his criticism of the court's verdict against Asia Bibi. Following the death of Taseer, who was a liberal Muslim, five hundred Islamic clerics voiced their support for his brutal murder and urged the public to boycott his funeral.[18] This took place during the governmental mourning period. It is crucial to underline that the greatest dangers came from non-state actors and from the societal arena.

When Taseer's assassin Mumtaz Qadri was executed in 2016, some twenty-five thousand people gathered to commemorate his death. This gathering led to countrywide violence by religious parties and groups that supported the killing of Taseer, and armed forces had to be called to disperse the protestors.

Another notable killing was also connected to the case of Asia Bibi. Shahbaz Bhatti—Pakistan's minister for minority affairs and the only Christian member of the nation's cabinet—was shot dead on March 2, 2011, two months after the assassination of Taseer. Ever since the ruling on Asia Bibi's case, Bhatti had been repeatedly threatened for his criticism of the blasphemy law. Before his death, he spoke about the death threats that he was receiving for his staunch support of Asia Bibi, and he knew that these threats came from the Taliban and al-Qaeda.

Seven months later after Salman Taseer's assassination, his son Shabaaz Taseer was abducted by a Taliban group and was only released after five years of captivity.[19]

While an accusation of blasphemy may lead to members of religious minorities suffering horrific legal consequences in countries like Pakistan and Afghanistan, false accusations of this sort can cause serious harm even to practicing Muslims. Two notable examples of this are the cases of the Muslim women Farkhunda Malikzada (from Afghanistan) and Mashal Khan (from Pakistan).

In 2015, Malikzada was falsely accused of burning some pages of the Koran. Malikzada, who was wearing an Islamic veil at the time of her death by mob lynching, was a religious studies graduate, and she had a teaching position at a local Islamic religious school. She had engaged in an altercation about other issues with the imam of a local mosque, and as a result, she was subsequently accused of burning the Koran. In response to this allegation, Malikzada stated that she was a Muslim and that Muslims do not burn the Koran. The accusation against her led to a rumor, however, and a large crowd formed to punish her; she was stoned and burned to death in an open street. At the time of the killing, there were police at the scene.

Shortly after this brutal killing, Hashmat Stanekzai, a spokesperson for the police chief's office in Kabul, wrote on Facebook, "This [individual] thought, like several other unbelievers, that this kind of action and insult would get her U.S. or European citizenship. But before reaching her goal, she lost her life."[20]

A similar example of how an accusation of blasphemy can be devastating for an individual is the case of Mashal Khan. On April 13, 2017, he was lynched by classmates on the campus of his university in Mardan, Pakistan. Prior to his death, Khan had criticized the university authorities for certain irregularities, and subsequently the authorities opened an investigation against him and two other students, accusing all three of posting blasphemous items on their Facebook accounts, evidence for which was never uncovered.

Khan's friends stated to the police that Mashal was a devoted Muslim. Eyewitnesses to his murder told the police that the mob had attacked him not because of an accusation of blasphemy, but because he had been accused of spreading the Ahmadi faith using Facebook. When he was killed, at least twenty police officers were present.[21] A subsequent police investigation cleared Mashal and two other accused students of all charges of blasphemy.

It's not always an individual who is the victim of an accusation of blasphemy. On September 29, 2012, an entire Buddhist community and six ancient Buddhist temples were burned to ashes at Ramu, Chittagong, in Bangladesh, after a Facebook rumor had been carefully orchestrated and spread far and wide, to the effect that certain groups of Buddhists had insulted the Islamic prophet.[22]

In many countries, *apostasy* (the renunciation of one's religion, or the changing of religious affiliation) is considered to be an even greater sin than blasphemy. In 2018, there were twenty-two countries in which apostasy was criminalized.[23] In twelve of these (Afghanistan, Iran, Malaysia, Maldives, Mauritania, Nigeria, Qatar, Saudi Arabia, Somalia, Sudan, United Arab Emirates, and Yemen), apostasy is punishable by death. Pakistan has a death sentence for blasphemy, a term so vague that it can mean almost anything.

In short, in thirteen countries around the globe today, you can be legally put to death for stating that you do not believe in God.

In 2010 in the Maldive Islands, in a question-and-answer session after a talk given by a Muslim speaker, one audience member said that he was "Maldivian but not Muslim." Other people in the audience attacked him, and he was carried off by the police. Later he was charged with apostasy and was threatened with a death sentence. According to Maldivian law, apostasy necessarily leads to the death penalty unless the accused, when given the chance, changes their mind and readopts the Muslim faith. It's pretty obvious that, under such circumstances, it's easier to lie about one's sudden change of heart than to go to meet one's maker. When merely having certain *thoughts* is a crime, then it's impossible for human rights to survive.

In some countries, religion isn't the only thing that is dangerous to criticize. Critical examination of New Age superstitions or traditional superstitions can also be highly risky. India is a prime example. In 1989, Narendra Dabholkar founded the organization Maharashtra Andhashraddha Nirmoolan Samiti (MANS), whose mission was to criticize superstitions of all sorts. On the morning of August 20, 2013, while he was taking his daily walk, Dabholkar was assassinated by two unidentified men. The murder was blamed on the fact that he had been campaigning for a law that would forbid the sale of New Age products of any sort in India.

In 2017, Indian secular activist and journalist Gauri Lankesh was shot dead in Bangalore. The killing of Lankesh followed the assassinations of several other outspoken secular humanists in the past few years, including scholar Malleshappa Kalburgi, Narendra Dabholkar (mentioned earlier), and Govind Pansare.

In 2018, Indian secular humanist and human rights activist Babu Gogineni was arrested by the Madhapur police in Hyderabad for "hurting religious sentiments," among other charges.[24]

ISIS

In recent years, a certain type of religious fanaticism has spread widely across vast regions of the earth. The organization al-Dawlah al-Islāmīyah fi al-Iraq wa-l-Sham, for short called "Da'ish" or "Daesh" in Arabic, and in English "ISIS" (Islamic State of Iraq and Syria) or "ISIL" (Islamic State of Iraq and the Levant), advocates systematic violence to an unparalleled degree. The organization, originally an Iraq-based offshoot of al-Qaeda, changed its name to "Islamic State of Iraq" in 2006 and at that time started to operate as an organization independent of al-Qaeda.

By 2011, it had expanded into Syria during that state's civil war, which started the same year. In 2014, the group proclaimed a worldwide caliphate.[25] By December 2015, ISIS occupied an area in Syria and Iraq that was larger than many countries. Fundamentalist Sunni groups in other states— for example, Nigeria's Boko Haram—began to pledge allegiance to the growing movement.

In July 2017, the group lost control of its largest city, Mosul in Iraq. Today ISIS has largely been beaten back, and it no longer controls nearly as much territory as it did at its peak. Nonetheless, its ideology still enjoys wide support, and a number of terrorist cells around the world still operate according to the ISIS vision of creating a new Islamic caliphate.

ISIS is the successor to al-Qaeda, and at its high point it was better organized and had greater financial resources than al-Qaeda ever had. In 2014, the leader of ISIS, Abu Bakr al-Baghdadi, declared an Islamic caliphate in those areas of Iraq and Syria that the movement then controlled. Abu Bakr al-Baghdadi has a doctoral degree in Islamic history and law from Baghdad University in Iraq, which shows, not too surprisingly, that higher education does not confer immunity to fanaticism. His motto, "We shall avenge all wrongs done to us," can be seen in many videos of executions of innocent people posted on the web.

The rapid advance by ISIS during the years 2014–2016 caught the world deeply by surprise. During its heyday, the organization's income was thought to lie in the range of $3 million to $7 million per day. Oil was shipped illegally to Turkey and Iran, where it was sold to willing buyers at half the normal rate. The revenue thus gained went directly to ISIS, and the money was used not only for the purchase of weapons but also for the support of social operations, such as schools and health care institutions. This was a clever strategy, because it led to ISIS being warmly welcomed as a savior by some people in dire need.[26]

Thanks to its skillful exploitation of YouTube and other social media, ISIS reached millions of viewers and lured people to join it, from as far as Europe and even the United States. Of course, most Muslims disown the goals and methods of ISIS, but there is no denying the fact that its political ideology is tightly linked to Islam, albeit in an extreme and grotesque interpretation thereof. And the notion that one will be rewarded in heaven for acts of martyrdom carried out on earth does much to facilitate the recruiting of people willing to die for the cause. A particularly vivid but well-known example is the idea cherished by each of the young men who, as a team, crashed fully loaded passenger planes into the World Trade Center and the Pentagon on September 11, 2001—the dream-vision that they would all go to heaven, where each one would be greeted by a harem consisting of seventy-two beautiful and eager virgins.

According to Jeffrey R. Macris, a professor of history at the United States Naval Academy, the ideological roots of ISIS can be traced back and linked to Muhammed ibn Abd al-Wahhab, the eighteenth-century founder of a movement today called "Wahhabism." Cambridge University historian David Motadel has also linked ISIS—especially its style of extreme violence—with Wahhabism.[27]

Muhammed ibn Abd al-Wahhab's book *Kitab al-Tawhid*[28] is the origin of many of the violent ideas practiced and adopted by ISIS, and its credo has strong similarities with the credos of many ISIS members, including the idea of accusing anyone who does not accept the god of Islam of apostasy.[29]

Kitab al-Tawhid is the origin of such ideas as the requirement for believers (that is, Muslims who subscribe to this school of Islam) to destroy any places or objects where idols are worshipped. An example of the execution of such abhorrent ideas took place when the statues of the ancient city of Palmyra in Syria (a World Heritage Site) were severely and intentionally damaged by ISIS.[30]

Another example was ISIS's wanton destruction of the mosque of the prophet Jonas in Mosul, Iraq, which was carried out since that mosque was considered to be polytheistic.[31] And there are hundreds of other churches and ancient religious sites that were destroyed by ISIS in its attempt to impose its extremist form of Islam all across the Middle East (and then, hopefully, the entire world).

While Muhammed ibn Abd al-Wahhab is the true source of many of ISIS's core beliefs, including its most violent behaviors, he is rarely referred to in the ISIS literature today, garnering only a few rare mentions in ISIS's magazine *Dabiq*.

Egyptian writer and ideologue Sayyid Qutb (1906–1966) was one of the most important ideological thinkers behind the scenes of the twentieth-century Islamic world. During the 1950s and 1960s, Qutb was among the leading intellectuals in the Muslim Brotherhood, as well as its propaganda chief. Today he is remembered for his key role in founding the militant Islamic movement, which was launched and grew in Egypt during the 1970s.

In his speeches and his writings, Qutb championed many bizarre notions, such as the conspiracy theory of Jewish people trying to conquer the world, and the innate intellectual inferiority of women. Swedish historian of religion Eli Göndör writes:

> [Qutb] was convinced that women, by virtue of their hypersensitive nature, are not suited for taking responsibility in such things as finances or the making of crucial decisions. As for Jews, according to Qutb, they have recently invaded disciplines such as history, philosophy, and sociology, with the goal of demoralizing the world while also taking over the realm of medicine, as well as the worldwide sales of coffee.[32]

These are clearly highly irrational, aberrant points of view. And yet Sayyid Qutb is revered by members of the Muslim Brotherhood, al-Qaeda, and ISIS. Qutb is one of the key contemporary Islamist thinkers whose ideas were key in launching the violent practices of ISIS.

Today, ISIS has mostly become a network of independent terror cells throughout the world, and it is no longer a threat to the nations of the Middle East; however, its ideology still inspires fanaticism and terror attacks, and the movement is probably the most extreme and dangerous ideology since the Nazis.

Nordic Terror

On July 21, 2011, Norway was hit by the worst terror attack in its history. In the district of Oslo that houses the Norwegian parliament, Anders Behrling Breivik set off a bomb that killed eight people. Then he systematically shot to death sixty-nine idealistic, socially liberal, and politically active teenagers on the island of Utøya, very near Oslo. Breivik was driven to these acts by his loathing of Muslims, of a multicultural society, and of the project of secular enlightenment. Before carrying out his brutal attack, he posted a fifteen-hundred-page manifesto (in English) on the web, which he called "A European Declaration of Independence." In it, he lays out his ideology and describes, among other things, "an unholy alliance" among secular humanists, Marxists, and the global capitalistic system as follows:

> Cultural Marxists, suicidal humanists or capitalist globalists are all multiculturalists. "Multiculturalist" is a label for individuals who support multiculturalism [. . .], the European hate ideology which was created to destroy

our European cultures, national cohesion and Christendom (in other words Western civilisation itself).[33]

Breivik saw secular humanism as a threat both to Christianity and to the notion of "national unity culture." In a certain sense, he was correct. Secular humanism strives for a nonreligious and cosmopolitan society where everyone is welcome, independently of their background and their set of beliefs. Of course, in such a society, the influence of Christianity is very limited, and Breivik saw this as a profound threat.

Further along in his manifesto, Breivik writes:

> Secularists, it seems to me, are also less keen on fighting. Since they do not believe in an afterlife, this life is the only thing they have to lose. Hence they will rather accept submission than fight.[34]

Breivik's ideology is a bizarre hodgepodge of religious fanaticism and extreme nationalism. He is a clear example of how dangerous fanaticism can be when it takes possession of an energetic personality.

Christian Extremism

Western media are constantly abuzz with reports about Islamist extremism, but there are numerous other extremists who are seldom mentioned in public forums. Their ideas and interpretations of beliefs are every bit as deranged as those of the most extreme Islamists, but their methods are not as brutal—at least not yet.

Among the world's highly developed countries, the United States is the most religious, even though its constitution is far more secular than the Swedish constitution. In the United States, religion is frequently cited as a force that strongly unites the American people (recall the pious statements issued by President Eisenhower after the Pledge of Allegiance was altered to include the words "under God"). But American people's religious tolerance decreases rapidly when systems of belief other than Christianity are brought up. This emerges clearly in a poll made in the United States by the Pew Forum.[35]

The poll revealed that 85 percent of the people surveyed considered that the United States was not ready for a president who is an atheist.[36] In some American intellectual circles, people speak of *atheophobia*, meaning a widespread contempt for atheists. Atheists have long been, and are still, discriminated against in the United States. There are six states—Arkansas,

Mississippi, North Carolina, South Carolina, Tennessee, and Texas—that have laws on the books that forbid atheists from working in the public sector. For example, the Constitution of the State of Mississippi (Article 14, Section 265) declares the following: "No person who denies the existence of a Supreme Being shall hold any office in this state." In Arkansas, atheists are not allowed to testify as witnesses in court. In 2014, an organization called Openly Secular Coalition was founded, whose aim was to eliminate such obsolete laws all across the United States.[37]

Luckily, the great majority of Americans see Christianity mainly as a sort of social glue, without any deeper meaning. But there are some Christian sects in the United States with members who, in their thoughts (but fortunately less often in their actions), can be compared to Muslim extremists. Christianity can be dangerous.

The Joshua Generation

A nine-year-old girl screams out with terror over the poisoning of American society by Satanic demons. As she prays to Jesus to save her, the tears run down her cheeks, and her whole body trembles. This girl is at an American summer camp for the worship of Jesus—a so-called "Jesus camp." Her parents believe that this is the key that may save the United States from the state of moral free fall in which it currently finds itself. Children who have been "saved" are the only path to turn a godless nation into a Christian one and to restore God's Kingdom.[38]

Between 1.5 and two million American children are never given the chance to attend a public school. They are taught at home by their parents to keep them from being exposed to the "godless education" and the "godless society" that are found outside the safe little cocoon of the Christian home. Evolution is of course not part of homeschooling, except for being mentioned as an example of the immoral teachings of modern science.

Homeschooled children are often called the "Joshua Generation" in the United States, in honor of Moses's military leader and disciple Joshua, who, according to the Old Testament, was responsible for the bloody recapture of the Holy Land.[39] Homeschooling is now perfectly legal in all fifty U.S. states, thanks to highly effective lobbying by Christian groups. And when children are old enough to study subjects that their parents don't know sufficiently well, then there are plenty of private Christian schools that are eager to take over the duty of educating them.

One of the more prestigious of such schools is Patrick Henry College in Virginia (https://www.phc.edu). The school offers a college education for Christian students who were homeschooled. Its founder, Michael Farris, is very clear about his goals; in an interview he candidly declares that he wants to educate a new generation of Christians whose task is to play significant political roles in society.[40] The school's credo can be found on its website:

> The Bible in its entirety (all 66 books of the Old and New Testaments) is the inspired Word of God, inerrant in its original autographs, and the only infallible and sufficient authority for faith and Christian living.[41]

In the classes at Patrick Henry College there is one simple and clear principle that holds: When the Bible and science arrive at differing conclusions, then science has clearly misinterpreted its data.

Another of the school's credos is the following:

> Satan exists as a personal, malevolent being who acts as tempter and accuser, for whom Hell, the place for eternal punishment, was prepared, where all who die outside of Christ shall be confined in conscious torment for eternity.[42]

Students at Patrick Henry College have frequently been solicited by electoral campaigns to help fundamentalist Christian candidates get elected to the Senate, such as Tom Coburn in Oklahoma and Jim DeMint in South Carolina. Coburn is an advocate of the death penalty for doctors working in private abortion clinics, and DeMint is against allowing single mothers to be teachers, as they are considered to be pernicious role models for youth.

Coburn was one of Oklahoma's two senators until he retired in 2015, and DeMint was a senator from South Carolina from 2005 to 2013 when he became head of the Heritage Foundation (www.heritage.org), a well-known conservative think tank. He resigned from this position in May 2017 after a unanimous vote by the foundation's board of trustees.

Waiting for the Second Coming of Christ

A significant number of conservative Christians in the United States are *premillennialists*—that is, they believe that Jesus will soon return to earth and restore God's reign. Many believe that this will happen during the current century. This scenario is described in *Left Behind* (www.leftbehind.com), a series of books garnering sales of more than sixty-three million copies. A "Christian apocalyptic thriller film" with the same name as the book series was filmed in 2014, with actor Nicolas Cage playing a major role in it.

The *Left Behind* books were written by Tim LaHaye, a preacher who was Michael Farris's mentor and teacher. *Left Behind* tells the story of fundamentalist Christians who are saved by Jesus and who disappear from the earth during the so-called "Rapture." These people literally go up in smoke, ascending to heaven, while chaos breaks out on earth: airplanes crash and cars drive off the road as their pilots and drivers simply vanish into thin air.

Remaining on earth are those whose belief in Christ was not great enough, or who did not lead a sufficiently virtuous life. They are the ones who are "left behind." After a while, the Antichrist returns to the earth, disguised as an angel of peace (in the book, the Antichrist shows up as Nicolae Carpathia, secretary-general of the United Nations). After a few years of increasing strife, Armageddon—that is, the final battle between Christ and the Antichrist—takes place, and Jesus comes out victorious and restores God's reign upon the earth. (According to the Book of Revelation [16:16], Armageddon is the location of this final battle between Good and Evil.)

The tale in *Left Behind* is based on the prophesies in the Book of Revelation in the Bible. There it states that the return of the Antichrist will deceive the people who remain on the earth, and that he will seem to them like an angel of peace. He will try to bring all nations together and set up a world government in the name of peace. All this takes place before he reveals his true, evil nature to the people. It is therefore no accident that the novel's Antichrist is the United Nations' secretary-general. Many fundamentalist Christians in the United States see the United Nations as a symbol for the Antichrist's world government and are highly critical of attempts by the United Nations to help resolve political problems on earth.

The book series *Left Behind* has also been issued, in a simplified adaptation, as a children's book, and even as a video game. In *Left Behind: The Game*, the players try either to save unbelievers or to kill them all. (A player earns more points for successfully saving an unbeliever than for killing one.)

Tim LaHaye also runs the PreTrib Research Center (www.pre-trib.org), a think tank that looks for signs of Jesus's return to earth and tries to prepare us for the severe deprivations that the Antichrist will subject us to. LaHaye's wife is also involved in pro-Christian activities in the United States. She founded Concerned Women for America (www.cwfa.org), an organization of conservative women that opposes such things as feminism, same-sex marriage, and liberal abortion laws.

Christian Taliban

A small number of fundamentalist Christians in the United States are *postmillennialists*, and with this concept we are coming closer to the true American Taliban.[43] To be sure, this group of people is vanishingly small in the vast throngs of American Christianity, but it is nonetheless very real.

A postmillennialist is someone who believes that Christians themselves bear the full responsibility for restoring a Christian kingdom on earth. Only when that goal has been achieved will Jesus come back and take over as leader. The difference between premillennialists and postmillennialists might seem marginal to some, but it is actually very significant, as we shall soon see.

In the United States, the postmillennialist movement is also called "Christian Reconstructionism,"[44] and it was developed in conservative Presbyterian circles during the 1970s. Its founder is usually considered to be Rousas John ("R. J.") Rushdoony (1916–2001). He wrote a number of books, of which the best known is probably *Institutes of Biblical Law*. Today, the movement is represented by such people as Gary North, head of the Institute for Christian Economics, and Gary DeMar, head of the lobbying organization American Vision (https://americanvision.org). On its website, American Vision states:

> We believe in the personal, bodily return of our Lord Jesus Christ at the consummation of history. The dead, consisting of believers and non-believers, shall be raised up in final judgment. Those who are saved shall be raised up unto everlasting life and those who have rejected Christ unto eternal damnation. We believe that it is the current duty of the Church (Christians) to expand and build the Kingdom of Christ wherein Christ shall be the head of all things unto all men.[45]

There are many people who believe that Marvin Olasky, a Presbyterian elder and professor of journalism, belongs to this same movement, although this claim has been disputed. In any case, his book *Compassionate Conservatism* (2000), with a foreword by former president George W. Bush, advocates a neoliberal form of economics, as well as, paradoxically enough, a sort of "social survival of the fittest," which is of course an allusion to Darwin's theory of evolution but adapted (invalidly) to a social context.

The theological basis of this movement has to do with the conditions for the return to earth of Jesus. Christians have taken it upon themselves to create a Christian seizure of power on earth and to establish the Old Testament as the law of the land. But only when a majority of the earth's people have been converted to Christianity and God's rule has been restored can Christ return to earth.

This theological viewpoint is often called *dominion theology*. Its teachings are drawn from the Old Testament, in particular from Genesis (1:28). In the King James translation of the Bible, the key passage runs as follows:

> And God blessed them [Adam and Eve], and God said unto them, Be fruitful, and multiply, and replenish the earth, and subdue it: and have dominion over the fish of the sea, and over the fowl of the air, and over every living thing that moveth upon the earth.

The majority of Christians today take these words to mean that humans have a special role in creation and are responsible for *managing* nature and animals, rather than subduing or subjugating them.

But dominion theology interprets these words differently: God is seen as urging Christians to dominate over all other beings and to unite all the world's societies under God's law. Christian Reconstructionists also consider that the 613 laws that are stated in the Pentateuch (the opening five books of the Old Testament) are binding for all humans, independent of their nationality, culture, or religion. These laws are given in the books of Moses—namely, Genesis, Exodus, Leviticus, Numbers, and Deuteronomy. Genesis is about creation and the earliest stages of history, including the Flood. Exodus states laws pertaining to the covenant between Yahweh and Israel. Leviticus states laws about sacrifice and cleanliness, and Numbers sets forth a number of laws about rituals. Deuteronomy stresses the importance of worshiping one god alone and contains a number of laws about the necessity of monotheism.

These laws can be divided into two main groups: ceremonial and moral. The Reconstructionists believe that Jesus freed Christians from having to follow all the ceremonial laws, since the New Testament gives new rules that replace them; however, the moral rules of the Old Testament still hold fully. Those laws should therefore, as a first step, be incorporated into the new Christian society in the United States, and as a second step, into the rest of the world. Only then can God's Kingdom be realized, and only then can Jesus return to the earth.

In political terms, all this implies that the Christian Reconstructionists are trying to get the U.S. federal government to play a subordinate role in society. They believe that Christian churches should take charge of the welfare and the education of the citizenry, and that the U.S. Constitution should be modified so as to be consistent with the laws stated in the Old Testament.

Such a radical change would entail many very serious consequences, among which are these. All homosexuals should be executed. Blasphemy, adultery, and idolatry should all be punishable by death. Any religious organization that does not accept the Old Testament's laws would by definition be guilty of idolatry. This of course would apply not only to all non-Christian religions but also to liberal branches of Christianity. In the Reconstructionist society, all members of such organizations would thus be subject to the death penalty.

Women's rights would be rapidly reduced to the level of the rights that slaves formerly held in the United States, and in fact some Reconstructionists want to reintroduce slavery as part of the structure of society, in conformity with the laws of Moses. All abortions would be forbidden, no matter for what reason. Doctors who carried out abortions would be sentenced to death. The prison system would be abolished, to be replaced by a system of "societal payback," with fines and forced labor being imposed on people who commit minor infractions, and death for others.

William O. Einwechter is a pastor who belongs to the Reconstructionist movement. In 1999, he wrote a notorious article called "Stoning Disobedient Children," in which he advocated the death penalty, by stoning, for recalcitrant children. It can be found in the *Chalcedon Report*, published by the Chalcedon Foundation.[46] In support of his position, Einwechter cites the following passage from the Christian Bible:

> If a man have a stubborn and rebellious son, which will not obey the voice of his father, or the voice of his mother, and that, when they have chastened him, will not hearken unto them: Then shall his father and his mother lay hold on him, and bring him out unto the elders of his city, and unto the gate of his place; And they shall say unto the elders of his city, This our son is stubborn and rebellious, he will not obey our voice; he is a glutton, and a drunkard. And all the men of his city shall stone him with stones, that he

die: so shalt thou put evil away from among you; and all Israel shall hear, and fear. (Deuteronomy 21:18–21)

In a later article, Pastor Einwechter softened his position, explaining that the term "child" in Deuteronomy can also mean a young man who refuses to obey his parents. He claims that death by stoning was never intended to apply to small children but, rather, to adolescent children or young adults who need to face the consequences of their behavior.

Jesus Shall Crush the Nonbelievers

The Christian Reconstructionists constitute a very tiny minority among American Christians, and luckily they have practically no influence on American politics today. But the conservative Christians who now have, or have recently had, a considerable influence can be just as extreme. A good example is the late television preacher Jerry Falwell (1933–2007).

In his book *Nuclear War and the Second Coming of Jesus Christ*, Falwell wrote that "those who fail to follow Jesus shall have their eyes melted and their flesh burnt" and "Jesus shall crush the nonbelievers."[47]

In 2004, Falwell founded the influential lobbying group called the Moral Majority Coalition (he himself had formed the group's predecessor, the Moral Majority, back in 1979). The organization does its best to whip up opposition to abortion and homosexuality and aims to set up censorship against "anti-family propaganda." Falwell was also very clear in his criticism of secular humanism. He once stated:

> We're fighting against humanism, we're fighting against liberalism . . . we are fighting against all the systems of Satan that are destroying our nation today. . . . Our battle is with Satan himself.[48]

Falwell also came out with the following notorious statement about AIDS: "AIDS is not just God's punishment for homosexuals; it is God's punishment for the society that tolerates homosexuals."[49]

Falwell's view of the role that religion should play in politics is quite clear: "The idea that religion and politics don't mix was invented by the Devil to keep Christians from running their own country."[50]

Another influential figure at the borderline between Christian Reconstructionism and more traditional fundamentalism is Pat Robertson, founder of the lobbying group called Christian Coalition of America (https://.cc.org). In 1992 in Iowa, to raise funds to oppose a state equal-rights amendment, Robertson wrote a letter to his supporters in which he says the following:

> The feminist agenda is not about equal rights for women. It is about a socialist, anti-family political movement that encourages women to leave their husbands, kill their children, practice witchcraft, destroy capitalism and become lesbians.[51]

Most people outside of the United States are completely unaware of the surging of Christian fundamentalism in the world around them. It is far easier for them to perceive, and be frightened by, fundamentalism in "alien religions," such as Islam. People who live outside of the United States are horrified by the Taliban in Afghanistan and the Islamist jihadists in Syria and Iraq, but what we have just seen is that even in "Christian lands" there are tributaries that can be just as frightening.

One essential thing distinguishes such frightening tributaries, however. The organized Christian fundamentalists in the United States, despite their fanatical desire to serve their angry God, do not go around killing school-children in terrorist attacks, or blowing themselves and other people up with bombs worn around their waists—at least not yet. Let us pray that will not come to pass.

~

INTERLUDE: ON HUMAN SUFFERING AND MEDICAL PRACTICE

How far should humans go in trying to control the course of nature? This question comes up over and over again today in discussions of such things as organic farming or genetically modified foods. In prior eras, though, the question most frequently arose in connection with the role of medicine in controlling human diseases.

Numerous Christian churches have often wished to slow down or stop the progress of medical research. For example, in the Middle Ages and the early Renaissance, the Catholic Church vigorously opposed the dissection of human bodies. It was believed that when Judgment Day arrived, there would be a physical resurrection, and if a person's body had undergone a post mortem, it would be hard, even for Almighty God, to restore the damaged body to its original pristine state. The "infallible" Holy See finally backed away from this stance, however, when it at last recognized how much knowledge medical science had gained through dissections.

Another historical example is certain Christian churches' view of smallpox. That disease, known ever since antiquity, spreads very rapidly, causing epidemics. (We now understand that smallpox is extremely contagious because the virus is carried by microscopic droplets that float through the air.) The mortality rate due to smallpox was very high in olden times, and people who survived it were often severely disfigured.

The first smallpox vaccine was developed toward the end of the eighteenth century by Edward Jenner in England, who fortuitously discovered that milkmaids who'd had cowpox were immune to smallpox. The liquid that can be extracted from the cowpox rash contains the cowpox virus, and it could thus be transmitted from animals to humans. This procedure came to be called "vaccination" (from *vacca*, Latin for "cow").

Despite this amazing boon, both the Catholic and the Protestant churches originally protested against the use of vaccination. Their argument was based on the thesis that mere mortals should never try to interfere with God's will. If someone comes down with smallpox and then dies (or survives), this is God's will. People should never try to interfere with God's control over life and death.

In the end, however, the churches bowed to reason and conceded, accepting the smallpox vaccination—but even so, they never admitted that they'd been wrong; they had merely buckled under popular pressure. After all, the advantages of the vaccine were blindingly obvious to everyone, and it was extremely simple to show that there was a far lower incidence of smallpox in areas where vaccinations had been carried out. In 1980, the World Health Assembly declared that, thanks to vaccination, the scourge of smallpox had been successfully eradicated from our planet.

A third historical example is the attitude of most Christian churches toward anesthesia during surgery. In 1847, chloroform was introduced as an anesthetic by British physician James Young Simpson. Now patients could be anesthetized or put to sleep during operations, which of course was a fantastic piece of progress. Previously, surgeons had used alcohol to drug their patients, or else they had simply tied the patient down to the operating table.

And yet, back then, Christian churches protested against the use of anesthetics in the case of childbirth. The reason was that the Bible describes an event in heaven that God was not happy about: Eve ate the apple of knowledge (since it gave the gift of knowledge) and then she tricked Adam into eating it as well. God punished Eve for this grave misdeed, saying to her (in Genesis 3:16), "I will greatly multiply thy sorrow and thy conception; in sorrow thou shalt bring forth children."

In England and Scotland, bitter battles were raging over whether women should have the right to use anesthesia in the case of very painful childbirth—in particular, in case of Caesarean sections. It was Queen Elizabeth I who put a halt to the church's cruelty in this case. Being a highly determined woman, as well as the mother to several children, she refused to bow to the church's declared position on these matters.

Today, we see similar tendencies in the fervent religious opposition to all sorts of old and new medical ideas, such as abortion, euthanasia, stem-cell research, and the detection of genetic defects in fetuses. Thanks to basic research, modern medicine is making fantastically rapid strides ahead, and this explosion of knowledge and of possibilities of course raises all sorts of brand-new ethical dilemmas, no matter what one's religious outlook is. For this reason, even in a completely secular environment, people will be forced to take stances on many thorny issues in medical ethics.

What should our attitude be toward, say, surrogate motherhood, or toward diagnosing genetic defects in a fetus when it is still in utero, or toward making DNA maps to help insurance companies recognize more vulnerable individuals so that they can charge them higher rates, or toward developing drugs that enhance intelligence, and so forth? No matter what our moral stance is, medical progress will force us to confront many new and difficult questions as time marches forward.

CHAPTER TEN

THE BATTLE OVER OUR ORIGINS

Concerning Evolution, Creationism, and Anti-science

We are here because an odd group of fishes had a peculiar fin anatomy that could transform into legs for terrestrial creatures; because the earth never froze entirely during an ice age; because a small and tenuous species, arising in Africa a quarter of a million years ago, has managed, so far, to survive by hook and by crook. We may yearn for a "higher answer"—but none exists.—Stephen Jay Gould[1]

The United States is today, as we have seen, a battleground where all imaginable species of religion are in constant mortal combat, mercilessly vying with each other to survive and propagate. Of course there is a flip side to this violent story as well: there are many socially committed altruistic organizations in the United States, some religious and some nonreligious, that are doing wonderful work in helping the needy and the sick—all this in a land where the economic gulfs are huge and where the social safety net is extremely fragile.

But it is also in the United States that the war between science and anti-science (as sponsored by religion) is being the most flagrantly and bitterly fought. And yet the United States was founded by settlers who clamored desperately for total separation of church and state. How did this paradox come about?

Texan Tom DeLay, who for two years (2003–2005) was the Republican Majority Leader in the House of Representatives, gave a remarkable explanation for the horrific massacre in 1999 at Columbine High School in Colorado (see Michael Moore's extraordinary documentary *Bowling for Columbine*), in which two seniors at the school coldbloodedly murdered twelve of their schoolmates and one teacher: "Our school systems teach our children that they are nothing but glorified apes who have evolutionized (a word coined by DeLay himself) out of some primordial soup of mud."[2]

DeLay was referring to the fact that Darwin's theory of the evolution of species, rather than the story of creation as related in the Bible, was taught in Columbine High School's biology classes.[3]

In the years 2001–2009, George W. Bush, a literalist Christian, was president. His successor, Barack Obama, gave the opposite impression. There is good evidence suggesting that Obama is not particularly religious, and in fact Obama's mother described herself as a secular humanist. Obama, in his autobiographical memoir *Dreams from My Father*, writes that his mother was "a witness for secular humanism."[4]

In practice, however, it is next to impossible for any American politician to express even the slightest doubt about the existence of God, and so Obama, once he was running for president, bowed to this convention and often expressed himself in pious terms.

But it wasn't always that way. The U.S. Constitution was drawn up at the end of the eighteenth century by a group of men (and no women) of whom several clearly expressed great skepticism about religion. In the Constitution, particular care was taken to separate religion from politics. According to the First Amendment to the Constitution (1791), the government must not take a stance either for or against any religion or religious idea. Specifically, it says: "Congress shall make no law respecting an establishment of religion, or prohibiting the free exercise thereof."

Four of these men eventually became the first presidents of the United States—namely, George Washington, John Adams, Thomas Jefferson, and James Madison. All of these myth-enshrouded political leaders were deeply critical of religion and wanted to limit its influence in the political sphere.

These presidents made pungent criticisms of religion that touched on politics, morality, and even the nature of knowledge. It is not hard to find quotes from these towering figures in American history that, were they to be uttered in today's America, would cause enormous uproars. Here are a few:

Religious controversies are always productive of more acrimony and irreconcilable hatreds than those which spring from any other cause.—George Washington[5]

Twenty times in the course of my late reading have I been on the point of breaking out, "This would be the best of all possible worlds, if there were no religion in it!"—John Adams[6]

The day will come when the mystical generation of Jesus, by the Supreme Being as his father, in the womb of a virgin, will be classed with the fable of the generation of Minerva in the brain of Jupiter.—Thomas Jefferson[7]

Religious bondage shackles and debilitates the mind and unfits it for every noble enterprise.—James Madison[8]

Enlightened ideas like these were, for many decades, very standard in American society. In today's America, however, this kind of attitude has lost its respectability. As American society has developed, there have been countless skirmishes in the 150-year-long battle between those who take the Bible literally and those who read its stories as metaphors and myths. Today in particular, a very serious battle is raging between those who believe in evolution's explanation of the origin of species and those who believe in the Bible's tale of creation. And as often happens in the United States, this battle has been partially fought in the courts.

Darwin's Legacy

One evening in December 1831, the HMS *Beagle* sailed out of the harbor in Plymouth, England, with a twenty-two-year-old man named Charles Darwin (1809–1882) on board. For five years, this ship sailed throughout the southern hemisphere, and during that whole time, young Darwin, wherever he got the chance, was avidly studying the flora and fauna that they encountered. Many years later—in 1859—he published his pathbreaking book *On the Origin of Species by Means of Natural Selection*, which drew its inspiration from that five-year voyage.

For the first time in the history of science, a theory had been presented that, in a self-consistent fashion, explained how animals and humans had come to be, and all without the intervention of any kind of higher power. A creator was possible, but not necessary, according to Darwin. No longer was any god—with or without a capital *G*—needed. The idea that humanity had stemmed from some mythical Adam and Eve in the Garden of Eden struck Darwin as highly unlikely. In fact, the most likely possibility, to his mind, was that humans were descended from apes.

This notion was seen by many religious people as deeply offensive and hugely threatening. In the early twentieth century, the book *The Funda-*

mentals: A Testimony to the Truth appeared in the United States. It was a reaction against Darwin's theory of evolution and against the growing tendency to embrace scientific ideas. A religious movement came out of this, called the fundamentalists, which called for a return to utter faith in the Bible's truth. For the fundamentalists, Charles Darwin came to symbolize the threat that science posed to religion and to the truth of the Bible's Holy Word. Darwin's evolutionary explanation of the origin of human beings was seen as incompatible with a literal interpretation of the Bible, and therefore as godless, destructive, and immoral.

The idea that humans were created exactly as is told in the Bible came to be called "creationism" (often spelled with a capital *C*, especially by its adherents), and today, throughout the world, there are numerous versions of the creationist teachings. The most classical and the closest one to the Bible, however, stems from Irish archbishop James Ussher, who, in the seventeenth century, scrutinized the Old Testament's texts, and from what he found in them, calculated that the earth was created in the year 4004 BCE. (In fact, his calculations were somewhat more accurate than that. He calculated that the earth had been created at exactly 8:00 p.m. on Saturday, October 22, 4004 BCE. If you were an early riser, it would have been pretty rough for you that day!) According to the Bible-believing fundamentalists, the entire human species derived from Adam and Eve and was only 6,000 years old. The opposing notion—that humans came from apes—was simply scoffed at, much as one would scoff at a ridiculous fairy tale.

But for people with a secular outlook on life and some basic knowledge of how science works, it's *creationism* that comes across as pure nonsense. Within Islam (whose story of creation is virtually identical to the Christian one) and within evangelical Christianity throughout the world, however, creationism retains a strong following, and most of all in the United States.

Monkey Trial, 1925

The first half of the 1920s in the United States was marked by major social changes and a deep shifting of values. Young people were dancing to jazz, experimenting with abstract art, and discussing not only Darwin's theories about the origins of humanity but also Sigmund Freud's theories about sexuality and the hidden drives lurking inside our psyches.

Traditionalists were horrified; they thought that the theory of evolution made God superfluous, thereby undermining morality and removing any reason to lead a virtuous life. To block such repugnant ideas from gaining any foothold in society, they tried to use legal means. Thus in March 1925, the state of Tennessee passed a law that forbade teachers from "teaching any theory that denies the story of divine creation as taught by the Bible and teaching instead that man was descended from a lower order of animals."[9]

In the summer of that year in the hamlet of Dayton, Tennessee, a twenty-four-year-old teacher of biology named John Scopes was indicted for having taught Darwin's theory of evolution. The trial of John Scopes was opened by the Honorable William Jennings Bryan, a renowned politician who not only had run for president three times but also was a fundamentalist Christian lawyer known far and wide for his fiery oratory; he spearheaded a crusade against the godless theory of evolution. About the Scopes trial, Bryan famously proclaimed, "If evolution wins, Christianity goes!" In response, Scopes's lawyer, the equally famous lawyer Clarence Darrow, declared, "Scopes isn't on trial; *civilization* is on trial!" It was thus with great excitement that the mass media followed the battle between modernity and traditionalism, and before long, the case against Scopes had been redubbed the "Monkey Trial."

And the Monkey Trial quickly grew into a huge media spectacle, not without its humorous sides. When one day, Bryan stood up in the courtroom and passionately complained that evolutionists claim that humans don't even come from *American* monkeys but from *Old World* monkeys, the journalists were delighted, even though they mostly sympathized with Scopes. And when Bryan was slickly holding forth, his nemesis Clarence Darrow would do his best to distract the jury by letting the ashes at the tip of his cigarette grow longer and longer without ever falling off (the wily Darrow had inserted a rigid wire inside his cigarette, which held the ashes suspended in the air, seemingly without any support). In a word, it was truly a circus. In the end, though, teacher Scopes was convicted and was ordered to pay a fine; but guilty or not, he had gained many converts to his side.

The Christian fundamentalists' crusade against Darwin's evolutionary theory rushed on apace. Between 1925 and 1928, they fought this battle over and over again, with varying degrees of success. After Scopes was found guilty, both Arkansas and Mississippi introduced legislation to forbid the teaching of evolution in their schools.

Evolution Scores a Couple of Victories

In 1968, the time was ripe for another legal trial in the ongoing battle between the theory of evolution and creationism, and this time the Christian fundamentalists were dealt their greatest setback ever, all thanks to the U.S. Constitution—and therewith, the early American presidents' viewpoint wound up triumphing.

The Arkansas state law forbidding the teaching of "theories that claim that man evolved from the apes" provoked Susan Epperson, a teacher in that state. She challenged the state's law, arguing that it contradicted the federal constitution. This lawsuit was considered by the U.S. Supreme Court, and Epperson won the ruling hands down. The court found that the Arkansas state law violated the First Amendment to the U.S. Constitution (*Epperson v. Arkansas*, 1968).[10]

The fundamentalists retreated and licked their wounds. Slowly a new strategy started to emerge. Instead of seeking to squelch the teaching of our descent from the apes, they launched the idea that evolution and creationism should *both* be taught in schools. And the phrases "creation science" and "scientific creationism" were coined, clearly suggesting that here one was dealing with a *scientific* theory, every bit as respectable as the theory of evolution. The reasoning was, if there were two rival scientific theories, then of course schools must teach them both.

In 1987, this new legal strategy was put to the test in Louisiana, and the upshot was a new law that proclaimed that evolution could not be taught in schools unless "creation science" also was taught. But when that law's constitutionality was tested in the Supreme Court, it was found to conflict with the Constitution, and it was repealed. Once again, the creationists lost a skirmish.

Intelligent Design

How does creationism manage to survive? Why do so many evangelical Christians and pious Muslims throughout the world so vehemently oppose the theory of evolution? It's not that evolution is filled with glaring holes or that it is deeply counterintuitive. There are far more counterintuitive scientific theories (such as quantum mechanics or cosmology) in which many questions remain unanswered.

The answer is that the theory of evolution casts serious doubt on one of the main arguments for belief in God. It shakes the very pillars underlying religiosity. If the recombination of genes, together with natural selection, can account for all the variety of all the flora and fauna on earth, including human beings, then there is no need for a divine creator who brought humans into existence. God simply becomes an entity unneeded by the laws of nature.

In 1990, in response to the setbacks of the 1970s and 1980s, a new Christian think tank called the Discovery Institute was founded in Seattle, Washington; its main purpose was to try to figure out new strategies for bringing creationism into U.S. schools. It is still trying to realize this goal today. On the institute's website[11] one can find passages defining the various missions of the institute, including these paragraphs:

> Scientific research and experimentation have produced staggering advances in our knowledge about the natural world, but they have also led to increasing abuse of science as the so-called "new atheists" have enlisted science to promote a materialistic worldview, to deny human freedom and dignity and to smother free inquiry. Our Center for Science and Culture works to defend free inquiry. It also seeks to counter the materialistic interpretation of science by demonstrating that life and the universe are the products of intelligent design and by challenging the materialistic conception of a self-existent, self-organizing universe and the Darwinian view that life developed through a blind and purposeless process.
>
> The worldview of scientific materialism has been pitted against traditional beliefs in the existence of God, Judeo-Christian ethics and the intrinsic dignity and freedom of man. Because it denies the reality of God, the idea of the Imago Dei in man, and an objective moral order, it also denies the relevance of religion to public life and policy.
>
> [W]hile modern discoveries in biology and ecology have given us a greater appreciation for the importance of other creatures, these same discoveries are sometimes misused to promote an extreme vision of "animal rights" that places animal welfare above the welfare of human beings. Our Center on Human Exceptionalism counters pseudo-scientific attacks on human dignity by defending the unique dignity of persons, what we call human exceptionalism, in health care policy and practice, environmental stewardship, and scientific research.

The Discovery Institute operates using a new strategy: since the U.S. Constitution forbids religious influences in schools, the questioning of evolution must be done in a way that would completely avoid references to the Bible or other religious ideas. And thus, if it were possible to cloak the arguments for creationism in purely scientific garb and to promote creationism as a genuine scientific theory rather than as a biblical teaching, it might be possible to slip it into school curricula without any conflict with the Constitution.

And therefore the term "creationism" was dropped and replaced by the new phrase "intelligent design." This was supposedly a strictly scientific theory about the origin of species, including human beings—a scientific alternative to Darwin's theory. The new theory asserted that complex biological organisms could only have been concocted by the mind of an intelligent designer—but whether that great being was God or some other mysterious power was not stated. All efforts were being made not to get ensnared in yet another fight with the Constitution.

The basic idea of intelligent design runs roughly as follows. Biological organisms are so complex and so ingeniously constructed that they could not possibly have emerged as the result of processes like those that Darwin described, involving constant genetic variation and natural selection among the variations. For life as we see it on earth to have come about via evolution is just too unlikely, and therefore all these complex functions and organisms must have been invented to serve some *purpose*. And therefore, there must exist some creator, some intelligent thinker, behind the scenes of earthly life.

The example most frequently wheeled out in this context is the *eye*, where a number of parts need to cooperate in order to allow vision to take place. The advocates of intelligent design claim that these parts must all have come into existence at the very same moment, in order to give rise to the faculty of vision. There is no way that random variation, combined with natural selection, could have resulted in a working eye—or so goes the argument.

Swedish researchers Dan Nilsson and Susanne Pelger, however, have convincingly shown how evolution can develop an organ as complicated as an eye.[12] Nilsson and Pelger show how a light-sensitive spot of skin can gradually evolve into a fully functioning eye possessing a lens, and in fact they show how such a transformation can occur relatively quickly. It turns out that it would take about four hundred thousand generations to get the job done, which may strike some readers as an astronomical number, but actually, in evolutionary terms, it is a mere "blink of an eye" (so to speak).

The Discovery Institute very efficiently runs two major campaigns advocating the theory of intelligent design: one directed at changing the mind of the public and the media, and the other directed at influencing politicians and school boards.

The movement claims that intelligent design is a scientific theory and also claims that there is an actively ongoing scientific controversy among researchers, pitting the theory of evolution against that of intelligent design. And from this it of course follows that innocent schoolchildren should have the right to be exposed to both sides of the question and to select the superior one of the two rival theories, since that, after all, is how modern science "evolves." Survival of the fittest theory!

Suggesting that schools should teach rival theories and allow students to choose between them sounds like a wise idea: offer a smorgasbord of scientific theories and tell the children that, as of yet, there is no definitive answer. But this idea is just the thin edge of the wedge, and it leads to a very slippery slope. Should geography classes offer an unbiased choice between a round-earth theory and a flat-earth theory? Should physical science classes in schools teach both astronomy and astrology? Both chemistry and alchemy? Both medicine and shamanism? Should Holocaust deniers be given equal time along with those who tell the true history of World War II? Where will it all stop? There is only one sane answer to such queries: Schools should teach only those ideas that have stood up under the harsh scrutiny of scientific research, rather than offering a random potpourri of ideas.

But creationists argue that no process of gradual evolution could possibly have led to the phenomenal kinds of complexity that we see in animals and humans. One of the most commonly trotted-out images is that of the design of a passenger airplane. The argument runs more or less this way:

> That the blind and purposeless forces of nature could give rise to hugely complex entities like ourselves is as improbable as a tornado sweeping through a scrapyard, picking up thousands of random pieces of junk, and then tossing a perfect Boeing 747 down onto the ground.

Unfortunately, this argument is based on a profound misrepresentation of how evolution works. This can be shown by the following thought experiment. Suppose that you have twenty dice in your hand, and you toss them all onto the table in front of you. What are the chances that you will get twenty 6s?

For any *one* die, the chance that you'll get a 6 is one in six, of course. The chance of throwing all *twenty* dice so that they show a 6 is infinitesimal—one in 6^{20}, to be precise. That is a tiny number: 1 in 3,656,158,440,062,976. If you were to throw all twenty dice in one second, and then were to repeat the process over and over again, you would have to throw your dice for more than one hundred million years to have a reasonable chance of getting all twenty dice to come up with 6 at once.

But that's not how evolution works. Imagine instead that you have a set of twenty dice and that every time you toss them, you can set aside all those dice that got 6s. On your first throw, perhaps you get *three* 6s out of twenty possible. So you set those aside, and now you toss the seventeen remaining dice. This time maybe you get just *one* 6. All right; you set it aside, and now you toss the remaining sixteen dice. Maybe this time, two more 6s. And so forth. It'll take you on the order of a minute or so until you have set aside twenty 6s—nothing like one hundred million years to reach your goal.

And this is in fact how evolution works: advancing by one bit at a time, taking advantage of each small bit of progress. Small variations are made, and advantageous ones are selected for. The image of the tornado randomly throwing pieces of junk together and coming up with an immaculate jet airplane is wildly off the mark. It is not only misleading but it is also *deliberately* misleading. It is intellectually dishonest. Indeed, it is intellectual dishonesty of a high order.

Intelligent Design in American Schools

In 2005, yet another legal case arose in the United States involving the battle between religion and science. It came out of a decision by the Dover, Pennsylvania, school district to teach intelligent design as an alternative to evolution in biology classes.[13] After a six-week trial, the verdict was handed down, and science was the victor.[14]

The Dover school board's action took on almost sectarian forms, involving threats, lies, and manipulation. Some members of the board went so far as to draw Charles Darwin's family tree on a large piece of paper and then to burn it in effigy.

In the trial, everything was cleared up. The judge concluded by saying that it was remarkable that so much good tax money and human resources had been wasted on a trial of this sort.

All attempts to wrap scientific clothing around the biblical tale of creation wound up being completely futile. The jury members themselves showed this in a very elegant manner, without taking sides in the question about whether an intelligent creator existed. They based their action on the following argument.

A fundamental need for any scientific theory is for it to be both testable and falsifiable. For a theory to be falsifiable means that there is a way to say what kinds of evidence might topple it; however, none of the advocates of intelligent design could explain what kind of scientific experiments, discoveries, tests, or logical proofs could falsify the idea of intelligent design. And that meant it was not a scientific theory. The jury members in Dover didn't take a position either for or against an intelligent designer, but they merely observed that the theory had no place in the teaching of biology.

In sum, the theory of intelligent design was determined not to be a scientific theory and therefore had no place in scientific education.

The intelligent-design movement also tried to reveal the existence of loopholes and flaws in the theory of evolution. One of their main attempts was based on the claim that there were holes or missing links in the transitions between species. Such an argument, however, was based on two fundamental misunderstandings.

The first of these was that it is not in the least surprising that there aren't in existence fossils and other kinds of physical evidence that document *every* single tiny stage of the evolution of one species out of another. If there were, it would be miraculous—like having a record of when every leaf ever fell from every tree on earth throughout the last billion years. We don't need such a detailed record in order to believe that all those trees' leaves actually fell.

The second misunderstanding was that the mere existence of gaps in the fossil record did not in any way confirm the theory of intelligent design. In fact, even if the theory of evolution were totally refuted, that wouldn't necessarily increase the likelihood of intelligent design. It would merely show that there must be some other way to explain the origin of species, rather than the way we believe today.

The Discovery Institute, which created the modern theory of intelligent design, has a very different agenda from the scientific one of explaining the origin of species. The Discovery Institute's hidden agenda is religious, and that is something that the U.S. Constitution will not permit in American schools.

The court in fact discovered, among the institute's internal documents, a clearly formulated strategy declaration—namely, that the mission of the intelligent-design movement is "to defeat scientific materialism and its destructive moral, cultural and political legacies."[15]

One could hardly formulate it more clearly. Darwin's theory of the origin of species is considered by its enemies to be unethical, immoral, and godless. Evolution is considered as god denying and atheistic.

But such reasoning is based on a deep misunderstanding: Darwin's theory doesn't deny the *possibility* of God's existence—it only denies the *necessity* of God's existence.

Creationism in Sweden

In Sweden, too, there are creationists—principally Muslims and Christians belonging to "Swedish free churches."[16] The Swedish creationist society Genesis, led by Anders Gärdeborn, includes a variety of creationist movements.

Another noteworthy creationist on the Swedish scene is Mats Molén, who runs a creationist museum in the northern city of Umeå, called "Den Förhistoriska Världen" ("The Prehistoric World"). At one time, Molén was also the leader of Genesis, but today he mostly travels around Sweden and gives talks about creationism in schools and other educational institutions.

On its website, the Prehistoric World Museum professes to being extremely open minded in the following manner (the following is translated from the Swedish website of the museum):[17]

Working with philosophy

First of all we want that the visit at our museum will start to think critically and not just embrace what various authorities say (including what is said at the museum). This is fundamental to all human socializing, and also the basis for democracy. That includes respect for others' opinions, even if you do not share them. (If we all have the same views, there's a risk that it is brainwashing!)

We therefore welcome a discussion on various issues. We oppose all discrimination and the "ivory tower" kind of reasoning (i.e., someone consider themselves superior to others and therefore do not want to discuss).

If you want to go deeper into these issues, we also welcome a profound discussion about different interpretations of Earth's history (this applies only if you want to discuss). Research related to today's issues is often directly applicable in practice, such as how best to build a bridge or make a good medicine, so you can't do too many mistakes. But regardless of how the ancient history took place we are here. Alternative explanations for the past and our origin therefore do not often provide immediate practical consequences. This can be seen for example of the discussion in the scientific literature.

But the question concerning origins is basically a religious/philosophical question, whether you're an atheist or believe in a creator. Therefore, that question, for various reasons is extra sensitive to many people and some even refuse to discuss their faith. We accept and understand the latter position, but we can bring up and discuss even these difficult questions, if it can be made openmindedly.

There are unfortunately many who name their faith or religious view with the label "fact" or "science." You can believe that your faith is science. However, there is a problem if you want to present dogmatically this faith and calling it fact or science, while at the same time you are not willing to let anyone critically examine and challenge this belief system. The museum is opposed to such an undemocratic and unscientific approach.

Tough questions

This museum, the Prehistoric World (Den Förhistoriska Världen) is run by a non-denominational Christian association which cooperates with all who wants to cooperate and who dares to discuss difficult questions. But discussing difficult questions isn't something which all dare to do—but we offer you help to get started!

We do NOT believe that life is meaningless, as many scientists and authors of textbooks has come to believe lately. The latter belief is often due to unconsciously confusing the scientific method (which is a method to explore the world and the universe) with a philosophy/religion that says that there is no real purpose in life.

It is difficult to understand why someone might be interested in bringing out a faith of meaninglessness. It is anyways pointless with a message that life has no meaning and it isn't making anyone happy. (Perhaps someone unwittingly finds a sense of just spreading a message, regardless of whether the message is good or bad, true or false?)

You can find information about this belief in futility in the scientific press (they do not always say so, but the basic idea is meaninglessness).

Many scientists have also stated that they are not interested in discovering what is true. They are only interested in explaining everything with their own beliefs/philosophies/theoiesy. (See for example a statement by Scott Todd, from the scientific journal *Nature*, here quoted in an article by creation scientists [https://creation.com/whos-really-pushing-bad-science -rebuttal-to-lawrence-s-lerner].)

The latter, to exclude all thoughts except what you believe in yourself, does NOT include how we address the issues at the museum. That is, we believe in openness and discussion, even about the most basic questions. (However, we do not always have the same views as the web sites we link to.)—The board of directors, the Prehistoric World

It is of course pleasing to me to see how strongly some of this philosophy echoes the principles suggested in this book, but when one looks at it more carefully, one sees that it does not adhere to those principles.

Very few people in Sweden take creationism seriously, but now and then one will see, in the newsletters put out by Swedish free churches, some note to the effect that a newly discovered fossil doesn't agree with the family tree of species so far worked out by researchers. It's easy to imagine the triumphant headlines: "New Discovery Shows Darwin Was Wrong!"

The flaw in such an argument is nicely revealed through a simple analogy. Imagine you've laid out the pieces of a jigsaw puzzle on your dining room table, and as you start putting it together, you soon recognize that it's a photo of old-time Swedish movie star Greta Garbo, famous for her line "I vant to be alone." When you've nearly finished, you're disappointed to see that a piece or two is missing, and then you pick up a piece that doesn't seem to fit anywhere in the puzzle. If you were a creationist, you might at this point exclaim, "Well, what do you know—a piece that doesn't fit in! I guess this means it *isn't* Greta Garbo, after all!"

For Darwin's theory of evolution to be correct as a whole doesn't mean that every tiny detail in the overall picture has to be exactly right at all moments, or that if even the slightest detail is missing, then the whole picture is invalidated. It's self-evident that, as time progresses, we will find ourselves adding new details and revising old details in the story that tells how all the different species are related to one another.

Indeed, in recent years, methods involving DNA samples have given us new clues about our origin. For example, renowned Swedish researcher Svante Pääbo has mapped out the DNA of the Neanderthals, and thanks to his work[18] we now know that the Neanderthals were not ancestors of ours (that is, not direct predecessors of *Homo sapiens sapiens*), but another type of human that lived in parallel with "us" (that is, with our direct ancestors) up until about thirty thousand years ago.

All new knowledge we gain about biology through our understanding of DNA points in the same direction—namely, that there is no intelligent design involved. The process of natural selection does the job all by itself.

But in a sense, intelligent design does exist! I recently had the chance to talk to the 2018 Nobel Laureate in Chemistry, Caltech's Frances H. Arnold. In 1993, she conducted the first directed evolution of enzymes, which are proteins that catalyze chemical reactions. She told me that the intelligent design movement tries to use her research as evidence for their very unscientific hypothesis, which made her very annoyed.

The truth is, Dr. Arnold is *herself* performing intelligent design, as a human and scientist. She uses science and her intellectual skills to direct evolution as an intelligent designer. Now that is *genuine* intelligent design!

~

INTERLUDE: ON THE NORMAL AND THE ABNORMAL

Is it abnormal to be homosexual? Well, sure—in the same way as it's abnormal to be born on January 1, or to be six feet three inches tall, or to have green eyes. Normality is a statistical concept, but it is often erroneously used as a value judgment.

The fact that somewhere between 5 and 10 percent of people are homosexual makes it slightly abnormal in a statistical sense, but in that same statistical sense, it is considerably more abnormal to be a comedian, a florist, a university president, a Jew, or a secular humanist.

Usually, "normal" is equated with "most frequent," so that unusual things are, by definition, abnormal. But no value judgment should be attached to that word. Why should unusual traits provoke negative reactions, such as fear, scorn, or verbal abuse?

Down through the ages, many of the "cosmic authorities" (such as priests) to which people have traditionally turned have cast a critical eye on belief systems that are different from their own—or on types of sexual behavior or social interaction that are not shared by the majority. And yet, the most reasonable stance toward such behaviors should be exactly the opposite, for in truth, a great diversity of human traits, experiences, and lifestyles is a source of enrichment, rather than a threat, to our existence on this planet.

If an abnormal type of behavior is harmful to ourselves or to others, we should clearly question it. But that holds just as much for normal types of behavior. Take smoking, for instance. Smoking was once extremely popular in the United States; it was considered perfectly normal to smoke. Eventually, however, the unhealthiness of smoking started to be suspected, and the reasonableness of smoking was called into question. Today, as a result of years of medical research, it's far more unusual in the United States to smoke than not to smoke. In other words, what yesterday was taken as normal is very abnormal today.

Unfortunately, most of us do not pay as much attention to behavior that we see all around us as we do to behavior that breaks the norms. While the latter often turns us off, the former is generally met with mere silence. But what matters about someone's behavior is not whether it is normal or norm breaking; what matters is whether it contributes to well-being and happiness. In sum, what society considers "normal" is just a matter of statistical distribution and should not be confused with value judgments.

CHAPTER ELEVEN

THE HISTORY OF IDEAS

Concerning the Roots of Secular Enlightenment

Dare to know!—Immanuel Kant, "What Is Enlightenment?" (1784)

The story of the development of the secular view of what it means to be human goes back a long way. We'll now take a look at how this viewpoint grew out of venerable philosophical traditions. Essentially, the secular view of human beings started to come into prominence as people's fear of mysterious supernatural events and their terror of arbitrary, harsh gods started to weaken.

Reason, Compassion, and Various "Golden Rules"

For millennia, humanity has scratched its collective head over such riddles as how gods could exist, or what would constitute a meaningful life, or what determines whether a particular act is morally right or not. This kind of incessant moral reflection is an intrinsic part of the human condition, and a central element in moral reflection is compassion (meaning "co-suffering" or "sharing of feelings").

The capacity to feel compassion for others is a deep human attribute, but contrary to what many people think, it is not a uniquely human attribute. Experiments with chimpanzees and other primates have shown that such animals can show empathy and compassion, and that they even have a sense of morality.[1]

The ability to "co-suffer" with another being evolved through natural selection as an important strategy for survival. In the Western world, empathy, compassion, and human rights were discussed as philosophical concepts long before Christianity formulated them as part of its traditions.

Already in the seven centuries stretching from 900 BCE to 200 BCE, philosophical thoughts about compassion had been formulated in Confucianism and Taoism in China, in Hinduism and Buddhism in India, in Greek philosophy, and in the monotheistic society in Israel. An example is a Buddhist principle taken from the *Udānavarga* (5, 18): "Do not harm others in a way that you yourself would not wish to be harmed." (The *Udānavarga* is an early collection of thoughts and utterances by Buddha and his disciples.)

In the sacred writings of Jainism (a religion that arose in India during the same historical period and in the same cultural environment as Buddhism), one finds the following: "The principle of good behavior is not to harm anyone."

In the philosophical movement called "Taoism" (or "Daoism"), one finds similar thoughts: "Consider your neighbor's gain to be your own gain, and your neighbor's loss to be your own loss." (This quote is taken from the *Tai-Shang Kan-Ying P'ien*, meaning "Treatise of the Exalted One on Response and Retribution.")

In some early philosophical texts from India there can also be found secular philosophies and currents of thought. Thus one of the *Upanishads* (the oldest philosophical texts written in Sanskrit, and whose title is traditionally rendered as "secret teachings transmitted from master to attentive pupil") calls into question the existence of the god Brahma. In roughly 600 BCE, there arose a philosophical school called Lokāyata or Cārvāka, which steered clear of holy scriptures and the transmission of venerated traditions. Instead, it taught that there exists only one world—the material world. Believers in Lokāyata therefore denied the existence of any sort of god, as well as the immortality of the soul. Lokāyata clearly illustrates the fact that a skeptical attitude toward religious belief is neither a recent development nor an exclusively Western phenomenon.[2]

Chinese thinker Confucius, who lived in the fifth century BCE, formulated his moral and political principles completely independently of any gods or supernatural notions. Although he did not explicitly deny the existence of gods, his teachings focused on human beings and their earthly existence. Confucius devised a version of the Golden Rule some five hundred years before the birth of Jesus:

> Zi Gong [a disciple of Confucius] asked: "Is there any one word that could guide a person throughout life?"
> The Master replied: "How about 'shu' [reciprocity]? Never impose on others what you would not choose for yourself."[3]

Jesus is the central figure of Christianity. There is no strong evidence about where or when he was born, but most likely someone with that name was born in Palestine in roughly the year 4 BCE, and this person was taken by people around him to be a prophet. After his death roughly thirty-three years later, Christianity became a widespread religion.

Christians tend to believe that the form of the Golden Rule found in the Bible is superior to the versions cited earlier. The Christian version—namely, "Do unto others as you would have others do unto you"—is concerned with what you should *do*, rather than what you should *not do*. And on first glance, it might well seem that this positive formulation, stemming from the Bible, is the best one. But the negative formulation also has some advantages. Swedish philosopher of science and ethics Birgitta Forsman, in her book *Gudlös etik*,[4] writes:

> It is clearly better for me to leave other people in peace and not to demand certain forms of behavior from them than to force them to do things that I personally think are good. Tastes, after all, can vary from person to person. . . . The worst case is captured in the clever saying, "A sadist is a masochist who follows the Golden Rule." This whole topic is closely related to the political and philosophical question about positive and negative rights. Negative rights, such as the rights not to be subjected to torture, slavery, or imprisonment, are usually considered to be more fundamental than positive rights.

The Philosophy of Ancient Times

In ancient Greece, many theories were put forth about human beings and their place in the cosmos. Greek society was sufficiently well organized that the struggle to survive was not people's main concern. At least for a few thinkers, there was time left over to reflect on life. These Greek thinkers gave birth to philosophy, science, and drama—diverse ways to explore the nature of humanity and of the world.

Here we can also sense the earliest traces of secular humanistic thinking. The Greek philosophers were the first people, as far as we can tell, to systematically ponder how the world was constructed and how it worked. Today, these philosophers are usually called "natural philosophers"—thinkers about the nature of nature. Three of the most important of them were Thales, Anaximander, and Anaximenes, all of whom lived in the sixth and fifth centuries BCE.

Thales speculated that the entire world was made out of just one basic substance, out of which other substances could be formed. He thus reflected about how the world's complex structures and phenomena could be reduced to simpler things.

Anaximander is often considered to be the founder of astronomy. But he also developed a theory, based on his study of fossils, of how people originated from creatures of the sea. His ideas about the roots of the human species thus anticipated Charles Darwin's theory of evolution by some twenty-three centuries.

Anaximenes imagined that everything in our world originated in an ocean of air, and that all other substances arose from the condensation or the dilution of air.

A most significant fact about all three of these natural philosophers is that they never resorted to explanations based on myths or religious ideas; rather, they strove to ground their ideas in observations and rational thinking.

Philosopher Democritus (whose dates are roughly 460–370 BCE) further developed the ideas of Thales, and most famously he devised the theory that the world consists of tiny indivisible particles, which he called *atoms*. (The prefix "a" means "without," while "tom" means "piece," so Democritus's word "atom," itself built out of two *semantic* pieces and four *alphabetic* pieces, means "without pieces.") Democritus's philosophy was thus naturalistic, in that it saw everything as being made of atoms following mechanistic laws of nature that hold throughout the entire universe. Democritus believed that even consciousness arises as a result of the laws of nature, and that it ceases to exist when we die.

Democritus is often quoted as saying, "Nothing exists except atoms and empty space; everything else is opinion." This is of course not true, but it very clearly shows his naturalistic approach to reality.

Protagoras (roughly 485–415 BCE) developed ideas in the same spirit as Democritus. Among his writings we find this statement: "Concerning the gods, I have no means of knowing whether they exist or not, nor of what sort they may be, because of the obscurity of the subject, and the brevity of human life." Thus, although Protagoras did not directly deny the existence of gods, he basically considered them to be irrelevant. He is probably best remembered for this statement: "Man is the measure of all things."

In the third century BCE, there lived the great Greek philosopher Epicurus. Like Democritus, Epicurus believed that when a thinking creature dies, the atoms constituting its body and its mind are scattered to the four winds, and that nothing remains of the creature. He also believed that the meaning of life lay in the creation of good things during our brief sojourn on earth, and that if we focus on life's small pleasures, we can attain inner peace and harmony. This viewpoint about life later came to be known as *epicureanism*. According to Epicurus, a person is a physical entity that has no immortal soul. And justice, for Epicurus, resided in the careful following of agreements and contracts that we human beings draw up, whose goal is to keep us from doing harm to one another.

Throughout the centuries, Epicurus was hardly ever mentioned in typical histories of Western civilization, as his theories were considered to be incompatible with Christian theology.

The most important contribution of all these early thinkers is not the precise ideas that they proposed about the world; instead, it is the fact that they thought about the *real world*, rather than thinking about a hypothetical god or society of gods. Unlike medieval philosophers—who devoted their lives to the pondering of such bizarre and futile questions as "How many angels can dance on the head of a pin?"—the much earlier Greek philosophers asked questions about visible, audible, and tangible phenomena. In this way, they saw the world as being natural, rather than supernatural.

Three Intellectual Giants

It is probably fair to say that Western thought derives more from three particular Greek philosophers than from anyone else. These three thinkers—Socrates, Plato, and Aristotle—all lived in the fourth and third centuries BCE.

Plato was a pupil of Socrates, and Aristotle was a pupil of Plato. Together, these philosophers demonstrated that it is possible to think logically about human beings and the world with the help of the tools of reasoning, guided by an attitude of careful investigation.

Socrates showed how a new method of writing, consisting of dialogues in which many questions were posed and various possible answers were explored, could be used to investigate and analyze all types of complex and subtle topics. Merely by asking a series of penetrating questions, Socrates could impart insights and knowledge to his hypothetical dialogue partner. This technique became known as the *Socratic method*.

Plato was a democrat in a certain sense, although not in today's sense. He believed that society should be governed by an enlightened elite, not by the citizenry. That is hardly democratic! As far as knowledge was concerned, however, he was a democrat—that is, he thought that it was always the best (i.e., the most logical) argument that would win in a debate. Success in debating did not have to do with who said what or with how debaters expressed themselves; truth and objectivity would always triumph over personality and rhetorical fanfare.

Aristotle's attitude about understanding the world was one of investigation. He also tried to formulate a type of morality rooted in reason, based on his studies of human nature. His main focus was on how to attain happiness and well-being in this life on earth, as opposed to an imagined life after death.

It goes without saying that all of these ancient philosophers were wrong about many things. Today we understand far more than they did. But they blazed the trail for a sort of thinking that ever since then has shaped humanity's collective pursuit of knowledge and attitude of curiosity.

It is interesting to note that many if not most of Plato's writings are still extant today, while practically nothing written by Epicurus survived. In all likelihood, this is a result of the fact that the monks who carefully copied out and translated the Greek texts, thus assuring their survival, considered Epicurus's ideas to be incompatible with their belief in God, and they thus did not wish to propagate them in any way. In this manner, Christian theological credos wound up having a huge effect on which ancient philosophers might influence our thinking today.

Humanity at the Focal Point

The Greek thinkers, rather than focusing on gods, placed humanity at the center. They saw it as every citizen's duty and obligation to learn to reason more reliably, to develop moral principles, to play a responsible role in society, and to live a sensible life in harmony with other citizens. This doesn't mean that these philosophers always denied the existence of the famous Greek gods. But the relation of people to gods was not found at the focal point of their attempts to understand the natural world and the human condition. Nor was morality seen as reliant on any kind of divine pronouncement or celestial authority.

Moreover, the Greek thinkers believed that people's abilities can be *improved*: that humanity has the potential to develop and evolve. In that way did ancient Greece set the stage for the style of education in the entire Western world. The ideals of these philosophers included an all-around development of people's intellectual and artistic capacities, which included, among other things, philosophy, logic, rhetoric, mathematics, astronomy, and drama.

The Greeks called this ideal form of education *paideia*. The Romans quickly adopted this notion for themselves, giving it the analogous Latin name of *humanitas*. From this Latin word comes the English word "humanism." The origin of humanism in ancient Greece thus had to do with the idea of humans as evolving creatures, independent of gods or other supernatural beings or forces.

One of Aristotle's pupils was Alexander the Great (356–323 BCE). Under his aegis, the Greek empire expanded until it included much of the world as it was then known to the Greeks. When the Romans eventually defeated the Greek army in the second century BCE, the Greek notions of education had already been fully absorbed into the Roman empire's culture. The Roman reign lasted for roughly six hundred years, all the way until the end of the fourth century CE. The Greek ideals of education were respected by the Romans for this whole time.

Roman philosopher Lucius Annaeus Seneca was born in what is now the Spanish city of Córdoba, within a few years of 0 CE, so he was a contemporary of Jesus. As a young man, he was sent off to faraway Rome to study philosophy and rhetoric. Seneca wound up playing a major role in Roman politics, but eventually he was forced into exile. Seneca was a harsh critic of religion. The following is one of his more famous cynical commentaries on the topic: "Religion is regarded by the common people as true, by the wise as false, and by rulers as useful."

The Middle Ages

After the fall of the Roman Empire, what we usually call the Middle Ages set in, and that period lasted nearly a thousand years. The "dark" Middle Ages were, in many respects, not as dark as today we tend to think, but the ancient Greeks' ideal vision of education, as well as their focus on human beings, took a back seat to the dominant and ever-growing church, and to the Christian-style theory of divinity.

Almost all artistic and intellectual efforts were oriented toward religion or theology. The church—basically the Catholic Church—did everything in its power to ensure that it had a total stranglehold on philosophical teachings and all other types of education. Anyone who cast doubt of any sort on religious ideas and rules was harshly punished, if not with hanging or beheading (or some other grisly type of execution), then with torture and lengthy persecution.

During this period, the first European universities were founded, but they limited themselves nearly exclusively to the study of Christian theology. Ideas about humanity itself, and about the rest of the cosmos, entered the picture only insofar as they related to humanity's relationship to God.

The natural philosophy of the Greeks was thus handed over, albeit unintentionally, to Arab philosophers in the Middle East, who continued to develop it themselves. We can therefore thank the Arab and Muslim worlds for the fact that much of Greek philosophy managed to survive through those centuries.

Arab philosopher Averroës (1126–1198 CE) was born in Córdoba in what was then the Islamic land of Spain, which at that time enjoyed a great deal of intellectual freedom.[5] Averroës wrote scholarly commentaries on all of Aristotle's writings and, in that way, helped ensure that Aristotle's ideas would be passed on to as yet unborn generations. In Europe, during the Middle Ages, when Christianity reigned supreme, Aristotle had been largely relegated to the dust piles of history and forgotten; however, thanks to Averroës's scholarship, Aristotle's philosophy came back into prominence and wound up being integrated, at least in bits and pieces, into Christian theology.

Averroës maintained, among other things, that whenever some piece of religious writing contradicted ideas that the natural philosophers had developed, then the religious text should always be interpreted either metaphorically or mythologically—not taken literally. This was a radical stance, and it cleared the way for natural philosophy and science to flourish, at the expense of literal interpretations of the Bible and the Koran. In this manner, Averroës laid down solid intellectual foundations on which the coming Enlightenment in Western philosophy would eventually be based.

The Renaissance

It was only in Renaissance Italy (the French word *renaissance* means "rebirth"), toward the end of the fourteenth century, that the humanistic ideals of education from ancient Greece and Rome were brought fully back to life. Humankind once again became the focal point. The Renaissance notion of a "universal man" (*uomo universale*) was, in many ways, a rediscovery of long-forgotten viewpoints concerning knowledge and the potential of humanity.

The study of humanity and of the human condition emerged, during the Renaissance, into the foreground—and in a wholly new fashion. University courses expanded so as to include subjects like astronomy, law, geometry, medicine, and art. The teachers of these subjects, along with their students, came to be known as *humanists*. The Renaissance humanists developed their own new philosophical tradition that started calling into question the church's way of dealing with the concept of "truth." In particular, they launched the idea that the church was no longer considered the primary source of knowledge, or even of morality. Knowledge could be gained in ways that were different from, and superior to, revelations supposedly emanating from God. One's own observations and personal experiences could be just as valid a pathway to knowledge. Independent thinking became a *virtue*, rather than a form of behavior to be feared or punished. And the mission of philosophy gradually came to be seen as the asking of good questions, not merely the deferential serving of Christian theology and the institutions of the church.

The humanists believed that the religious teachings of the Middle Ages had rendered human beings subservient to God, and they wished to restore humanity to a state of self-respect and autonomy.

Criticisms of the Catholic Church came not only from humanists but also from representatives of the ongoing reformation of the church, led by independent thinkers like Martin Luther (1483–1546). In 1517, Luther published his revolutionary tract "95 Theses on the Power of Indulgences." The very recent development of movable type by Johannes Gutenberg allowed Luther's ideas to be rapidly and widely distributed all over Europe, and this led first to the splintering of the Christian church, then to the founding of Protestantism.

Toward the end of the sixteenth century, the ideas of the natural philosophers began exerting an influence on people's thinking in general. Out of this came the first inklings of modern science. Science quickly made such rapid strides that old-fashioned Christian theology had a hard time keeping up with and coexisting with the new types of knowledge about the world.

In the sixteenth century, English philosopher and diplomat Francis Bacon called into question the church's religious teachings and its claim to be a source of knowledge (let alone *the* source of knowledge), and his ideas had a more durable impact than earlier criticisms had had. In place of the church's teachings, Bacon advocated philosophy and natural science as tools to understand the world. He believed that reason, not revelation, is the true source of knowledge.

In the seventeenth century, natural scientists such as Italy's Galileo Galilei and England's Isaac Newton developed a brand-new scientific way of looking at the world, which contradicted the church's teachings and which suggested that the entire universe was ruled by timeless mathematical laws of nature.

During these years of rapid change, it was risky for thinkers of any sort to publicly question the reigning Catholic doctrines. Many people, in fact, had warned Galileo against asserting that Copernicus's heliocentric model of the cosmos was literally true. (A *heliocentric* worldview states that the earth rotates around the sun, rather than the reverse. The Catholic Church stuck for a very long time with the *geocentric* viewpoint, according to which the sun goes around the earth.) Galileo nonchalantly dismissed all such warnings, and as a result, with his bold theories he brought down on himself the wrath of the Catholic Church. During the Inquisition, he was arrested and threatened with torture and execution; however, after he had confessed his sins, his punishment was reduced to a jail sentence, which then turned into house arrest.

Many people have wondered whether Galileo actually gave the officials of the Catholic Church sufficient evidence for his heliocentric views. Today, of course, we know that he was right, but it's possible to imagine that he was right for the wrong reasons; therefore, one might, in some sense, maintain that the church acted "in good faith" when it rejected the heliocentric position. On the other hand, it's clear that the church was perfectly ready to torture or kill any scientist who dared to contradict the literal interpretation of the Bible. In his earlier-cited 2011 book *Humanism: A Very Short Introduction*, British philosopher Stephen Law writes:

In 2000, Pope John Paul II publicly apologized for, among other things, the Church's trial of Galileo. However, in 1990, Cardinal Ratzinger—the former Pope Benedict—quoted philosopher Paul Feyerabend, seemingly with approval:

"At the time of Galileo the Church remained much more faithful to reason than Galileo himself. The process [meaning "the trial"] against Galileo was reasonable and just."[6]

Precisely what Ratzinger meant by this remark is a matter of debate, but it obviously raised a few humanist eyebrows.[7]

How far the church was prepared to go is shown by the case of Giordano Bruno (1548–1600). Bruno was a monk in the Dominican order, in a monastery near Naples. His studies of science and mathematics led him to formulate ideas about life possibly existing in other parts of the universe—ideas that clashed profoundly with the Catholic Church's doctrines. As a result, he was accused of heresy. He had little choice but to abandon the monastery and his monk's garb, and he moved first to France, then to England, and thence to Germany. In those lands, he was highly respected for, among other things, his vast scientific knowledge and his original mnemonic techniques for memorization. Finally, however, the Catholic Inquisition caught up with him, and he was thrown into jail in various venues over a period of eight years, at the end of which he was put on trial for heresy, found guilty, and finally burned at the stake in Rome's famous Campo dei Fiori ("Flower Meadow") on the ignominious date of February 17, 1600. This is the kind of respect the Catholic Church had for thinkers in those days.

Rationalists and Empiricists

During the seventeenth century, mathematical and analytical tools were developed that helped thinkers of the day explore the universe much further—and in a systematic fashion.

Thanks to the discoveries of such brilliant minds as French philosopher/mathematician René Descartes (1596–1650), Dutch philosopher Baruch Spinoza (1632–1677), and German philosopher/mathematician Gottfried Wilhelm von Leibniz (1646–1716), later thinkers of the Renaissance acquired key tools for the deeper analysis and comprehension of the laws of nature.

Among other things, Descartes pondered what kinds of knowledge we can truly be sure of. He tried to analyze the basis of knowledge by asking himself whether all of his beliefs might not be the result of an "evil demon" that gave him misleading perceptions. How could one be sure that this is not the case?

He concluded that at least his own existence was beyond doubt, because in order to be tricked, he at least had to exist. This was the idea that lay behind his celebrated motto *Cogito ergo sum* ("I think, therefore I am"). Descartes was also a dualist who believed that body and soul are essentially different substances, a body being a physical structure in three-dimensional space (*res extensa*) that cannot think, and a soul being a thinking entity (*res cogitans*) that has no physical extent and no location in space. In subsequent centuries, Descartes's dualistic vision had a profound influence on Western thought.

Sweden's Queen Christina (1626–1689) was fascinated by Descartes's ideas and asked him to come to Stockholm to be her teacher. Unfortunately, Descartes was not prepared for the harshness of Swedish winters, and after having spent only a few months in the Royal Palace in Stockholm, he died of pneumonia.

Baruch Spinoza, who by birth was Jewish, is considered as one of the most outstanding rationalists of his day. He maintained that our faculty of reason, not our sensory perceptions, is the source of our knowledge. His definition of God departed strikingly from traditional conceptions; he identified God with nature, a viewpoint that some people later called *pantheism*. Many of his contemporaries thought of him as an atheist, and for that reason, already at the tender age of twenty-four, he was barred for life from belonging to any Jewish congregation in Amsterdam.

Gottfried Wilhelm von Leibniz contributed to the development of some of the primary tools that are used today in mathematical descriptions of nature, especially calculus, which he discovered simultaneously with Isaac Newton. He made many technical discoveries, one of which was a "mathematics machine" that, in a certain sense, was a forerunner of today's computers.

What these three rational philosophers had in common is the belief in reasoning as the basis of all knowledge, whereas the *empiricists* were inclined to see knowledge as originating in sensory perceptions.

The most important representative of the empiricist point of view was English philosopher John Locke (1632–1704), whose contributions were made mostly in the latter part of the seventeenth century. Aside from being a salient figure of the Enlightenment in the British Isles, Locke inspired political thinking about freedom and equality. He is often considered to have been the first empiricist. He denied the existence of innate ideas and believed that when a baby is born, its mind is a *tabula rasa*, or "blank slate," which then gets filled up by sense impressions during the course of life. In his book *An Essay concerning Human Understanding* (1689), Locke tried to explain the limits of knowledge.

Two other key figures in empiricism were Irish philosopher George Berkeley (1685–1753) and Scottish philosopher David Hume (1711–1776).

Hume would also play a central role in the coming Enlightenment, which heavily criticized religion, as we shall soon see.

The Enlightenment

Scientific progress, new mathematical tools, and the rational turn in philosophy opened the door to a profoundly new attitude toward the world. Could it be that we human beings might somehow be able to understand ourselves in this world we inhabit? Could it be that, with the help of human reason and scientific education, we might feel comfortable with a world utterly lacking in divine guidance?

Trains of thought of this sort gave rise to a new intellectual movement called the Enlightenment, which was launched in the middle of the 1700s and had its center in France. The key idea of the Enlightenment was a belief in humanity and its power of reason. Blind faith in rulers and in the authority of the church was rejected by Enlightenment thinkers. Rather, the Enlightenment's linchpin was the idea that society consisted of human beings, all of whom had equal worth.

Denis Diderot (1713–1784) was one of the leading philosophers of the Enlightenment in eighteenth-century France. He was the principal editor of *L'Encyclopédie*, the first universal reference work, which consisted of thirty-five volumes, and which took thirty years (from 1751 to 1780) to complete. It was the largest encyclopedia ever published, and Diderot himself penned several thousand of its roughly sixty thousand articles; he also read, discussed, and corrected most of the rest. He was also an essayist, philosopher, lawyer, and critic of art, literature, and music. On top of all that, he was an outspoken advocate of women's rights.[8]

Diderot was a committed atheist, and his *Encylopédie* contains many radical and naturalistic ideas. In one of his articles, Diderot described "the enlightened thinker" as someone who "trampling on prejudice, tradition, universal consent, and authority—in a word, on all that enslaves most minds—dares to think for himself." (As for that final word "himself," well, those were the days when women were totally eclipsed behind men. Even Diderot wasn't *fully* enlightened, alas.)

Diderot slipped several articles into the *Encyclopédie* that criticized the church and its theology, even though doing so was very controversial and was opposed by the state-level censor. His writing rooms were often searched for articles that were critical of religion, of the state, or of the reigning morality. The *Encyclopédie* was banned even before it was published, and Diderot himself was arrested and thrown into prison on numerous occasions. In his *Philosophical Thoughts* (1746), he wrote that "[s]kepticism is the first step towards the truth."

Diderot did his best to support himself through his writing, but he also had the good fortune of being helped financially by Catherine the Great (1729–1796), the empress of Russia. Thanks to her support, Diderot traveled to Russia and helped her develop a plan for a future Russian university.

Another important philosopher of the French Enlightenment was François-Marie Arouet, better known by his adopted pen name, Voltaire (1694–1778). He was deeply critical of the Catholic Church and made valiant efforts to defend human rights and the freedom of speech. A famous line often attributed to Voltaire and frequently cited in discussions about the freedom of speech is this: "I disagree with what you have to say, Sir, but I will defend to the death your right to say it."

Apparently Voltaire never made this exact statement, but it was put into his mouth 128 years after he died by English writer Evelyn Beatrice Hall in her 1906 book *The Friends of Voltaire.* Although it is not an exact quote, it certainly captures his attitude eloquently.

Voltaire fought against all forms of religious dogmatism and superstition, and he strongly believed in the power of human reason to help shape a better world. His most famous work is the novel *Candide* (1756), an ironic satire about the naïve Doctor Pangloss, who is convinced that we live in "the best of all possible worlds." Voltaire, who rejected Christianity, viewed religion as a way for the ruling classes to hold ordinary people in thrall.

In another famous work, *Traité sur la tolérance* (A Treatise on Tolerance), dating from 1763, Voltaire describes, in a powerful attack on the Catholic Church, the "violent passions that are aroused by the dogmatic spirit of Christian religions." In it, he declares that only philosophy and tolerance can in the end defeat religious fanaticism, and as a case in point, he describes the Ulster Rebellion in Ireland:

A more populous and wealthier Ireland will no longer see its Catholic citizens, over a two-month period, sacrificing its Protestant citizens to God, burying them alive, hanging their mothers on scaffolds, tying their daughters to the necks of their mothers, and watching them expire together; or slicing open the bellies of pregnant women, pulling out the half-formed embryos, and tossing them to dogs and swine for food; or putting swords in the hands of prisoners and pointing these weapons toward the breasts of the prisoners' wives, fathers, mothers, and daughters, hoping in this way to cause them to murder all their closest kin, and not just murder them but also ensure that they will suffer eternally in hell.[9]

The "holy war" that Voltaire here describes is strongly reminiscent of the war raging today between different brands of extremist jihadists ("holy warriors") in the name of Islam, especially in Iraq and Syria.

A prominent contemporary of Voltaire and Diderot was German-born French philosopher Paul-Henri Thiry, baron d'Holbach (1723–1789), who was deeply committed to the Enlightenment, wrote a number of books, and contributed to the *Encyclopédie*. The most important of his own books was *Système de la nature* (1770), which was burned on account of its objectionable content.

Holbach was an atheist, and he argued that Christianity not only clashed with rationality but also was in fact a system designed to oppress humanity. He believed that everything in the world could be reduced to matter and the laws governing it, and he denied the existence of gods and of the soul. He was also critical of the French monarchy and of the social inequality that it gave rise to.

Secular Philosophy, Morality, and Politics

The French Revolution, in 1789, was profoundly inspired by the ideas of the philosophers of the Enlightenment. One of the Enlightenment's great thinkers outside of France was Scottish philosopher David Hume, mentioned earlier. Hume made significant contributions not only to the theory of knowledge and the philosophy of religion but also to moral and political philosophy.

In his first major book, *A Treatise of Human Nature* (1739), Hume formulated his fundamental criticism of religion, which he had been developing in earlier writings. As a result of his ideas, which were correctly seen as advocating atheism, he was denied a position as professor of philosophy. His greatest work concerning the philosophy of religion, *Dialogues concerning Natural Religion*, was first published in 1776, and in it he stated his position with full force.

Hume criticized many claimed testimonies of miracles and wonders (these terms meant violations of natural law), and his criticisms had a strong influence on thinkers of his day. He analyzed in a very general manner the circumstances under which we ought to believe in personal reports, and he concluded that there are far better explanations for strange and astonishing reports of alleged miracles than simply accepting the claims that such miracles actually took place.

The ideas of the Enlightenment spread rapidly throughout Europe during the eighteenth century. Scientists, philosophers, and writers joined together and further developed the ideas of the Enlightenment.

German philosopher Immanuel Kant (1724–1804) exerted a major impact on the ideas of the Enlightenment. In his famous essay "Beantwortung der Frage: Was ist die Aufklärung?" (An Answer to the Question: What Is Enlightenment?), published in 1784, he describes enlightenment in the following manner:

> Enlightenment is the emergence of human beings from their own self-imposed immaturity. Immaturity is the inability to use one's understanding without relying on guidance from another person. This immaturity is self-imposed when the reason for reliance on others is not that one lacks knowledge, but rather, that one lacks the determination and the courage to rely on oneself and to be independent of others. Sapere aude! Dare to know! Have the courage to rely on your own understanding![10]

Kant's essay captured the Enlightenment's belief in the ability of humanity to develop through philosophical reflection, critical reasoning, scientific research, and political change. His expression "*Sapere aude!*" ("Dare to know!," or "Have the courage to use your own rationality!") could even be taken to be the motto of the entire Enlightenment.

Perhaps Britain's foremost Enlightenment thinker during this period was lawyer/philosopher Jeremy Bentham (1748–1832). Toward the end of his life, he wrote *Analysis of the Influence of Natural Religion on the Temporal Happiness of Mankind* (1822), in which he described religion as irrational (although perfectly explicable as a natural phenomenon) and, moreover, as harmful to society. Bentham is usually considered to be the founder of *utilitarianism*, the modern philosophy based on his notion of usefulness. According to Bentham, the principles of morality do not have a supernatural origin but, instead, can be completely derived from just one basic premise: "The greatest happiness for the greatest number of people." Bentham was a strong advocate of education, seeing it as a source of greater well-being and a better society.

Bentham also had pioneering ideas concerning the philosophy of animal rights. For instance, in *An Introduction to the Principles of Morals and Legislation* (1789), he writes:

> The day may come when the rest of the animal creation may acquire those rights which never could have been withheld from them but by the hand of tyranny.
>
> It may one day come to be recognized that the number of the legs, the villosity of the skin, or the termination of the *os sacrum* are reasons equally insufficient for abandoning a sensitive being to the same fate.
>
> What else is it that should trace the insuperable line? Is it the faculty of reason, or perhaps the faculty of discourse? But a full-grown horse or dog is beyond comparison a more rational, as well as more conversable animal, than an infant of a day or a week or even a month old. But suppose they were otherwise, what would it avail? The question is not "Can they reason?" nor "Can they talk?," but "Can they suffer?" Why should the law refuse its protection to any sensitive being? The time will come when humanity will extend its mantle over everything which breathes.[11]

In Great Britain, the Enlightenment was also represented by, among other philosophers, writer and feminist Mary Wollstonecraft (1759–1797). She was the mother of Mary Shelley (1797–1851), who wrote the novel *Frankenstein* in 1818. Mary Shelley's husband, Percy Bysshe Shelley (1792–1822), also belonged to the group of Enlightenment thinkers who were punished for their skepticism and their refusal to accept religious claims about truth. Before he married Mary, already as a nineteen-year-old, Shelley published *The Necessity of Atheism* (1811). For this act, he was expelled from Oxford University. Only a few years later, in 1822, a journalist at the *London Courier*, upon learning of Shelley's having perished at sea, took a cheap shot at the late poet's atheism by writing "Shelley, the writer of some infidel poetry, has been drowned; *now* he knows whether there is a God or not."

The Enlightenment as a Movement for Freedom

The Age of Enlightenment and its philosophy were understood by many as a freedom movement. Its philosophy raised people's awareness of the fact that they did not need to accept being oppressed, either by religion or by politics. New ideas of this sort came to play a central role in the American colonies' fight for freedom from Great Britain—and in the creation of the United States of America.

A central figure and political powerhouse in the fight for freedom in the British colonies in the eighteenth century was Enlightenment thinker, writer, scientist, and politician Benjamin Franklin (1706–1790). Originally a book printer in Philadelphia, Franklin soon became an important figure in the formation of opinions and in the cultural life of the British colonies in North America. Later he also became famous for his careful studies of electricity (he was the first person to recognize that it came in two varieties—positive and negative) and for his invention of the lightning rod. Franklin founded the American Philosophical Society, which contributed to the spreading of the Enlightenment philosophy from European to American soil.

In the conflict between Britain and its colonies, Franklin served for a while as a neutral go-between, but then he decided to throw his support behind the colonists' quest for independence. He became part of the committee that drafted the Declaration of Independence, and on July 4, 1776, the thirteen colonies declared their independence from Great Britain. That date became the national holiday of the United States. Working with Thomas Jefferson (1743–1826), Franklin helped create an American nation that was based on the philosophical ideals of the Enlightenment. Jefferson subsequently became the country's third president in 1804.

Another important Enlightenment thinker in the eighteenth century was British writer Thomas Paine (1737–1809), who in 1774 traveled to North America, where he advocated the colonies' liberation from British rule, and in fact Paine was the person who came up with the name "the United States of America." He was one of the first voices to call for the abolition of slavery and the death penalty. He also pushed for women's rights and came up with an idea resembling today's Social Security payments to elderly people. In 1793, in his book *The Age of Reason*, Paine writes:

> I do not believe in the creed professed by the Jewish church, by the Roman church, by the Greek church, by the Turkish church, by the Protestant church, nor by any church that I know of. My own mind is my own church. All national institutions of churches, whether Jewish, Christian or Turkish, appear to me no other than human inventions, set up to terrify and enslave mankind, and monopolize power and profit.
>
> Whenever we read the obscene stories, the voluptuous debaucheries, the cruel and tortuous executions, the unrelenting vindictiveness with which

more than half the Bible is filled, it would be more consistent that we call it the word of a demon than the word of God. It is a history of wickedness that has served to corrupt and brutalize mankind; and, for my part, I sincerely detest it, as I detest everything that is cruel.[12]

Up until the publication of this work, Paine had been acclaimed as a great hero, but this kind of antireligious writing turned many people against him.

Today, ironically, the United States has become, as we have already noted, a very religious country, despite its secular founding, and the Christian conservatives and the fundamentalist movement exert a powerful influence on American politics. An example of this troubling tendency is that Thomas Paine, many years after his death, was described by the usually open-minded and famously progressive American president Theodore Roosevelt as "a filthy little atheist."[13]

The Nineteenth Century

The 1800s were the century when science truly came into bloom. Great steps forward in medicine and technology rapidly changed the conditions of life of many people for the better.

An important thinker of the period was England's John Stuart Mill (1806–1873), whose father, James Stuart Mill, was a friend of Jeremy Bentham. John Stuart Mill became the leading figure of nineteenth-century British liberalism. The two works of his that have had the greatest lasting impact are *On Liberty* (1859) and *The Subjection of Women* (1869), whose ideas he developed with his wife Harriet Taylor Mill. In them, he argues for the inviolable freedom of the individual and for complete equality of the sexes. His influence on Enlightenment thinkers was particularly strong during the latter half of the nineteenth century.

One could not write about the Enlightenment in Britain without mentioning the great naturalist Charles Darwin (1809–1882) and his colleague Thomas Henry Huxley (1825–1895). Neither of the two was an Enlightenment philosopher in the strict sense of the term; rather, both were scientists. As we saw earlier, however, Darwin's theory of natural selection and the origin of species made it possible to bypass the idea of a god that, using an inconceivably vast intelligence, carefully designed humans and animals. Darwin's *magnum opus*—his book *On the Origin of Species*—appeared in 1859 and immediately sparked a great uproar. Quite a number of years later—in

1871—he finally published *The Descent of Man*, which specifically described how human beings are a product of natural evolution.

As was noted earlier, Darwin's theory of evolution was, and still is, rejected by religious fundamentalists of the Christian, Jewish, and Islamic faiths, but in serious scientific circles there is no doubt whatsoever that the theory of evolution is the correct way to describe the development of all biological species, including human beings.

Darwin's friend Thomas Henry Huxley was a physician who specialized in anatomy. In 1863, he showed, in his book *Evidence as to Man's Place in Nature*, the great similarities that exist between the brain structures of people (not only of men, despite his book's title!) and apes. Huxley soon came to be the most important advocate of, and popularizer of, Darwin's theory; in fact, his performances earned for him the droll nickname of "Darwin's bulldog." Huxley not only had a profound passion for science but also was a very eloquent orator and could skillfully muster powerful arguments against religious viewpoints. It was he who coined the term "agnosticism," and moreover he was the first person to suggest that birds had evolved from dinosaurs.

One of Huxley's great rhetorical performances took place at Oxford University on June 30, 1860. The occasion was a debate between Huxley and Bishop Samuel Wilberforce (1805–1873), a preacher who believed in interpreting the Bible in the most literal manner possible, and whose "greasy" demeanor had earned him the nickname of "Soapy Sam." Charles Darwin would have attended, but he was very ill at the time and had to stay at home in bed. Bishop Wilberforce opened with half an hour of remarks using language that was "now strictly logical, now witheringly dismissive, always flamboyant"—and with this type of language he tore at the theory of evolution as fiercely as he could, concluding snidely by asking Huxley whether it was on his [Huxley's] *mother's* side or his *father's* side that he was descended from apes. It is said that Huxley replied more or less as follows (no exact transcript of the debate exists):

> I am not in the least ashamed at having an ape as an ancestor; on the other hand, I feel deep shame at the fact of being related to a human being who abuses his great talents by attempting to hide the truth behind smoke-screens of empty verbiage.

Vivian Green, in her book *A New History of Christianity*, reports Huxley's retort somewhat differently but in greater detail, as follows:

A man has no reason to be ashamed of having an ape for his grandfather. If there were an ancestor whom I should feel shame in recalling, it would be a man, a man of restless and versatile intellect, who, not content with an success in his own sphere of activity, plunges into scientific questions with which he has no real acquaintance, only to obscure them by an aimless rhetoric, and distract the attention of his hearers from the real point at issue by eloquent digressions, and skilled appeals to religious prejudice.[14]

Whatever the two opponents actually said, it was a monumental occasion, and the words on both sides were very heated, as it was the first time that evolution had ever been publicly debated. There is no consensus as to which of the two was the "winner" of the debate, but the battle had certainly been launched.

In Germany, philosopher and theologian Ludwig Feuerbach (1804–1872) proposed that the idea of God was a projection of humanity's own inner nature. Specifically, in his 1841 book *The Essence of Christianity*, he tried to show that the Christian concept of God was in actuality the concept of an ideal human being, a notion that every person carries inside their mind. (Incidentally, Feuerbach's older brother Karl Wilhelm was a profoundly gifted mathematician who at age twenty-two discovered one of the most beautiful theorems in all of Euclidean geometry, called simply "Feuerbach's theorem." Tragically, he suffered from a mysterious mental illness and died very young.)

German philosopher Friedrich Nietzsche (1844–1900) was a contemporary of Ludwig Feuerbach's. In 1869, he became professor of Greek at the University of Basel in Switzerland, but soon he had to leave his post because of illness. He was strongly critical of German antisemitism and nationalism. In his 1882 book *The Gay Science*, Nietzsche famously declared "God is dead," and in *Thus Spake Zarathustra* (published in 1883–1885), he returned to this theme.

Nietzsche was an early advocate of cosmopolitanism (the idea of being a citizen of the world, as opposed to nationalism, which is essentially the belief in nations and borders), and he advocated a united and border-free world. He wrote that the Christian concept of morality was a "slave morality"—that is, a life-denying attitude rooted in self-denial, self-deprecation, and a sense of rage—and that the state of being religious was a neurotic condition. As he grew older, Nietzsche grew psychologically more and more unstable, and finally, in 1889, he suffered a serious mental breakdown. After Nietzsche's death, his sister, Elisabeth Förster-Nietzsche, managed her brother's archives until she herself died in 1935. Unfortunately, though, she fell victim to the scourge of Nazism, and as a result, she did her best to paint her brother as a great Nazi thinker—an absurdity, since he had expressly rejected all notions of nationalism and antisemitism.

In Sweden, too, during the nineteenth century, there were Enlightenment thinkers who criticized Christianity and the church's power. In 1888, Viktor Lennstrand (1861–1895) founded the Swedish Utilitarian Society and became its first president. As a youth, Lennstrand had originally been a deeply believing evangelical Christian, but at around age twenty, he broke away from all religion and became a lecturer and writer focusing mainly on the criticism of religion. For this "crime," he served numerous prison sentences.

In 1889, Viktor Lennstrand launched a newspaper called *Fritänkaren: Organ för Sveriges fritänkare* (The Freethinker: Organ for Sweden's Freethinkers), and here is how he explains, on the front page of the newspaper's first issue, the reasons that such a newspaper was needed:

> Whereas there will soon no longer remain any place in our nation where free thought is allowed to thrive and where honest truth-seekers can gather for the purposes of exchanging ideas and agitating for the enlightenment of the people and for liberation from medieval superstitions; and whereas the reactionary and conservative press not only approves of but openly advocates the violent suppression of every movement that aims at the liberation of reasoning and thinking and the fostering of awareness for its own sake . . . it thus seems to us that there is only one choice left: to found our own journalistic medium, acting as a guardian for all true freethinkers of all stripes; a medium in which we, before a thinking public, can present the language of truth, and in which we can exonerate ourselves from unjust accusations and present our case for the judgment of the public court.[15]

This editorial text, in Swedish, was printed on the front page of the very first edition of *Fritänkaren*, on June 1, 1889. Ironically, the new newspaper was edited and printed in a building whose address was Kyrkogatan 9A—that is, 9A Church Street.

As a result of lectures that he gave, bearing titles such as "Christian Morality and Rationalistic Morality," "Is There a Life after This One?," and "What Do We Utilitarians want?" Viktor Lennstrand was prosecuted numerous times for blasphemy. He was also prosecuted for violation of freedom of the press when he printed his "Six Theses of Örebro" (named after the Swedish town in which he stated them in a lecture), which ran as follows:

1. The books of the bible were written in the same way as we write books and letters today.
2. The god of the Old Testament is a sinner.
3. Christianity's gods have no greater reality to them than do Odin, Jupiter, or Serapis [an ancient Græco-Egyptian deity].
4. Nature, human experience, and history all are evidence against the existence of an omnipresent heavenly realm and an all-powerful god.
5. There is no new Jerusalem, nor is there a fire-and-brimstone hell, nor is there a personal consciousness beyond the grave.
6. The highest form of morality and ethics has ties neither to theology nor to religious dogma, and a life of service is the best possible sort of life.[16]

Freethinkers and Atheists

We have seen that secular thinking sprang out of ancient Greek roots, then developed further during the Renaissance in Italy and during the Enlightenment in England and France, until it reached its stage today. A common thread running through all these stages was free thought: the right (and the duty!) of human beings to think without dogmas but with the help of the reasoning faculty that is our innate heritage. That is why, in the present book, the term "free thinking" has been used almost interchangeably with the idea of a nonreligious viewpoint and, particularly, with the idea of a godless world.

Certain freethinkers were able to imagine a god who (or "that"), being an "original creator," had launched the universe, setting it into motion but subsequently playing no role in the course of events that ensued. Such an attitude about such an abstract god is called *deism*. Deists think that even if there is (or once was) a divine entity that created the world, we humans today no longer have any relationship with that entity. And thus, religion and the church are human inventions, and they do not have legitimacy coming from any higher powers.

Other freethinkers were outspoken atheists. But what freethinking deists and freethinking atheists had in common was the view that human problems should be solved through rational analysis and careful study—and with the help of a morality rooted in human values, rather than through a blind faith in some sort of invisible god. The freethinkers' views thus went hand in hand with the ideas of the Enlightenment as they were developing.

A modern form of deistic belief might be that some sort of divine being caused the Big Bang (the legendary cataclysmic event postulated by cosmologists to be the moment our universe came into existence), but aside from that, this being has nothing to do with human beings or the world we inhabit. Such a deistic viewpoint might be compatible with secular values and a secular notion of how society should be organized and should develop, but it is certainly *not* compatible with the secular-humanistic belief that the world is rationally comprehensible and explicable. After all, if the existence of the universe is natural and potentially explicable, then there cannot be any supernatural and "inexplicable" god that created it.

∼

INTERLUDE: ON FREE WILL

Do we human beings have free will?

Either our choices are determined by the past, or they are independent of it. Either we make decisions freely, or we don't. But which way is it?

The question is very tricky, particularly for those of us who have a naturalistic view of reality. The matter is simpler for people who believe in some sort of god; after all, for them, a person consists of a body and a soul, and the two are separate phenomena. From their point of view, God gave each one of us a free will so that we could choose our actions. That makes free will an autonomous part of reality, end of story.

But how is it for those of us who don't accept that notion? What do we—we who hold no scriptures sacred, and who refuse to buy into the idea of eternal souls divinely imbued with free will—think about the notion of free will?

We know that all physical processes obey certain precise mathematical laws—the laws of nature. The classical laws of nature (those that physicists believed in until the mid-1920s) were deterministic, and thus all events were in principle predictable. Quantum mechanics, which superseded the classical laws, says that what is deterministically predictable is a set of probabilities rather than actual outcomes, and that the outcomes are a result of the "dice" that Einstein hated. Though it is no longer deterministic in the way classical physics was, quantum mechanics doesn't include or allow intervention by any kind of "will." The idea of "will" is not part of the laws of physics.

And so, how can human beings have free will? If consciousness is an outcome of matter and of physical laws governing it, don't these laws also apply to phenomena involving consciousness? Doesn't the fact that physical matter—whether inanimate or animate—always obeys natural law, whether it's classical or quantum mechanical, throw free will completely down the drain?

It feels to me as if I have free will. I can choose, for example, between ravioli and risotto when I go to my favorite Italian restaurant (and I could just as easily have chosen to go to the Indian restaurant); I can rapidly surf all the channels on my television and then decide whether I want to watch the Japanese cooking show or the old Ingmar Bergman movie. Are all my choices predetermined, or are they all random results of microscopic dice throws?

Indeed, could the randomness at the heart of quantum mechanics be the source of my free will? At first it sounds appealing, but unfortunately it doesn't make any sense. After all, it hardly felt as if my choice of risotto tonight, rather than ravioli, was made at random, nor did it feel as if I chose randomly to watch *Smiles of a Summer Night* rather than *Iron Chef*. To the contrary: it felt, in both cases, as if I first thought things through, and then made a judgment call on the basis of my just-made reflections. What would randomness have to do with a carefully thought-out judgment call?

If I think through the alternatives and their consequences with great care and in great detail, but then if my choice-making mechanism rudely ignores my thoughts and simply takes the flip of a coin or the toss of some dice as the basis for its choice, what purpose did my careful ponderings serve? None at all!

Modern neuroscience sheds fascinating light on these ancient enigmas. There are experiments that demonstrate that the brain activates decision-making processes before we consciously make up our minds. Neuroscientist J. D. Haynes placed a subject in front of two buttons, one on the left and one on the right. Subjects were allowed to decide when to push either button, and which button to push. And while they were doing so, their brain activity was being recorded by an fMRI device (functional magnetic-resonance imaging). Haynes discovered that sometimes the decision had already been made as long as ten seconds before it was actually carried out.

As if that weren't enough, different parts of the brain were seen to be activated, depending on whether the upcoming decision was going to be to choose the lefthand button or the righthand button. Thus the "free" decision as to which button to push was completely determined long in advance.

From experiments of this sort (and there have been many), it seems as if the impression that we all have—namely, that of making conscious choices—is simply an illusion. But can free will truly be nothing but an illusion?

Douglas Hofstadter, in chapter 23 of his book *I Am a Strange Loop*, gladly accepts the idea that "will" is a real (although emergent) phenomenon inside a brain, but he argues that it's a grave error to think that one's own will, or anyone else's will, is "free":

> I am pleased to have a will . . . but I don't know what it would feel like if my will were free. What on earth would that mean? That I didn't follow my will sometimes? Well, why would I do that? In order to frustrate myself? I guess that if I wanted to frustrate myself, I might make such a choice—but then it would be because I wanted to frustrate myself, and because my meta-level desire was stronger than my plain-old desire.
>
> Thus I might choose not to take a second helping of noodles even though I—or rather, part of me—would still like some, because there's another part of me that wants me not to gain weight, and the weight-watching part happens (this evening) to have more votes than the gluttonous part does. If it didn't, then it would lose and my inner glutton would win, and that would be fine—but in either case, my non-free will would win out and I'd follow the dominant desire in my brain.
>
> Yes, certainly, I'll make a decision, and I'll do so by conducting a kind of inner vote. The count of votes will yield a result, and by George, one side will come out the winner. But where's any "freeness" in all this? . . .
>
> Our will, quite the opposite of being free, is steady and stable, like an inner gyroscope, and it is the stability and constancy of our non-free will that makes me me and you you, and that also keeps me me and you you.[17]

I am very sympathetic to Hofstadter's stance, which essentially claims that the free will riddle is just a pseudo-problem caused by an illusion, but even though I feel he is on the right track, I would hesitate to say that anyone today has definitively solved the free will mystery. We're surely getting closer, but we're not all the way there yet. So in my view, the free will question remains central to modern consciousness research and to the discipline called "neurophilosophy."

Whatever free will is (or isn't), one thing is true for sure: It is crucial that we organize our society and our lives as if our wills were free. It would be unthinkable for us to do otherwise, since the moment that we cease to believe that people really do make up their minds, and that people really do frame their own personal decisions, then we will no longer have any grounds for any sort of moral accountability—and that would destroy society as we know it.

There's no doubt that the human brain can learn moral rules and values that make it possible for us to all live together in harmony. This is a great thing about human nature. No matter how free or unfree we are, we can and must welcome universal moral standards.

CHAPTER TWELVE

SECULAR VOICES IN OUR DAY
Concerning Awe, Politics, and Religion

What can be asserted without evidence can also be dismissed without evidence.—Christopher Hitchens[1]

It was during the twentieth century that the form of enlightenment known as "secular humanism" came into full bloom. As we will see, although it stirred up controversy even after so many centuries, it acquired early and strong advocates who collectively helped forge a Western society leaning toward secularism.

Perhaps the foremost advocate for this tradition in Europe during the twentieth century was British mathematician, philosopher, and social theorist Bertrand Russell. (Russell's godfather was John Stuart Mill, who died only one year after little Bertie was born. As he grew older, Russell's thoughts were deeply affected by his godfather's thoughts.)

Among Russell's most influential works were the three massive tomes called *Principia Mathematica* (1910–1913), which he coauthored with eminent British philosopher Alfred North Whitehead (1861–1947). In this grandiose work, the two authors valiantly strove to develop all of mathematics on the sole basis of pure axiomatic logic. Two decades later—in 1931—the two brave thinkers' noble dream, however, was suddenly and forever shattered by Austrian logician Kurt Gödel's revolutionary discovery of the fundamental and irremediable *incompleteness* not only of *Principia Mathematica* but of *all* axiomatic reasoning systems aiming at capturing all of mathematical truth. Russell never fully recovered from this devastating blow.

In his famous (and also infamous) book *Why I Am Not a Christian* (1958), Russell describes how he had come to his skeptical point of view concerning religion; the publication of this book deeply shook up the British cultural establishment. Russell was also a political activist, who, among many other things, fought for the abolishment of nuclear weapons and, in his later years, stridently opposed the Vietnam War. Together with French philosopher Jean-Paul Sartre (1905–1980), among whose most famous works are *Being and Nothingness* (1943) and *Existentialism and Humanism* (1946), Russell in 1967 set up the International War Crimes Tribunal (often simply called the "Russell Tribunal") in Stockholm, whose declared purpose was to investigate and evaluate U.S. foreign policy and military intervention in Vietnam. Many prominent international figures participated, and Russell explains the rationale behind the tribunal by quoting Robert H. Jackson, the chief prosecutor at the Nuremberg War Crimes Trials:

> If certain acts and violations of treaties are crimes, they are crimes whether the United States does them or whether Germany does them. We are not prepared to lay down a rule of criminal conduct against others which we would not be willing to have invoked against us.[2]

When the issue, however, was raised as to whether alleged war crimes committed by North Vietnam would also be investigated by the tribunal, Russell's spokesperson haughtily replied: "Lord Russell would think no more of doing that than of trying the Jews of the Warsaw Ghetto for their uprising against the Nazis."[3]

This was no doubt a clever retort via mockery, but one sometimes has to wonder about double standards of this sort, which are occasionally held even by great thinkers. What's sauce for the goose is, after all, sauce for the gander.

As might be expected, Bertrand Russell's intense involvement in secular humanism gave rise to problems. In the United States, he was subjected to an intense campaign of slander, and although in 1939 he was offered a professorship at the City College of New York, he was viciously attacked for his atheism and his "immorality," and eventually he wound up being prevented from taking up the offer. In 1950, Russell was awarded the Nobel Prize in Literature "in recognition of his varied and significant writings in which he champions humanitarian ideals and freedom of thought."

Another major twentieth-century advocate of secular humanism was Austrian philosopher of science Karl Popper, whom we earlier met in chapter 4. Popper contributed to A. J. Ayers's anthology of essays *The Humanist Outlook* (1968) with an essay called "Emancipation through Knowledge," in which he described the key role of knowledge in liberating humanity from religious shackles.

For a while, Popper was associated with the British Humanist Organization. In his famous book *The Open Society and Its Enemies* (1945), he eloquently argues for a free and open democratic society and against totalitarian and fundamentalist points of view. Popper also argued for a naturalistic understanding of the universe and for an objective conception of truth. In his 1976 book *Unended Quest: An Intellectual Autobiography*, he declares: "My conviction [is] that there is a real world, and that the problem of knowledge is the problem of how to discover this world."[4]

Popper had nothing but contempt for truth-relativism (the notion that truth is observer-dependent) and for postmodern ideas to the effect that we "create our own realities." Again in his autobiography, he writes: "My sense of social responsibility told me that taking such problems seriously was a kind of treason of the intellectuals—and a misuse of the time we ought to be spending on real problems."[5]

The First Humanist Society in the United States

In the year 1929, Unitarian minister Charles Francis Potter (1885–1962) founded the First Humanist Society of New York City. This daring act was the culmination of a long spiritual trek; Potter had started out as a Baptist minister, then had left the faith and converted to Unitarianism, but eventually he abandoned Unitarianism as well because, in his opinion, even such a liberal pulpit did not afford him the necessary freedom of expression.

In 1923 and 1924, Potter held a series of highly publicized radio debates with Dr. John Roach Straton, a fundamentalist Baptist pastor and a highly colorful character in his own right. (For instance, Straton was such a great admirer of Uldine Utley, a fourteen-year-old preacher, that he invited her to preach in his Calvary church in New York City.) The Potter-Straton radio debates were soon published in four volumes provocatively titled *The Battle over the Bible*, *Evolution versus Creation*, *The Virgin Birth—Fact or Fiction?*, and *Was Christ Both Man and God?*

When Charles Potter founded the First Humanist Society in 1929, he felt that humanism itself was a radical new sort of religion, describing it this way:

> Humanism is not the abolition of religion but the beginning of real religion. By freeing religion of supernaturalism, it will release tremendous reserves of hitherto thwarted power. Man has waited too long for God to do what man ought to do himself and is fully capable of doing . . .

> Humanism will be a religion of common sense; and the chief end of man is to improve himself, both as an individual and as a race.[6]

Taking advantage of his new humanist "pulpit," Potter became a vocal advocate for many sorts of social reform, vigorously opposing capital punishment and campaigning for both birth control and women's rights. Also, a few years later, in 1938, he founded the Euthanasia Society of America.

Among the early members of the First Humanist Society's advisory board were such illustrious intellectual figures of the day as Julian Huxley, John Dewey, Thomas Mann, and Albert Einstein. Huxley remained connected to the First Humanist Society for a long time, eventually going on to found the International Humanist and Ethical Union (IHEU), which today acts as an umbrella organization for roughly a hundred secular humanist groups throughout the world.

The Myth of Albert Einstein's Religiousness

Albert Einstein, one of Potter's illustrious advisory-board members, is often trotted out by religious people as a prime example of how religious and scientific thinking are totally compatible. Eager to claim the godlike Einstein as a member of their club, they point to the oft-cited quote "God does not play dice" as if this statement proved that Einstein believed in the existence of God.

The claim, however, does not hold water. Einstein was not in any sense religious. The phrase "God does not play dice" was simply a pointed Einsteinian criticism of contemporary quantum physics. And in fact, what Einstein said was *not* "God does not play dice." The actual German quote comes from a letter that Einstein wrote in 1926 to his fellow physicist Max Born, and it runs as follows, first in German and then in English: "Jedenfalls bin ich überzeugt, dass der Alte nicht würfelt" ("In any case, I am convinced that the Old One does not play dice").[7]

By using the humorous expression "the Old One" in his letter to his friend, Einstein was jocularly referring to the unknown source of the laws of the universe. In particular, he was objecting to the brand-new quantum thesis, due in part to his own work on radiation in 1916–1917 and in part to Born's very recent work, that said that elementary particles have an intrinsic uncertainty to their behavior, and that certain processes can only be predicted probabilistically, rather than with the exactitude of the earlier, or *classical,* laws of physics. Although Einstein's own ideas had played a key role in the evolution of this probabilistic viewpoint, Einstein never accepted the conclusion that others (especially his friend Max Born) had drawn. He was convinced that all phenomena in the universe obeyed exact mathematical laws and that there was no room for any kind of intrinsic randomness. Therefore, in telling Max Born that "the Old One does not play dice," Einstein was vividly, metaphorically, and wittily expressing his skepticism about quantum mechanics as it then stood; his wit, however, did not in any way imply or reveal an underlying belief in any kind of god—whether Christian, Jewish, Muslim, or otherwise—or even in an Old One.

Today, the predictions of quantum mechanics, though mind bending, have been confirmed by all sorts of subtle experiments, often to an astonishing ten or twelve decimal places of accuracy. Although quantum mechanics is without a shadow of a doubt the most exact and complete theory in all of science, it has not yet proven possible to unify it with our *intuitive* understandings of the nature of reality. The interpretation of quantum mechanics is therefore still a source of lively controversy today; however (as was discussed earlier, especially in chapters 2 and 9), its controversiality to experts does not mean that quantum physics provides support for various religious or New Age–inspired ideas. Unfortunately, though, quantum slogans are still often cited by religious and New Age groups as if they were the ideal support for the groups' unscientific views.

In contrast to what some religious people would like, Albert Einstein's involvement in the First Humanist Society suggests that he would have described himself as a secular humanist, in today's sense of the term. His metaphorical use of the term "the Old One" should be understood as representing his supreme sense of awe and wonder for the universe. It was his personal poetic fashion of talking about the laws of nature.

In "The World as I See It," an essay written around 1930, Einstein states:

> The most beautiful experience we can have is the mysterious. It is the fundamental emotion that stands at the cradle of true art and true science. Whoever does not know it and can no longer wonder, no longer marvel, is as good as dead, and his eyes are dimmed. It was the experience of mystery—even if mixed with fear—that engendered religion. A knowledge of the existence of something we cannot penetrate, our perceptions of the profoundest reason and the most radiant beauty, which only in their most primitive forms are accessible to our minds: it is this knowledge and this

emotion that constitute true religiosity. In this sense, and only this sense, I am a deeply religious man. . . . I am satisfied with the mystery of life's eternity and with a knowledge, a sense, of the marvelous structure of existence—as well as the humble attempt to understand even a tiny portion of the Reason that manifests itself in nature.[8]

In this statement, Einstein makes no concession whatsoever to *traditional* religious thought, nor to any form of mysticism or belief in supernatural phenomena. Indeed, he had no patience with the idea of a god or a life after death. For Einstein, the term "mystery" denoted his deeply felt sense of awe for the rule-following patterns of the universe—which it is the researcher's duty to try to discover and comprehend, using the pathways of reason. As he puts it:

I cannot imagine a God who rewards and punishes the objects of his creation, whose purposes are modeled after our own—a God, in short, who is but a reflection of human frailty. Neither can I believe that the individual survives the death of his body, although feeble souls harbor such thoughts through fear or ridiculous egotisms.[9]

When Einstein applied for a residence permit in Zurich (something he did several other times later in his life, in other countries), he filled in the blank labeled "religion" with "no religious affiliation." At that time, making such a statement was highly controversial (just as it is in many parts of the world even today), and his bold but honest act caused him quite a lot of additional trouble.[10]

Another quote from Einstein that is often dragged out by advocates of organized religion as a "proof" of the great scientist's religiousness is this: "Science without religion is lame; religion without science is blind."

With these provocative words, Einstein was expressing neither a belief in God nor a respect for ungrounded faith. Rather, he was talking about people's primordial drive to understand the universe they inhabit, and about his own personal faith in the possibility of reaching such understanding. Here is the wider context of the phrase in question, taken from Einstein's revealing memoir *Out of My Later Years*:

Science can only be created by those who are thoroughly imbued with the aspiration toward truth and understanding. This source of feeling, however, springs from the sphere of religion. To this there also belongs the faith in the possibility that the regulations valid for the world of existence are rational, that is, comprehensible to reason. I cannot conceive of a genuine scientist without that profound faith. The situation may be expressed by an image: Science without religion is lame, religion without science is blind.[11]

As is clear, Einstein here is referring to the fact that the traditional religious image of a god who created the universe in a comprehensible manner can be seen as a historical precursor of scientific theories and of humanity's deep-seated goal of coming to understand "God's works." This of course doesn't mean that there really is such a god, or that Einstein thought that there was one.

Indeed, throughout time, Einstein grew increasingly annoyed by deliberate and religiously motivated distortions of his words. In July 1953, he received a warmly appreciative letter, rife with biblical quotations, from a Baptist pastor, who asked Einstein whether he had reflected about his immortal soul and its relationship to the Creator, and whether he felt assurance of everlasting life after death. In the margin of this letter, Einstein wrote, in his own hand, and in English: "I do not believe in immortality of the individual, and I consider ethics to be an exclusively human concern with no superhuman authority behind it." It is not known whether he actually sent an answer to the pastor or not.[12]

Einstein's growing frustration with being considered conventionally religious emerges very clearly in his correspondence. For instance, in 1954, an atheistic man who had read a claim that Einstein was religious wrote Einstein from Italy to ask him whether or not this was true, and Einstein answered him as follows:

It was, of course, a lie what you read about my religious convictions, a lie which is being systematically repeated. I do not believe in a personal God and I have never denied this but have expressed it clearly. If something is in me which can be called religious, then it is the unbounded admiration for the structure of the world so far as our science can reveal it.[13]

The same year, only one year before he died, Einstein wrote a letter to the philosopher Eric Gutkind where he states the following:

The word "God" is for me nothing more than the expression and product of human weaknesses; the Bible a collection of honourable but still primitive legends, which are nevertheless pretty childish. No interpretation, no matter how subtle, can (for me) change this. . . . For me, the Jewish religion, like all other religions, is an incarnation of the most childish superstitions.[14]

Few would dispute that Albert Einstein was the twentieth century's foremost scientist, and for this reason, any religious group would have considered it a great coup to be able to claim that such an outstanding mind belonged to their camp. But as we have just seen, Einstein's supposed religiosity is sheer myth. To be sure, there do exist conventionally religious scientists here and there (although one finds that their proportion diminishes rapidly as one moves up the ladder of scientific achievement)—but Einstein was not among them.

Albert Einstein was awarded the Nobel Prize in Physics for 1921, although it was only given to him in 1922—and it was not for his celebrated theory of relativity but for his discovery of the little-known "law of the photoelectric effect." There is a great irony in the way the Nobel citation was worded, revealing that Einstein's greatest contributions were not understood by the Nobel committee.

The strange saga of the "photoelectric effect Nobel Prize" began way back in 1905, when the totally unknown and freshly minted twenty-five-year-old PhD Albert Einstein (working in a lowly patent office in Switzerland) wrote an article claiming, on the basis of a subtle analogy that he had devised, that light was made of particles ("light quanta," as he called them)—not waves. This audacious claim from the young patent clerk third class ran smack in the face against everything that physicists of the day held sacred about light, and it flat-out contradicted James Clerk Maxwell's profound equations for electricity and magnetism, which were considered by physicists the world over to be as solidly confirmed as any law of physics ever had been or would be. Consequently, no one took the brash upstart Albert Einstein's light-quantum hypothesis in the least seriously for many years.

In his 1905 article, Einstein not only made a tradition-smashing guess about light's nature but also cleverly *exploited* his guess to predict how electrons in metals struck by light would be ejected. This was the "photoelectric effect," and though it was known a little bit by a few specialists, it had not been studied deeply and was very little understood at that time. Einstein's calculation of the intensity of the stream of escaping electrons (as a function of the color of the incident light) was a daring prediction that went way out on a limb in its mathematical precision, and this was his gentle hint to his colleagues that checking out the behavior of the little-known photoelectric effect in detail would be a good way to test his hypothesis of light quanta.

Around 1912, Einstein's photoelectric-effect prediction was at last put to the test by Robert Millikan, a famed American physicist at Caltech. After several years of painstaking work, Millikan published his results, showing that Einstein's prediction had passed the test with flying colors: the prediction was fully and perfectly confirmed. This was a great triumph both for Millikan and for Einstein; however, Millikan then added some very strange comments. He said that it was remarkable that Einstein's *prediction* had been verified, since the *underlying idea* from which it had sprung—namely, the hypothesis of light quanta—was clearly nonsensical, and he concluded by saying that the mysterious behavior of the photoelectric effect was still awaiting a plausible scientific explanation. Millikan even had the gall to claim (without any basis in fact) that Einstein himself had finally dropped his belief in light quanta.

Several years later—in 1921—there *still* was no physicist on the face of the earth, other than Einstein himself, who believed in light quanta. But ironically enough, physicists all had accepted the correctness of Einstein's photoelectric-effect prediction, while paying no attention to its origin in Einstein's radical vision of light. Therefore, the Nobel committee's choice to award that year's prize solely for Einstein's "discovery of the law of the photoelectric effect" was exceedingly timid, since Einstein's truly *great* discovery, his truly *deep* discovery, was that light is made of particles—this was a *revolutionary* discovery!—whereas his law predicting the detailed behavior of the photoelectric effect was a quite minor discovery, and in any case it came straight out of the light-quantum idea. This attitude on the part of physicists seems almost incomprehensible today.

Einstein always referred to his hypothesis of the existence of light quanta as "the most revolutionary idea I ever had," but his Nobel Prize was certainly not given for that idea, but for a nonrevolutionary idea that followed trivially from it. What an irony.

And—as a final footnote—in 1923, when Arthur Holly Compton studied the behavior of light bouncing off of orbiting electrons in atoms, he found mysterious behavior that couldn't be explained by Maxwell's equations but that perfectly agreed with Einstein's light-quantum idea. The "Compton effect" was reproduced and extended in many laboratories, and this development rapidly reversed the opinions of physicists the world over. Soon all the experts had jumped on the light-quantum bandwagon.

In 1926, the catchy word "photon" was coined, and Einstein's picturesque term "light quantum" became a relic. Today everybody has heard of photons, and no one bats an eyelash when the idea of "particles of light" is mentioned. Such is scientific progress: what once was so radical an idea that virtually no expert anywhere believed it gets confirmed, becomes widely accepted, and finally winds up being taken totally for granted even by tiny schoolchildren.

Secular Voices Today

Today, secular humanism is represented on the international stage by such well-known philosophers as Rebecca Goldstein, Steven Pinker, Daniel Dennett, Peter Singer, Anthony Grayling, Michel Onfray, and Stephen Law; by such eminent scientists as Steven Weinberg, E. O. Wilson, Richard Dawkins, Sam Harris, Jim al-Khalili, and Brian Cox; and by such illustrious writers as Umberto Eco, Barbara Ehrenreich, Caroline Fourest, Salman Rushdie, Alice Walker, Joyce Carol Oates, Wole Soyinka, Taslima Nasrin, Philip Pullman, and Ayaan Hirsi Ali. And of course there are many, many more.

Physician and writer Taslima Nasrin, born in Bangladesh in 1962, has fought for many years for women's rights and for secular principles in the world of Islam. She worked in her native land until 1994, when repeated death threats forced her to flee. Today she lives under the protection of bodyguards, but despite the risks, she constantly writes and travels, giving lectures on her secular vision.

Several of her books deal with the oppression of religious minorities in countries with Muslim majorities. When she wrote newspaper articles on this topic in Bangladesh in 1993, she stirred up the wrath of certain Muslim fundamentalists, who called for her execution. Ever since then, she has felt she has no choice but to live in exile, at times in Sweden. In recent years, Nasrin has lived in India, but there she continues to be threatened by fundamentalists.

American professor of philosophy Daniel Dennett, born in 1942, has given a careful analysis of religion as a purely natural phenomenon, due to specific sorts of mental processes that evolution has selected for. He starts from the following simple question: "Why does religion exist?"

The answer to Dennett's simple question is of course very simple—as long as you grant that there *actually is* a god who demands submission, obedience, and unquestioning reverence from humans. (It is perfectly possible to imagine a god who created the universe but who then let the creation run its own course without meddling with it in any way, or revealing how it was created. But the gods that humans tend to relate to are always mixing themselves up in our affairs, coming down (or up?) to earth in order to set up their embassies—churches, mosques, temples—and also producing a son or some other type of heir. . . . Could this very mundane humanlike behavior be sufficient proof of the fact that the gods in question are human inventions?)

Of course, it's equally important to try to answer Dennett's question in the *other* case—namely, the case where there is *no* god. In that case, what could explain the existence of religion in the world? To answer this question is the central mission of Dennett's 2006 book *Breaking the Spell: Religion as a Natural Phenomenon*. Dennett writes:

> I might mean that religion is natural as opposed to supernatural, that it is a human phenomenon composed of events, organisms, objects, structures, patterns, and the like that all obey the laws of physics or biology, and hence do not involve miracles. And that is what I mean. Notice that it could be true that God exists, that God is indeed an intelligent, conscious, loving creator of us all, and yet still religion itself, as a complex set of phenomena, is a perfectly natural phenomenon.[15]

American professor of psychology Steven Pinker, born in 1954, often writes about the need for a new enlightenment, and he worries about the flourishing postmodernist and anti-science movements that have arisen in certain areas of the academic world and in debates about culture. Advocates of scientific thinking are often raked over the coals by humanists for advocating "scientism" (a pejorative term that, for the critics, denotes an exaggerated, even dogmatic faith in science's power to yield knowledge about everything, and to solve all the world's problems). But, in an article titled "Science Is Not Your Enemy," Pinker warmly embraces the notion of "scientism," casting it in a positive light rather than a negative one:

> Scientism, in this good sense, is not the belief that members of the occupational guild called "science" are particularly wise or noble. On the contrary, the defining practices of science, including open debate, peer review, and double-blind methods, are explicitly designed to circumvent the errors and sins to which scientists, being human, are vulnerable. Scientism does not mean that all current scientific hypotheses are true; most new ones are not,

since the cycle of conjecture and refutation is the lifeblood of science. It is not an imperialistic drive to occupy the humanities; the promise of science is to enrich and diversify the intellectual tools of humanistic scholarship, not to obliterate them. And it is not the dogma that physical stuff is the only thing that exists. Scientists themselves are immersed in the ethereal medium of information, including the truths of mathematics, the logic of their theories, and the values that guide their enterprise. In this conception, science is of a piece with philosophy, reason, and Enlightenment humanism.[16]

Pinker's criticism of the religious worldview is very harsh (again taken from the same article):

To begin with, the findings of science entail that the belief systems of all the world's traditional religions and cultures—their theories of the origins of life, humans, and societies—are factually mistaken. We know, but our ancestors did not, that humans belong to a single species of African primate that developed agriculture, government, and writing late in its history. We know that our species is a tiny twig of a genealogical tree that embraces all living things and that emerged from prebiotic chemicals almost four billion years ago. We know that we live on a planet that revolves around one of a hundred billion stars in our galaxy, which is one of a hundred billion galaxies in a 13.8-billion-year-old universe, possibly one of a vast number of universes. We know that our intuitions about space, time, matter, and causation are incommensurable with the nature of reality on scales that are very large and very small. We know that the laws governing the physical world (including accidents, disease, and other misfortunes) have no goals that pertain to human well-being. There is no such thing as fate, providence, karma, spells, curses, augury, divine retribution, or answered prayers—though the discrepancy between the laws of probability and the workings of cognition may explain why people believe there are. And we know that we did not always know these things, that the beloved convictions of every time and culture may be decisively falsified, doubtless including some we hold today.[17]

In other words, the worldview that guides the moral and spiritual values of an educated person today is the worldview given to us by science.

In his recent book *The Better Angels of Our Nature* (2011), Steven Pinker also tried to show that the human inclination toward violence and cruelty has in fact diminished throughout the centuries, rather than the reverse, keeping pace with secularization, education, and the development of prosperity.

British writer and evolutionary biologist Richard Dawkins, born in 1941, is considered by many to be our day's harshest critic of religion. In his book *The God Delusion* (2006), he mounts a severe attack on religious ideas and their consequences in the world. Dawkins also wrote *The Magic of Reality: How We Know What's Really True* (2011), in which he lists a series of myths and gives scientific explanations of those phenomena. In an interview in 2013 with the magazine *Sans*, he describes the phenomenon of religion, using an evolutionary-biology perspective:

I think that we should see it [religion] as a byproduct of certain psychological tendencies. I don't think that one can say that religion is a human trait that we have developed because it has evolutionary advantages. That might be true—it might be the case that religious people live longer because they have less chance of developing ulcers, as they are not as worried as other people are—but I don't think that such a theory of religion would be enthusiastically received. It's more likely that being religious has advantages in terms of certain psychological tendencies, such as a tendency to obey authority figures. One can easily see why natural selection would favor an inbuilt rule of thumb that says, "Believe what your parents tell you!" since the world is dangerous and human children are very vulnerable. If, in Africa, a child's parents say, "Don't pick up snakes!," that's a very good piece of advice for survival. And then when a parent later says something else, such as "Make sure to sacrifice a goat, so that we won't have a bad harvest," the child's brain has no way of distinguishing between sensible and silly advice. Since the child's brain has been programmed to believe whatever a parent says, it is intrinsically susceptible to parasitic attacks of nonsensical ideas—just as a computer is vulnerable to virus attacks coming from the outside. A computer is a machine that blindly obeys the instructions it receives, and that's how it should be. But for that very reason, it is intrinsically vulnerable to a virus that says, "Destroy this person's hard disk and copy this virus program to another computer!" And I see religion as being analogous to computer viruses.[18]

Today, Richard Dawkins devotes himself almost exclusively to working for secularization and the advancement of science. In 2006, he founded the Richard Dawkins Foundation for Science and Reason, which is an online meeting place for people who are Enlightenment oriented and who wish to increase the visibility of secular ideas.

Another contemporary critic of religion is the late writer and journalist Christopher Hitchens (1949–2011). His book *God Is Not Great: How*

Religion Poisons Everything (2009) is one of the most influential of recent criticisms of religion.

Hitchens's orientation toward religion, which, in his view, infiltrates and poisons many of the world's political conflicts, earned him the label "aggressive neo-atheist" from numerous opponents. In his book, he writes:

> A week before the events of September 11, 2001, I was on a panel with Dennis Prager, who is one of America's better-known religious broadcasters. He challenged me in public to answer what he called a "straight yes/ no question," and I happily agreed. Very well, he said. I was to imagine myself in a strange city as the evening was coming on. Toward me I was to imagine that I saw a large group of men approaching. Now—would I feel safer, or less safe, if I was to learn that they were just coming from a prayer meeting? As the reader will see, this is not a question to which a yes/no answer can be given. But I was able to answer it as if it were not hypothetical. "Just to stay within the letter 'B,' I have actually had that experience in Belfast, Beirut, Bombay, Belgrade, Bethlehem, and Baghdad. In each case I can say absolutely, and can give my reasons, why I would feel immediately threatened if I thought that the group of men approaching me in the dusk were coming from a religious observance."[19]

Later he describes what he witnessed in such places. For example, he writes this about Belfast, the capital of Northern Ireland:

> In Belfast, I have seen whole streets burned out by sectarian warfare between different sects of Christianity, and interviewed people whose relatives and friends have been kidnapped and killed or tortured by rival religious death squads, often for no other reason than membership in another confession.[20]

And about Beirut, the capital of Lebanon:

> Israel's irruption into Lebanon that year [1982] also gave an impetus to the birth of Hezbollah, the modestly named "Party of God," which mobilized the Shia underclass and gradually placed it under the leadership of the theocratic dictatorship in Iran that had come to power three years previously. It was in lovely Lebanon, too, having learned to share the kidnapping business with the ranks of organized crime, that the faithful moved on to introduce us to the beauties of suicide bombing. I can still see that severed head in the road outside the near-shattered French embassy. On the whole, I tended to cross the street when the prayer meetings broke up.[21]

To read the writings of Christopher Hitchens is very disturbing. His orientation toward religion is wild and unreasonable. Much of his professional life was devoted to criticizing religion and totalitarian ideologies of all forms.

Hitchens died of esophageal cancer in 2011 at the age of sixty-two. He didn't describe himself as an atheist ("someone without god") but, instead, as an *antitheist*—an opponent of god. In his final book, *Mortality* (2012), he paints a vivid picture of his battle against cancer. With his usual sarcasm and black humor, he writes:

> Rabbi David Wolpe, author of *Why Faith Matters* and the leader of a major congregation in Los Angeles, said the same. He has been a debating partner of mine, as have several Protestant evangelical conservatives like Pastor Douglas Wilson of the New Saint Andrews College and Larry Taunton of the Fixed Point Foundation in Birmingham, Alabama. Both wrote to say that their assemblies were praying for me. And it was to them that it first occurred to me to write back, asking: Praying for what? . . .
>
> Pastor Wilson responded that when he heard the news he prayed for three things: that I would fight off the disease, that I would make myself right with eternity, and that the process would bring the two of us back into contact. He couldn't resist adding rather puckishly that the third prayer had already been answered. . . .
>
> An enormous number of secular and atheist friends have told me encouraging and flattering things like, "If anyone can beat this, you can"; "Cancer has no chance against someone like you"; "We know you can vanquish this." On bad days, and even on better ones, such exhortations can have a vaguely depressing effect. If I check out, I'll be letting all these comrades down. A different secular problem also occurs to me: What if I pulled through and the pious faction contentedly claimed that their prayers had been answered? That would somehow be irritating. . . .
>
> One of Christianity's most cerebral defenders, Blaise Pascal, reduced the essentials to a wager as far back as the seventeenth century. Put your faith in the almighty, he proposed, and you stand to gain everything. Decline the heavenly offer and you lose everything if the coin falls the other way. (Some philosophers also call this Pascal's Gambit.)
>
> Ingenious though the full reasoning of his essay may be—he was one of the founders of probability theory—Pascal assumes both a cynical god and an abjectly opportunistic human being. Suppose I ditch the principles I have held for a lifetime, in the hope of gaining favor at the last minute? I hope and trust that no serious person would be at all impressed by such a

hucksterish choice. Meanwhile, the god who would reward such cowardice and dishonesty and punish irreconcilable doubt is among the many gods in which (whom?) I do not believe.[22]

A more lighthearted attitude is held by comedian Ricky Gervais, who often pokes fun at religion in what for many people is a provocative manner. On Twitter, he comments: "It's almost as if the Bible was written by racist, sexist, homophobic, violent, sexually frustrated men, instead of a loving God. Weird."[23]

In another context, Gervais poses the following witty multiple-choice question: "God doesn't prevent terrible things because: (1) He can't; (2) He doesn't want to; (3) He causes them; (4) He doesn't exist. Please vote now."

Many other notable cultural figures, actors, and artists, such as comedians Bill Maher, Stephen Fry, and Eddie Izzard, as well as magician James Randi and musician Tim Minchin, have spoken out in favor of the secular perspective in various venues. (Tim Minchin is an Australian pianist, singer, composer, and artist, who made a remarkable video about religion, the New Age movement, and pseudoscience, which can be viewed on my website.[24])

It's clear that during the first years of the twenty-first century, this cause has become more urgent and pressing. Since the attacks on the World Trade Center on September 11, 2001, more and more people have realized the urgency of defending secularism. A lengthy list of well-known people who support secularism can be found on my website.[25]

It is still common to find secular people who believe that there is a necessary connection between religion and moral values. A particularly clear example of this is the fact that in Sweden, the teaching of religion and ethics at both the elementary and secondary levels still takes place in one and the same class, which is called "religious education." I will return to this topic in a later chapter.

Ethics researcher Birgitta Forsman, in her book *Gudlös etik* (*Godless Morality*, 2011), describes the independence of morality from religious precepts in the following way:

> Religiousness, the feeling of sharing, and perhaps the love of music may all have played roles in helping humans to survive before society developed. Sharing held groups together, so that our ancestors were able to defeat their hominid competitors [e.g., the Neanderthals]. Religiousness, on the

other hand, may not have furthered our evolution; all that was needed was that it wasn't so harmful as to actually threaten survival. But now the time may have come for things to change. In a highly technological world with atomic weapons and laser rays, it may be that fantasies about gods that order us to carry out certain actions are genuinely threatening our survival.[26]

And the well-known Swedish actor Stellan Skarsgård, when I asked him if he had ever grappled with the question of God's existence, said the following:

No. I see the whole question as absurd and totally devoid of interest, because if there is a god that is good, then he will be able to watch all my behaviors and judge them, and if he is so vain that he wants to be prayed to, then he simply isn't worth praying to. I will believe that God doesn't exist until the opposite is proven to hold. To me, God is just as plausible as Santa Claus. I don't feel any need to grapple with the question of God's existence, but that doesn't mean that I believe that the only things that exist are things that we can see. Our ability to perceive reality is in fact enormously limited; it evolved during a time when humans were hunters and gatherers living in small groups of roughly sixty people altogether.

Many animals see only what they need to see in order to survive. And so it is also with our sense organs—we certainly can't perceive everything. When I told Milos Forman [Czech American film director] that I was an atheist, he said, "But life is so poor if one foregoes all mystical feelings!" I replied that it is religion that foregoes mystical feelings. That's the whole problem with religion. Everything that is incomprehensible and astonishing about our world is shrunk down to a two-thousand-year-old myth centered on a Bronze Age god. What is that, if not denial of mysteriousness? I am fascinated by unanswered questions. We should of course try to understand as much as we can. But to jump to the idea of God the moment that you don't understand something—well, that is just running away from mystery.[27]

For those who are interested in secular voices from the nineteenth, twentieth, and twenty-first centuries, I strongly recommend the book *Icons of Unbelief: Atheists, Agnostics, and Secularists*, edited by S. T. Joshi (2008), in which portraits are given of such notable people as Sigmund Freud, H. P. Lovecraft, Ayaan Hirsi Ali, Richard Dawkins, Carl Sagan, Mark Twain, and many more.

~

INTERLUDE: ON TELEPORTATION
AND THE FEAR OF DEATH

You probably remember *Star Trek*, don't you?

On board the USS *Enterprise* there was a teleporter—a machine that could send you (or any person) from one spot to another by disassembling you at an atomic level and then reassembling your atoms somewhere else.

We know that objects consist of atoms, and that all atoms of a specific element are identical. If we were able to map an object out completely, one atom at a time, then we could precisely reconstruct that object using identical atoms, but in a different place. We would then have a truly identical object to the original. There would be no way for us, or for anyone, to distinguish between them.

But what if the reconstructed object is a person? If we have a philosophically materialistic viewpoint about human beings, then what we just said would have to hold for people as well. But would a teleporter really be able to transport people? Can we humans actually reconcile ourselves with the claim that we are nothing but material objects? To believe such a thesis is easier said than done. Consider the following thought experiment.

Imagine that someday in the future you can be safely teleported between two spots. The way this happens is that all the atoms in your body are first recorded in a huge computer memory (their types and their exact locations, as precisely as is necessary), and then they are all destroyed. After that, identical atoms in a different place are brought together and assembled in exactly the same fashion, thus making an identical copy of your body in the new place. It's a very quick and remarkably efficient way of transporting a person from one place to another.

As you are lying there in the teleportation machine, with all your atoms just about to be obliterated, might there be any grounds for you to feel a slight fear of dying?

Perhaps not. After all, you're convinced that a human being, like everything else, is made of atoms, so you're quite calm. You also know that before you, thousands of people have been teleported in exactly the same way. It worked for them, so it ought to work for you, too.

But let's just add one extra piece to the scenario. You're lying there in the machine, waiting to be teleported, but nothing happens. Then you hear a knock at the door, and the operator of the machine steps in and politely asks you to climb out of it. She says, "Sorry—there's been a little glitch. You were delivered safe and sound at your destination, so everything is fine at that end. But something went wrong at this end, so at the moment there are two copies of you—the brand-new one, far away where you were headed, and this old one, still right here. Of course this is unacceptable, so we'll now have to eliminate the copy of you that's here. So please follow me down this hallway, to the little room at the end . . ."

As you approach the room, you notice a small sign hanging just above the door with the words "Doom Room" on it, and just underneath the sign there's a cute smiley face. The operator laughs and says, "Isn't that a delightful name for the room? Just about everybody likes it. Hop on in!"

So how does it feel now? In principle, everything should be just fine, since an intact and perfect copy of you has just been constructed somewhere else. But still and all, does everything feel just right to you, as the operator ushers you into the cute little Doom Room?

This thought experiment shows how unaccustomed we are to thinking of body and soul as merely two different facets of a single entity. Nonetheless, science certainly tells us that body and soul are one and the same. But sometimes, one's intuition and scientific facts are not so easy to reconcile with each other.

CHAPTER THIRTEEN

ENLIGHTENMENT

Concerning Freedoms, Rights, and Respect

Those who can make you believe absurdities, can make you commit atrocities.—François-Marie Arouet (aka Voltaire), *"Questions sur les miracles" (1765)*

Now is the time for us to look toward the future and to offer some solutions instead of just listing problems. How should we act in this world that is so profoundly drenched in irrationality, in various forms of extremism, in intensely believed religious dogmas, and in democracies that are falling to pieces left and right?

The secular enlightenment provides a clear vision of a society in which all people, no matter what their cultural background or philosophy of life is, can coexist and have equal standing. Such a society would draw a sharp distinction between religion and politics. The utterances of ancient prophets would not be considered relevant when it comes to the making of political and legal decisions. The basis for such decisions will be the best thinking that humanity can offer, not the claims supposedly made by gods and prophets.

In such a secular democracy, the religious beliefs of any individual can of course influence how that person votes, and politicians can of course use their own religious beliefs as sources of inspiration. A member of parliament would be free to claim that the parliament should decide a certain issue on the basis of its being supported by some deity or other; this is a right that is guaranteed by freedom of speech; however, even the most pious of politicians in such a society would realize that in order to be taken seriously by others, they would have to present *nonreligious* arguments to convince others of their stance.

Today, we have the secular means at our disposal to try to move toward that vision of secular enlightenment. All the tools of democracy, education, science, and technology are available to us to further the causes of human happiness and freedom, as well as the goal of improving the living conditions of people all around the globe.

Universal Rights and Human Liberation

A key aspect of the project of secular enlightenment is the idea of the *universality* of human rights. All people, whatever their origin or their culture might be, should be granted the same basic rights. In theory this sounds good, but what does it mean in practice?

First of all, we have to point out that this is not a scientific question but, instead, a matter of values. Of course one might be able to demonstrate empirically that such a policy would lead to less suffering, greater freedom, and more equality, but whether all this is *desirable* is a question of values. Fundamentally, it's all about the belief in people's right to self-determination. Or to put it another way, borrowing Immanuel Kant's terminology, enlightenment is all about people's "emergence from their self-imposed immaturity." The key issue is whether people are free of external constraints, so that they can act on the basis of their own goals and values.[1]

The principle of universality of values and rights is in direct conflict with *cultural relativism*, which denies that there are any universal rights. Those who advocate this point of view claim that no cultural value that bears on morality can possibly be "better" than any other cultural value. Their viewpoint is that basic biological facts are the only things that link human beings to each other, and that all significant differences between people are defined by the various cultures that they belong to. Thus they would say that the human race can be separated into diverse categories, and the category to which you belong determines your needs and desires, and by extension, it also determines your freedoms and rights.

This is a dangerous pathway down which to tread. If one has such a viewpoint, one can in principle justify any crime or transgression whatsoever. History has shown how tragic it is to single out one group of people as being fundamentally different from one's own—for instance, the Jews in Germany, the Gypsies in contemporary Europe, the Tutsi people in Rwanda, and the Muslims in the Balkans.

Fortunately, cultural relativism seems to be losing ground—at least from a scientific point of view. To be concrete, research in neuroscience, sociology, linguistics, evolutionary psychology, experimental economics, and evolutionary anthropology all points toward the same conclusion: it is correct to speak about humanity as one uniform species, cutting across all different cultures.

Sometimes the universalist credo is criticized for being "imperialistic" or "Western centered." But this is an incorrect view. All cultures and traditions, throughout history, have advocated human freedom. Belief in this principle can be found everywhere in the history of ideas. To be sure, in various eras and places, that idea has been undermined and suppressed, but the idea itself is certainly not limited to Western cultures.

A prerequisite to the realization of universal rights is a deep notion of *equality*. Thus we need to overcome stereotyped views of women and men, and other unjustified categorizations of people. In his book *A Theory of Justice* (1971), American social philosopher John Rawls (1921–2002) formulated the "First Principle of Justice" as follows: "Each person is to have an equal right to the most extensive total system of equal basic liberties compatible with a similar system of liberty for all."[2] (Rawls is best known for the above-mentioned book, in which he develops his own alternative to the "consequentialist" tradition, closely related to utilitarianism, in political philosophy.)

For human rights to become universal is merely the first step toward an enlightened society. Human rights must also be taken seriously, not merely be eloquent pronouncements having a fine noble ring to them.

Identity as a Trap

It is a grievous error to think that an individual human being is defined by a handful of simple labels. Of course we can point to certain traits that *contribute* to one's identity, such as biological gender, sexual orientation, age, skin color, ethnic origin, or citizenship. But it is just too tempting to think that, solely on the basis of these few criteria (or similar ones), we can predict all of someone's *other* properties. The fact is, people who happen to have the same sexual orientation or the same ethnic origin are, as a rule, completely different from each other in other regards.

Throughout history many names have been given to people's ways of categorizing other people. Race biology, phrenology, and astrology all testify to our human desire to place one another in neat little boxes. Why do we humans seem to need such simplistic notions of identity? Apparently it's because we believe that if we can break our complex and chaotic world up into simple building blocks, then we can deal with it more easily. In truth, though, such simplistic labeling only makes life more difficult—and, unfortunately, more dangerous.

Philosopher and economist Amartya Sen has pointed out that the need for labeling has been increasing in our globalized world, unfortunately with very worrisome consequences. Sen, who was awarded the 1998 Nobel Memorial Prize in Economics, is best known for his studies of individual and collective rationality. His book *Identity and Violence: The Illusion of Destiny* published in 2006, deals with the concept of identity and with the danger of assigning too much meaning to it. In the book, he describes the trap of thinking that someone's superficial identity reveals all the hidden facets of their character.

The truth is that any individual can have many diverse identities, for example, a Muslim or a Christian can be white or black, lesbian or straight, conservative or radical, secular or orthodox, socialist or libertarian, feminist or feminist-basher, a lover of heavy metal or of pre-baroque choral music, a chess player or badminton player, or theoretical physicist or illiterate homeless person. And for a particular individual, none of these identities is necessarily more important than any of the others.

To be sure, certain aspects of one's identity can be rooted in traits such as skin color, sex, sexual orientation, or age. Others can be rooted in one's attitudes or beliefs, such as religious convictions or political values. And yet others can be rooted in circumstances, such as where one happened to be born, or random factors, such as one's native language or the cultural traditions that one grew up with. Traits, attitudes, and random factors don't have to be mixed together—but that's what constantly happens in politics and in public discourse. It happens even in the first paragraph of the Swedish law on discrimination! Check it out:

> The purpose of this law is to oppose all forms of discrimination, and to support equal rights and possibilities for all, regardless of a person's sex, sexual identity, or way of expressing oneself, ethnic affiliation, religion or other type of belief, handicaps, sexual orientation, or age.[3]

In this passage, *traits* (such as a person's age, gender, or genetic handicaps) are mixed together with *attitudes* (such as a person's religion or other points of view). Of course one can validly talk about discrimination within any or all of these areas, but traits, attitudes, and random factors are deeply different and should not be treated in the same way. For instance, one can criticize someone else's beliefs but not their innate traits. After all, beliefs and attitudes (unlike innate traits) can always, at least in principle, be reconsidered or changed; therefore, they should also be open to being questioned. If someone's religious beliefs result in negative consequences for other people, then that person should bear the responsibility for these consequences.

If a body of laws were to treat a person's religious faith as a trait—that is, as a fixed, unchangeable element of their personality—then, in effect, religious people would be seen, under the law, as being incapable of thinking for themselves. In such a scenario, you couldn't be held responsible for your religious beliefs (or any of their consequences) any more than for your gender, your skin color, or your age. If you committed a crime on the basis of, say, a sexist doctrine enunciated in a sacred text, you would simply have to be pardoned, since your act would have been derived from the (presumably fixed and immutable) trait of your religion. And conversely, an objectively justifiable criticism of a religion would be seen, in such a scenario, as an illegal act of discrimination. Such a scenario is quite absurd.

Islam and the Criticism of Religion

A highly inflammatory topic in Sweden today is Islam. When an "alien" religion begins to show its face in Swedish society, Swedes suddenly take notice of the fact that something called "religion" exists. After all, most Swedes tacitly assume that our Swedish tradition is either no religion at all, or some form of Christianity. Some Swedes see Muslims as a danger to our "Christian values"; others are frightened to death about criticizing viewpoints defined as Muslim. Reluctance to criticize thoughts and values connected with Islam stems from a fear of being labeled "Islamophobic."

The truth is, though, Muslims come from all sorts of countries and have all sorts of interests, educational backgrounds, sexual orientations, professions, and values. They also have diverse opinions about how Islam should be understood. Apart from the fact that they label some of their beliefs and attitudes as "Muslim," they don't have particularly much in common with each other.

It is no more valid to generalize about "all Muslims" than to generalize about "all Christians," "all Jews," "all Hindus," "all secular humanists," or all members of *any* group of people who happen to share a certain viewpoint about life. One can survey any group of people about their values and norms by researching such questions as "What proportion of the members of this group (be they Muslims, Jews, Christians, secular humanists, or what have you) have a certain attitude about, say, women or homosexuality?" Obviously, such questions can be answered by doing careful social research.

And if such research is done, it will help us understand what might be called "mainstream Islam," "mainstream Christianity," or "mainstream humanism." Today, however, there is a resistance to generalizing in this sort of statistical fashion, especially about Islam. There are even people who claim that no statements of any sort can be made about Islam or any other religious belief. The goal of such a well-meaning stance may be to undercut prejudices about Muslims, but the unfortunate consequence is that even *objective* and *justified* criticisms of one religion or another are thereby rendered highly suspect. Religion is thus exempted from all criticism, since there is no precise definition of what it is. This is not a promising avenue toward healthy dialogue.

By contrast, a much more important insight is this: Even if one can make a well-founded statistical summary about the beliefs and outlooks of members of Islam (or any other belief system), one can't draw any firm conclusions about an isolated individual having those beliefs. We can speak in general about different philosophies of life and their relationships with each other. We can—and we should—analyze their consequences and critically analyze their content. But we cannot ever predict what an individual Muslim (or individual Christian, Hindu, or secular humanist) will think about a specific issue, unless we ask that person directly.

Essentialism and Viewpoints on Humanity

In the Swedish elections of 2014, the Swedish Democrats more than doubled their previous performance, pulling down 13 percent of the votes. They quickly exploited their new position of power, creating parliamentary chaos in Sweden. Similarly, populist and nationalist forces are increasing in strength in many parts of Europe. This is an ominous sign, and it is a reminder of Europe's dark history during the twentieth century. Underlying these new currents is an irrational viewpoint about humanity that is clearing a pathway for the flourishing of false ideas, such as those on which populism and extremism are based.

I am referring to the essentialist viewpoint of human nature, which forms the basis of religious conceptions of humanity, and which also sadly permeates populist, extreme right-wing, and racist viewpoints—meaning both purely biological racism and racism in an extended sense, which is to say, ethnic, cultural, and religion-based identity politics.

Essentialism is the thesis that certain entities *necessarily* possess certain properties, while other properties are *contingent*, or merely *possible*. A common type of essentialist thinking involves stereotypes of "masculinity" and "femininity." This viewpoint implies that there are fundamental differences between men's and women's thinking, and that these differences are irrevocably linked with gender, coming out of the biological nature of males and females. This would imply that the differences between men's and women's thinking cannot be influenced by such things as culture.

The opposite viewpoint starts from the idea that each person is defined by their thoughts, values, actions, and choices. These are the things that result in someone being the person that they are. Or, to quote Jean-Paul Sartre, "Existence precedes essence." First we come into existence, and then we create ourselves and our "essence." This view stands in direct opposition to the religious dogma that every person is created in God's image and carries a "divine spark" in them.

Both religious dogma and nationalistic dogma imply that all people are born with a preexisting essence. Either that essence is pre-given by God, or else it is dictated by one's "race," one's culture, one's religion, or one's ethnicity. A sexist viewpoint presumes one's gender to be a fixed, determining essence of one's nature.

In Swedish society, some essentialist forms of thinking are viewed as more acceptable than others. Remarkably enough, although many Swedes today seem to be keenly aware that one should never say such things as "All black people have the following properties . . .," they seem to be utterly unaware of the analogous prejudice inherent in statements such as "All women have the following properties . . ." How often do you hear remarks such as "Women who go into business have to imitate men in order to succeed"? Just think how unlikely you would be to hear a statement such as "Black people who go into business have to imitate white people in order to succeed"—and how deeply offensive it would sound.

Another example of essentialism is the racist view that still persists in many extreme-right movements in Europe. It is not only morally reprehensible but also scientifically fallacious.

DNA analyses show, for example, that East Africans and Europeans have a greater genetic overlap than East Africans and West Africans. It was

East Africans who, roughly seventy thousand years ago, emigrated from Africa, first through the Arabian Peninsula and then into Europe. This is why East Africans and Europeans are genetically closer than East Africans and West Africans are. This creates a puzzling dilemma for racists.

Research has shown that some thirty thousand years ago, Neanderthals and humans of our species were both alive, but then the Neanderthals died out; however, modern humans mated with Neanderthals, so that today, somewhere between 2 and 4 percent of our DNA comes from Neanderthal chromosomes. This holds for all humans except for pure Africans. They are the only people now alive who don't have a single trace of Neanderthal DNA.

Gender Essentialism and the Lens of Gender

The feminist movement has given us new insights into gender—specifically, into gender as a social construct. Looking at society through the lens of gender means paying attention to how prejudices, expectations, and sex-linked norms are projected onto each one of us throughout our entire lives. A form of behavior that could give a man a high social status can have entirely the opposite effect on a woman who exhibits the identical behavior (and, of course, vice versa). The social codes for "proper" masculine and feminine behavior are often unspoken, and we pick them up through our interactions with others, or they sneak silently into our minds via the mass media and subliminal messages in advertisements. There can be no doubt that this type of pervasive social influence plays a central role in how we members of a culture are shaped into "men" or "women" over and above our biological membership in a particular sex.

Unfortunately, some feminists and writers about gender hold science in contempt and consider that it is harmful, if not downright evil, to carry out scientific research on biological or cognitive differences between men and women; however, scientific research should not be controlled by ideological or political goals. The true problem is not the research or its findings; it is the irrational conclusions that far too often are drawn from such research. If some sort of biological differences are found between the sexes, these should not be cited as arguments in favor of traditional gender roles or in favor of treating boys and girls (or men and women) differently.

Suppose that after considerable research, there turned out to be a difference between the average performance of men and that of women in some type of intellectual activity (the usual claims cited for such differences being error ridden). Even if that were the case, the differences between diverse individuals belonging to a single gender would be far larger than the difference between the two averages. Thanks to this enormous intra-gender variability, the whole idea of special treatment based on gender falls completely apart.

A subtler logical trap is the very idea of comparing just men and women, ignoring other potential ways of dividing people up. If, for example, we were to compare righthanded and lefthanded people, or short and tall people, we might very well discover equally great, or even greater, differences between the averages of such groups. Would we want social policies to be based on such discoveries? Most of us would simply scoff at such a suggestion—and yet our lifelong tendency to think in terms of gender stereotypes prevents us from having an unbiased attitude, as we would for groups of people having a specific hair color, handedness, height, and so on.

I believe that in order to develop a view of people that is free from preju-dice, we have to totally abandon essentialism. This is a crucial ingredient in the project of a new enlightenment.

What Is Religious Freedom?

Religious freedom is written into Swedish and U.S. law, but what does this really entail? Of course, it means that anyone can believe whatever they wish; this should be an absolute and unrestricted right. A secular state should, however, be able to place limitations on religious freedom in terms of how religious beliefs can be expressed (in the sense of determining people's ac-tions). People should not be allowed to do just anything they please in the name of their religion. This is where political conflicts and controversies come into the picture.

In article 18 of the United Nations (UN) Universal Declaration of Hu-man Rights, we find the following statement: "Everyone has the right to freedom of thought, conscience, and religion." In a commentary in 1993, the United Nations committee on human rights clarified this statement as fol-lows: "Article 18 protects theistic, non-theistic and atheistic beliefs, as well as the right not to profess any religion or belief."[4]

The Swedish constitution also states that we all have the right to speak without restriction, to transmit and receive information without restriction, to gather without restriction, to demonstrate without restriction, to organize groups without restriction, and to practice any religion without restriction.

Historically, religious freedom was motivated by the need to react against various forms of coercion or persecution of religious minorities. Consider that, as recently as 1951, it was illegal to leave the Swedish Lutheran Church unless one simultaneously became a member of some other religious denomination. There are still lands today in which anyone who renounces Islam for Christianity or a lifestyle without any god is subject to imprisonment if not capital punishment. This is why the UN declaration is important.

Every person should have the right to express their beliefs, to gather with others, and to form groups based on common beliefs. Similarly, the right to follow one's own customs, whether they are religious or cultural, should be protected—but only as long as those customs do not conflict with other laws.

Religious freedom, as it is formulated in the UN Declaration of Human Rights, is actually logically inconsistent, in certain cases. Take, for instance, the case of the Roman Catholic Church. Catholicism requires Catholic parents to raise their children in the Catholic faith, so that they too will become believing and practicing Catholics. The UN declaration, through its respect for religious freedom, gives these parents the right to carry out this religious "duty." On the other hand, according to the UN Convention on the Rights of the Child, all children have their own rights. In particular, children have their own right to religious freedom, and this means that they cannot be coerced to participate in the Catholic faith or its rites. We thus see that the very notion of "religious freedom" involves two logically incompatible principles.

In practice, Swedish society accepts very few things in the name of religious freedom that are not covered by other laws. Adult Jehovah's Witnesses, just like adults of any other belief, can refuse to accept a blood transfusion for themselves, but they cannot legally prevent their children from receiving blood transfusions. This might be seen as a limitation of their religious freedom, but the secular law that safeguards children's lives takes precedence.

Nor does Swedish law permit the slaughter of animals without anesthesia by people whose religion would insist that this is how slaughter should take place. It is precisely this dilemma that is reflected in the debate over whether animal slaughter according to kosher (Jewish) or halal (Islamic) conventions should be allowed in Sweden. It turns out that in Sweden it is still legal to import meat that is produced abroad in accordance with halal or kosher conventions; this has to be seen as an inconsistency in our body of laws. Of course, laws should be applied consistently. Animals' rights should always be respected, whether the issue is religious slaughter methods, or recreational hunting for elk or ptarmigans.

Another question concerns which belief systems are interpreted as religions, and are thus treated with equal merit. People who claim the right to certain customs or modes of dress, arguing that these traditions are rooted in a well-known religion such as Christianity, Judaism, or Islam, are usually granted a fair degree of respect. But someone who cites their belief in an old Norse religion or who professes a reverence for nature will be looked upon with skepticism or will be seen as slightly nutty.

And which criteria should apply in passing judgment on the seriousness of different (claimed) religions, especially since we are supposed to respect religious freedom? Is it the number of believers? Or the belief system's historical roots? Or the literary quality of the belief system's scriptures? It seems pretty clear that such judgments are, to a large extent, arbitrary. The very notion of "religion," as it is treated by the law, is filled with capricious and inconsistent practices, since there is no clear definition of the word "religion." Is a belief in some sort of god a necessary criterion for a collection of beliefs to be considered a religion? In that case, traditional Buddhism ought not to be counted as a religion, since it is a godless philosophy of life. Or how about an irrational worship of something, which brings together a group of people? In that case, we would have to consider a group of passionate soccer fans to be members of a religious faith.

It is hard to find any objective reason that a belief system that includes a belief in God should have the right to special legal protection. It would be more reasonable to broaden the scope of rights so that all people have a right to freely practice their existential beliefs, whether or not those beliefs involve the existence of some kind of god.[5]

Pacifism—the ideology that rejects the use of violence, even in cases of self-defense—gives a clear example of this kind of thing. Someone who refuses to bear weapons or to use violence can of course have a religious reason for doing so. But this kind of refusal can equally well be rooted in a secular belief system or a philosophical conviction. If a country has obligatory military service that includes the use of weapons, but also accepts the idea that *religious* pacifists do not need to serve, then the same exemption should apply to a pacificist whose pacifism comes from nonreligious grounds (such as a humanist set of ethics). Any other legal position would be unreasonable.

Example: Sikhs and Knives

According to Swedish law, it is forbidden to carry knives or other dangerous objects in public places, in school areas, or in vehicles that enter public places. But according to the religion known as Sikhism, founded in sixteenth-century India, an initiated Sikh should always have five characteristics: unshorn hair (and for men, a beard); a comb; a *kirpan* (a sword or dagger); a steel armband; and knee-length underwear.

This gives rise to a conflict between religious freedom and the law proscribing the bearing of knives in public places. This conflict was put to a legal test in a very well-known case in Canada, where it is illegal to carry a knife in a school. Balvir Singh Multani, a Canadian citizen and an orthodox Sikh, went with his son to the Saint Catherine Labouré School in Quebec. The father wanted his son always to be carrying a *kirpan* (a knife with religious meaning), which symbolized honor and nobility in the Sikh religious tradition.[6]

The school administration refused to allow the knife on campus, since no weapons were to be allowed on school grounds. The administrators felt it was their responsibility to assure the safety of all the schoolchildren. Their view was that if the knife was carried around by the boy, then anyone at all might seize it and use it violently.

The highest court in Canada ruled in favor of the Sikhs and stated that the exclusion of all knives in the school violated their freedom of religion; however, the prohibition against knives was ruled as still applying to all *other* students who might wish to carry knives onto the school grounds. Only for *religious* reasons could this law have exceptions.

Now suppose that this boy had not been a Sikh but was a member of a family with a rural tradition on Canadian soil going back many generations. There are many cultures that have rituals of passing that are supposed to take place "when a boy becomes a man." A central feature of such a ceremony occurs when the boy's father presents his son with a special knife that has been handed down from one generation to the next. Usually these kinds of rites take place when the boy is fourteen or fifteen years old. The boy's identity as an adult and his link to previous generations are symbolized by his carrying this knife on him at all times.[7]

Neither in Canada nor in any other country would this hypothetical boy be permitted to carry his knife onto school grounds, even though in many lands the Sikh boy would have very good chances of being granted permission to have the knife with him in school.

Here we clearly see how the notion of religious freedom gives rise to an inconsistent and thus (in the long run) untenable view of human behavior, ethical values, and attitudes. It goes without saying that a moral attitude isn't "stronger" or "worth more" simply because it is based on religious notions or traditions. Attitudes and actions can be more or less strongly motivated and meaningful for individual people, completely independently of what they are rooted in.

For Ethical Independence Rather than Religious Freedom

It may be time to replace the notion of religious freedom in legislation with a more inclusive and neutral concept. We might, for instance, make use of the concept "ethical independence," according to which every citizen has the right to independence relative to the state and relative to the general public. This idea was suggested by British American professor of legal philosophy Ronald Dworkin (1931–2013), who defines it in his book *Religion without God* (2013):

> Ethical independence means that government must never restrict freedom just because it assumes that one way for people to live their lives—one idea about what lives are most worth living just in themselves—is intrinsically better than another, not because its consequences are better but because people who live that way are better people.[8]

One key principle of ethical independence would thus be that the state has no right to coerce people to follow a certain set of ethical rules. This means, for example, that a state cannot prohibit the use of drugs simply because drugs are known to be harmful to one's health. On the other hand, the state can become involved in the life choices of individuals for other reasons—for example, in order to prevent acts that harm innocent people. This might mean, for instance, passing legislation against the use of some drug if it is determined that the use of said drug can cause harmful effects to society at large; such a law could also help reduce the cost to society of taking care of misusers of the drug. Dworkin suggests that we should abandon the idea that religious freedom has a special status in terms of rights. Instead, we should apply the same attitude of freedom to all thoughts and beliefs, whether or not they have religious, secular, or philosophical roots. In other words, equal treatment for *all* beliefs.

We of course presume that if the notion of religious freedom were removed from the law and were replaced by the right to ethical independence, other freedoms would remain, such as the freedom of speech, the freedom of information, the freedom to gather, the freedom to demonstrate, and the freedom to hold meetings. These freedoms, taken together, protect all people's right to believe as they wish, and to practice their religious faith, either alone or together with other people, as long as doing so does not harm other people.

Religion and Medical Ethics

These days, new medical discoveries and inventions are appearing on the scene at a very rapid pace. DNA analysis, surrogate motherhood, prenatal tests, and stem-cell research are just a few examples that raise important ethical questions, and these questions are by no means simple. Even in a purely secular society, there would be difficult ethical dilemmas of this sort, which would need to be thought through very carefully, but when religious dogmas are allowed to stand in the way and block medical research and progress, something has really gone haywire. A crystal-clear example is research into embryonic stem cells.

British developmental biologist John Gurdon (1933–) is considered to be the pioneer of stem-cell research. In 1958, he was the first to clone an animal—he cloned a baby frog from a cell taken from an adult frog. For this work, he was awarded the Nobel Prize in Medicine in 2012. Today, stem-cell research is one of the most promising methods to find cures for such terrible scourges as Alzheimer's disease, Parkinson's disease, Lou Gehrig's disease (amyotrophic lateral sclerosis, ALS), and many other currently uncurable diseases. But in many parts of the world today, stem-cell research is illegal or highly restricted for religious reasons. The religious belief that underlies this opposition is that the human soul is allegedly created by God at the very moment of conception. On the other hand, there is no *secular* reason to forbid stem-cell research.

In stem-cell research, embryonic stem cells are taken from a blastocyst—that is, from a small group of cells that have had a few days after conception to develop. These cells are collected when the fertilized egg has divided several times, thus giving rise to roughly three hundred cells. For an idea of the size of such a cell group, consider that the brain of a housefly contains roughly one hundred thousand cells.

At that early stage of development, the embryonic cells have not yet started to specialize; any one of them has the potential to develop into any kind of specialized cell—for example, into a bone-marrow cell, a liver cell, or a kidney cell. Embryonic stem cells are thus very different from cells in a human fetus—and in any case, stem cells are never taken from human fetuses.

Today, many fertility clinics the world over use the technique known as *in vitro* fertilization (formerly known as "test-tube fertilization") to help couples who otherwise cannot have babies. In this technique, a few fertilized ova are grown externally for a few days, and then one or more of them are surgically inserted into the woman's uterus. The remaining eggs are either discarded or are frozen, to serve in possible future attempts to impregnate the woman, in case the first attempt fails. It is from such fertilized ova that stem cells can be taken.

Stem-cell research is a direct outcome of medical science's goal of saving lives and giving the gift of life. Since secular ethics tries to minimize human suffering, it encourages stem-cell research as a remedy to impairments and as a source of hope to people with diseases that today are incurable.

These days, new medical discoveries are constantly forcing us to face new kinds of ethical dilemmas, and in debates on such topics it is critical to respect the distinction between religion and science.

On Secularism and Social Health

Is it possible that believing in God is healthy, whether or not God actually exists? Could it be that the placebo effect works well, even if God doesn't exist? Might it be the case that people are kinder, more honest, and more generous if they believe in God? Such a causal link might possibly exist in certain *individual* cases, but the question we wish to examine here is how things are at the level of a whole *society*. In other words, is there a correlation between the proportion of religious people in a given society, and that society's overall well-being and health?

Roughly 5 percent of U.S. citizens define themselves as atheists. But according to Columbia University psychologist Melanie Brewster, only 0.09 percent of the prisoners in American jails say they are atheistic.[9] In other words, in comparison with the overall population, atheists are underrepresented in American jails by a factor of fifty. Of course this statistic on its own doesn't imply that atheists are kinder, more honest, or more generous than religious people, but it is certainly thought provoking. It might just be a statistical correlation that doesn't represent any sort of causal link. But in any case, this statistic doesn't lend any support at all to the *opposite* conclusion—namely, that believers in God are morally better people than atheists.

We know that secular countries with a higher proportion of genuine nonbelievers (as contrasted with people who are coerced into stating they are nonbelievers) are among the healthiest, best educated, and most open societies in the world. Such countries also have a lower infant mortality rate and a higher level of equality between women and men.

According to opinion polls conducted by the European Union, the percentage of nonreligious people in Europe is increasing. In Great Britain, for example, the percentage of people who call themselves atheists grew from 14 percent in 1963 to 42 percent in 2012.[10] In Sweden, only 29 percent of the population describes itself as religious, and in Japan the percentage is even lower: only 16 percent.[11]

In the United Nations' 2018 Human Development Report,[12] 187 countries were ranked by their Human Development Index (a measure jointly developed by Pakistani economist Mahbub ul Haq and Indian economist Amartya Sen). This index is based on life expectancy, per capita income, literacy (both for reading and for writing), educational level, infant mortality rate, and so forth. The ten countries having the highest HDI were the following:

1. Norway
2. Switzerland
3. Australia
4. Ireland
5. Germany
6. Iceland
7. Hong Kong, China
8. Sweden
9. Singapore
10. Netherlands

The ten countries having the lowest HDI were the following:

180. Mozambique
181. Liberia
182. Mali
183. Burkina Faso
184. Sierra Leone
185. Burundi
186. Chad
187. South Sudan
188. Central African Republic
189. Niger

All of the countries occupying the top twenty-five slots in this list had a strikingly high proportion of nonreligious people, with two exceptions—namely, Ireland and the United States. But although Ireland is traditionally strongly Catholic, in 2018 it voted to allow abortion and it also overturned a medieval blasphemy law; moreover, it has a gay prime minister of Indian origin. In all these countries, nonreligious views on life exert a strong influence, according to opinion polls that have been conducted in them.

The countries occupying the twenty-five lowest ranks have a very low proportion of nonreligious citizens, and in countries like Bangladesh, Nepal, Afghanistan, and Pakistan, essentially everyone in the entire population is a believer in some sort of god.

The countries with the highest murder rates are ones that have a very high percentage of believers, while the lowest murder rates coincide with the highest levels of nonbelievers. The seventy-five countries with the highest infant mortality rates are all deeply religious countries.[13]

British organization Save the Children comes out annually with a "maternity index," which lists the best and the worst countries in which to be a mother. Many parameters go into making such determinations, such as childbirth care, infant mortality rate, and maternal (and paternal) leaves from work. The top ten countries in the list are all largely secular, while the bottom ten are all deeply religious.[14]

And what about peacefulness? Which are the most peaceful lands? The organization Vision of Humanity annually publishes a *Global Peace Index* based on a number of variables, such as the subjective feeling of safety and security, the rate of violent crimes, wars engaged in, and access to weapons. The pattern is the same in this dimension as well. The most secular countries are the most peaceful, while the least peaceful are all highly religious.

What conclusions can we draw from all this? To start with, let us mention some conclusions that we *cannot* draw. Although there is a clear statistical correlation between the degree of well-being and the degree of secularism, we cannot presume that this is a *causal* link. The figures don't prove that secularism is the *reason* for well-being or that religiosity is the *reason* for poverty and disease. Nonetheless, the statistical connection is undeniable. Secular societies and secular attitudes go hand in hand with health and well-being. From a historical perspective, it's also clear that secularism has been one of the main driving causes behind social progress in many ways, not least concerning the rights of women and homosexuals.[15]

Many religious conservatives, whether they are Christians, Muslims, Jews, or Hindus, claim that to create a peaceful and prosperous society, it is necessary for all citizens to serve God and to submit to God. This claim is made equally strongly by orthodox rabbis on the West Bank in Israel/Palestine, by Wahhabi sheikhs in Saudi Arabia, by Salafists in Iraq and Syria, by the pope in the Vatican, and by the Christian Right in the United States. They all claim that God's will forms the core of social life, and that only if we bow down to God can we be assured of having a moral, successful, healthy, and safe coexistence. As a rule, these authorities also believe that when many people in a society deny God's existence, society's stability will be undermined. All the evils in a society—for example, crime, poverty, poor education, AIDS, and so on—can thus be pinned on the falling number of God-fearing citizens.[16]

Well, if this were indeed the case, then one should expect that the least religious (or most atheistic) countries in the world would be the hardest hit by crime, poverty, and illness. And by the same token, the most religious countries would be paragons of social stability. And yet exactly the opposite is the case. Those countries with the highest proportion of nonreligious citizens are the most peaceful, most stable, most free, most well-off, and most prosperous, while the most religious countries are the most unstable, violent, oppressive, and poor.

An obvious conclusion from all these statistics is that secularism does not lead to immorality or to the collapse of civilization. To be sure, a belief in God may be a comfort for an individual, but societies that have a secular style seem clearly to be the healthiest ones—no matter what religious authorities may claim.

Is Religious Belief Rising or Falling in Today's World?

There can be no doubt that the influence of religion on politics is rising in many countries around the globe. One often hears about "the return of God" in today's world. It is also clear that certain sorts of religious pronouncements are becoming ever stronger and more extreme in all the world's main religions. Conservative interpretations are sprouting everywhere in Christendom, Judaism, and Islam, not to mention Hinduism and Buddhism. (The Swedish Lutheran Church is an exception to this trend; in its liberal and even secular interpretation of Christian scriptures, it is at the extreme other end of the spectrum.) And yet paradoxically, it seems that the percentage of people who call themselves religious is actually declining with time.

In the United States, somewhere between 20 and 30 percent of the people state that they don't belong to any organized religion. By contrast, in 1990, only 10 percent of U.S. citizens made that statement. The shift is even clearer among young Americans: 32 percent of Americans under thirty years old claim to belong to an organized religion. In the 1980s, among people under thirty years old, there were twice as many evangelical Christians as there were people without any religious affiliation. Today the ratio is exactly the opposite: there are twice as many nonreligious people under thirty as there are evangelical Christians in that age group.[17]

According to a poll taken by Gallup International, there are many countries in which all of these striking changes have taken place in just the brief period between 2005 and 2012. Table 13.1 shows the percentage of people who describe themselves as "religious" or as "nonreligious." For thirty-seven countries throughout the world, the percentage of self-described "religious people" in 2005 is shown side by side with the percentage in 2012. The countries are listed in the order of *most religious* to *least religious* (in the year 2012). The final column shows the percent change between 2005 and 2012.[18]

As can be seen in table 13.1, in many countries, people are increasingly moving away from religion: in twenty-four of these countries, the proportion of believers went down, while in eleven of them it went up.

The claim of "the return of God" is thus correct in only one aspect—namely, religion seems to be growing more and more dangerous in many parts of the world, reverting to a far older style, such as when the Catholic

Table 13.1. Percentage of Self-Reported Religiosity by Country

Country	2005 (%)	2012 (%)	Change
Ghana	96	96	0
Nigeria	94	93	−1
Macedonia	85	90	+5
Rumania	85	89	+4
Kenya	89	88	−1
Peru	84	86	+2
Pakistan	78	84	+6
Moldavia	78	83	+5
Colombia	83	83	0
Cameroon	86	82	−4
Malaysia	77	81	+4
India	87	81	−6
Poland	85	81	−4
Serbia	72	77	+5
Italy	72	73	+1
Argentina	80	72	−8
Ukraine	70	71	+1
Ecuador	85	70	−15
Lithuania	75	69	−6
Bosnia and Herzegovina	74	67	−7
South Africa	83	64	−19
United States	73	60	−13
Bulgaria	63	59	−4
Iceland	74	57	−17
Russia	57	55	−2
Finland	51	53	+2
South Korea	58	52	−6
Spain	55	52	−3
Germany	60	51	−9
Switzerland	71	50	−21
Canada	58	46	−12
Netherlands	42	43	+1
Austria	52	42	−10
France	58	37	−21
Vietnam	53	30	−23
Czech Republic	22	20	−2
Japan	17	16	−1

Church carried out public burnings at the stake and other kinds of horrendous tortures. But as far as the human inclination to believe in supernatural
powers or in an almighty creator is concerned, that seems to be slowly fading
out of existence.

INTERLUDE: ON GRATITUDE AND FEELINGS OF GUILT

Should we feel thankful? If so, then to whom or to what?

It is healthy and sensible to feel gratitude to a friend who does you a favor, to a companion who consoles you, or to a friend who stands up for you in a time of crisis. We also can and should be thankful for being healthy, for the fact that the people we love are in good shape, and even for the basic fact that we are alive.

There is, however, another type of thankfulness, a forced and unnatural thankfulness—the thankfulness that is demanded by religion. I am talking about being thankful to a god.

When we start to feel gratitude toward an external power, an omnipotent being, things become problematic. It becomes a contingent thankfulness, which by its very nature assumes that this being could have chosen to let something else happen—something worse.

Why was I the only one who survived the accident, when no one else had that good luck? Why am I in good health, when so many of my friends are suffering from serious illnesses? Or why did I get hit by cancer when so many of my friends are fine?

A priest asks a group of six-year-old children, "Children, what would you like to thank God for today? Lisa, what do you want to thank God for?"

"I want to thank God for making my parents feel good and be happy," says Lisa.

"Sam, what do you want to thank God for?"

"I would like to thank God for keeping my sister healthy."

Perhaps this all sounds harmless. But how does Lena feel, little Lena who is sitting there knowing how deeply unhappy her parents are? They may soon be getting divorced. Or perhaps one of her parents is critically ill. What is Lena thinking?

"Why doesn't God want my parents to be happy? Why can't my daddy be healthy? What did I do to not deserve God's caring? Why do I not count, in God's eyes?"

To direct one's thankfulness toward a cosmic being, whether we call it God or something else to which we attribute intentions, powers, and the desire to affect people, might seem perfectly reasonable, so long as only positive things happen in our own little lives, and so long as we aren't worried about other people's lives.

But that's not how life really is. When we are suddenly hit by adversity, our prior feelings of thankfulness toward a higher power can in a flash turn into feelings of shame or guilt, or even feelings of betrayal or hatred.

We should not teach children to feel shame or guilt before a divine being. And for this same reason, we should not teach children to feel thankfulness toward a god. A more human form of thankfulness resides in humbly recognizing the fact that good things have happened to us and in being aware that we could easily have been hit by misfortunes instead. It resides in being thankful for our health, for the good things that we've had the luck to encounter in life, and for the deep bonds we have with our loved ones.

It's best to avoid feelings of gratitude to mysterious unearthly powers, but we should certainly cultivate an indefinite, undirected sense of gratitude for what's good in life, without contaminating it with a belief that supernatural powers were involved therein.

CHAPTER FOURTEEN

WHAT SHOULD WE
TEACH OUR CHILDREN?

Concerning Faith, Science,
and How Schools Talk about Life

> Governmental authorities are fooling themselves very badly if they think that it is useful for schools to teach children that God is the ultimate authority. Today you can't even get schoolchildren to believe in such rubbish.—Hjalmar Söderberg (one of Sweden's greatest authors), *Hjärtats oro* (1909)

If we are ever going to see a new dawning of enlightenment, it will have to start with our children. In a world burgeoning with fanaticism and superstition, schools have an unprecedentedly crucial role to play in teaching children the art of clear thinking and of developing an independent style of critical thought. Education is the nourishment that enlightened thinking needs in order to grow. In many parts of the world, children—especially girls—have very limited opportunities to go to school. We in the Western world live in privileged countries where going to school is taken for granted.

But going to school is not enough. The content taught in schools also has to be up to snuff. As far as existential and scientific questions are concerned, there are serious lacks in Swedish education, and they are similar to those in almost all Western lands.

When we speak about children, we tend to categorize them from an external point of view in a manner that we would never do with something like political ideologies. Without any careful reflection, we casually speak of Muslim, Christian, and Jewish children—but in fact, there are no Muslim, Christian, or Jewish children; there are only children with Muslim, Christian, or Jewish parents. We would never describe six-year-old Lisa as a "Democratic" little girl, or seven-year-old Carl as a "Republican" little boy. We would simply say that Lisa and Carl have parents who are Democrats or Republicans. In my opinion, we should learn to see children as individuals who are born without any kind of religious faith.

Let Children Discover the World as It Truly Is

In Sweden, children are required to go to school for at least nine years, and soon it may be ten or even more. A fundamental question is therefore what we should teach our children—more specifically, what underlying principles should determine the topics and ideas that are taught. It's the state's duty to make sure that every child receives a rational outlook on the world. Schools should thus not teach children to believe in one particular religion but, instead, should teach the scientific facts that humanity has discovered about the world and how it is made. According to the United Nations convention on children's rights, all children have this right to the freedom of religion: "States Parties shall respect the right of the child to freedom of thought, conscience and religion."[1]

Accordingly, we should consider it unacceptable for a school to teach creationism, theories of hellfire and damnation, or other religious notions. All teachings should be religion free. Of course a school can teach *about* religious philosophies of life, but it should not *transmit* one. This principle should hold both for the mandatory period of schooling and at higher levels of education.

The strange fact is that the United States is one of only two countries in the world (the other being Somalia) that have not ratified the United Nations Convention on the Rights of the Child. In 1995, President Clinton signed the convention, but antagonistic pressure groups lobbied on the U.S. Senate not to ratify it. These groups argued that adopting the agreement would allow children to choose their own religion instead of their parents' and would allow children to protest and to legally oppose parental decisions. Such groups were also fearful that ratification of the convention would coerce the United States to spend more on children's welfare than on national defense. As a result of such fear-mongering, the convention still remains unratified by the U.S. Senate, even after more than twenty years.

Children are vulnerable. They possess neither the intellectual nor the emotional tools to be able to figure out where they themselves stand relative to the confusing clamor of voices in the adult world. For this reason, it is crucial that a schoolroom should be a neutral space where neutral, objective methods are used to explore and analyze reality. Schools should also provide an underlying system of values such as the one voted into law by Sweden's Parliament. This means that they should convey a basic democratic viewpoint in which all people's equal rights are stressed, regardless of their skin color, gender, sexual orientation, philosophy of life, or ethnicity. Nor should any religion's viewpoint about morality or ethics be favored.

Some years ago, Nicholas Humphrey (1943–), a noted British professor of psychology who focuses on the development of intelligence and the nature of consciousness, gave a controversial lecture at Oxford. The talk was arranged by Amnesty International and dealt with religious free schools. Humphrey argued that parents should have freedom, but not arbitrarily great freedom, to make choices concerning their children. Of course, today's laws already place certain limitations on parents' rights to bring up their children as they wish. In fact, Sweden was the first country in the world to prohibit caning or whipping of children by their parents (1979), and today most European countries have followed suit. Sexual exploitation of children has also been criminalized.

Humphrey's point, however, was that we have to protect children not just from *physical* acts of aggression but also from aggressions of a *psychological* nature. We cannot let schools teach children that the positions of planets in the sky control our lives, or that premarital sex is punished in hell, or that women should be subjugated to men, or that one has the duty to kill people whose religion is not one's own. Nor should schools teach that the Bible is literally true, or that humanity was divinely created some six thousand years ago. Schools are responsible for giving children the reasoning tools and analytical tools that are needed to examine and understand our world. This is a school's primary duty.

Magical and religious notions have a strong tendency to survive down through the generations. This is probably a result of the fact that children grow up strongly imprinted with their parents' ideas. Humans, in contrast to all other animals, survive and evolve thanks to the passing on, both orally and in writing, of knowledge accumulated by earlier generations. Children have a much greater chance of surviving if they depend on their parents' description of the world and its dangers. "Don't pick up a snake on the ground!" "Don't try to pet a tiger!" "Don't swim in a river where there are crocodiles!" Young children, however, cannot distinguish between wise advice (e.g., "Don't go anywhere near that tiger!") and useless advice (e.g., "To make rain fall, sacrifice a goat when the moon is full").

Unfortunately, a child's placement of faith is unreliable, and it can be abused. A parent who tells their child, "You will wind up in hell unless you pray to God whenever you think unclean thoughts" is very likely to damage the child for the rest of the child's life.

For many religious groups and sects, it is crucial that members not be allowed to be influenced by outsiders, and for this reason, a central feature of the *modus operandi* of such groups is to prevent members from mixing with other groups. Any group member who criticizes the group is said to have been possessed by demons or by Satan. Being submissive to the group is considered to be a virtue, and not asking questions about the world is seen as spiritual purity. Indoctrination is the most successful when it's the converts themselves who are gung-ho to build high walls against external thoughts.

Indoctrination at an early age can be very hard to break away from. For that reason, psychological acts of aggression against children can be just as harmful as physical acts of aggression. Such acts can be definitive and irreversible. It is thus crucial, in certain cases, that schools be able to protect young children and teenagers from their parents—from indoctrination and from attempts to block children's chances to have open, two-way contacts with the outside world.

Sectarian Schools in Sweden

In Sweden, schools run by religious sects are legal. In the prelude, I wrote about the Christian doomsday sect known as the Plymouth Brothers, located in Hyltebruk, in southern Sweden. This sect runs its own school, attended only by children whose parents belong to the sect. The congregation consists of about four hundred people all told, and it is probably the most extreme sect in Sweden today. Their religious views diverge considerably from traditional Christian ones.

In the 1960s, the Plymouth Brothers' then leader announced that doomsday was nigh, and that the Antichrist had taken over the world. The group's members were therefore told to cut off all contact with the outside world. Ever since then, total isolation from "infidels"—that is, from those who do not belong to the group—has been central in the sect's daily life and practice. Children would supposedly be saved in this way, spared from the fate of burning forever in hell's "sea of fire." Members of the sect are not allowed to eat with or share their living quarters with any outsider. They cannot live in any building owned by anyone outside the sect, nor live in a building in which any nonmembers live. They cannot even sign a contract with anyone outside the sect.

These people see the outside world as evil, as a work of the devil. The young members of the sect are instructed to steer clear of academic studies, so as not to seem prideful in God's eyes; moreover, married women are forbidden from working outside the home. Men are seen as women's superiors. Homosexuality is the work of demons, and God's rightful punishment for it is death. The sect's interpretation of the Bible is literal and therefore creationist; they believe that humanity was created some six thousand years ago. The theory of evolution is seen as a piece of nonsense that is simply carrying humanity away from God.

Children who grow up in the Plymouth Brothers' sect are not allowed to have friends outside of the sect. They don't play soccer in the local youth soccer league, nor do they make music with children in the local music school. And when a child in the group reaches the age of twelve, he or she is given a choice: "Are you with us or against us?" Any child who chooses to question the sect is thrown out of the community and no longer can even eat meals with their own family.

Some time ago, for such children, the local public school was a window onto the world, an opening toward normal relations with other children of the same age. But in 2007, the sect was legally permitted to run a school solely for its own children, and the school, just like other schools in Sweden, was provided some state funding. This state of affairs is nothing less than a tax-financed act of psychological aggression on an already disadvantaged group of children.

Religious Education in Schools

In 2011, the Swedish National Agency for Education established a new curriculum for religious education in elementary and secondary schools. The new curriculum was indisputably better than the previous one: broader, more objective, and more inclusive. And in it was included, for the first time ever, a guarantee that pupils would be taught about secular humanism. This would imply that, if the curriculum were followed, children would come to understand that a moral philosophy of life doesn't need to rest on stories of supernatural events, or on the existence of one or more gods, or on any kind of magical forces or powers. In fact, if the legislators took their own law seriously, then the very subject matter known as "religious education" would have to be renamed "education about philosophies of life" (or something along those lines); however, they didn't dare go that far with their curriculum reform. (As we have seen, all religions could be described as "philosophies of life," but not all philosophies of life are religions. Thus any religion would

315

be a valid candidate for inclusion in a class on philosophies of life but not vice versa.)

In the nationally established curriculum for religious education, pupils in the seventh through ninth school years would learn about the following topics:

- the main ideas and scriptures of Christianity and the outstanding features that characterize Christianity's three main branches (Protestantism, Catholicism, and the Eastern Orthodox Church);
- the main ideas and scriptures of other major world religions, such as Islam, Judaism, Hinduism, and Buddhism;
- various interpretations and practices, in the world's major religions, in today's society;
- an outline of the history of the world's major religions;
- new religious movements, new types of religiousness, and personal religiousness, as well as how such developments have arisen; and
- secular points of view on life, including secular humanism.

The Swedish National Agency for Education also published a document with more detailed comments about the curriculum. In it one finds the following:

> In grades 7 through 9, the curriculum mandates coverage of secular philosophies of life. Such philosophies are often based on a scientific worldview and they do not see ethical values as having a divine origin. Instead, they promote the idea that ethical principles can be completely justified by thoughts that human themselves create. . . . This characterization applies to such well-known philosophies of life as secular humanism . . .
>
> Atheism, for example, implies that one denies the existence of any god. Starting from such a basis, one can then form other conceptions about many things, ranging from the origin of the universe to ethical issues, but having an atheistic worldview per se does not imply unique answers to other types of questions.[2]

When this new curriculum was accepted, I was naturally very curious as to how the secular perspective would be presented in schoolbooks. The result was not very encouraging, alas.

Several of the books that I checked were quite old, but I also looked at ones that had been updated after the new curriculum had been adopted. These texts were often riddled with conceptual confusions and blatant errors of fact, but most important, secular philosophies of life were generally described inaccurately and from a religiously colored perspective.

To be sure, there are some schoolbooks that succeed in presenting the topic in an exemplary manner; however, ones that do a poor job outnumber those that do a good job by a very large factor. I very carefully checked out (among others) several textbooks written by educator and priest Börge Ring. Ring's priestly aura shines through everywhere in his writings, which is unacceptable in schoolbooks that are supposed to be totally neutral with respect to philosophies of life. I have chosen to single out this particular writer as an example, but the problem recurs almost everywhere in the majority of texts that I examined.

In Ring's book *Religion—helt enkelt* (Religion, Simply Presented), secular humanism is discussed only in the last chapter, side by side with Jehovah's Witnesses, the Mormon Church, creationism, and Satanism! Börge Ring's obsession with devil worship is a running theme in many of his books, and it does not correspond in any way to the very sparse—in fact, vanishingly small—distribution of Satanism in the world.

The sloppiness in such texts is not limited to those passages that describe the secular philosophy of life. For example, in a passage dealing with evolution deniers and creationists, Ring says: "[C]reationism and science are in opposition to each other, and the discussions about them contain many obscure arguments concerning physics."[3]

Of course it is perfectly correct that science and creationism are incompatible, but the controversy concerns biology, not physics. Later, Ring writes: "The number of followers [of creationism] cannot be accurately stated, but it is only in the United States that the movement has had much influence."[4]

This is not quite right. Although creationism (along with its cousin, "intelligent design") has so far not had a very large influence in Sweden, it is frequently cited not only in the Swedish free churches but also in the evangelical Christian movement throughout the world—and that is the branch of Christianity that is most on the rise at the present time. Moreover, creationism is also the dominant belief in Islam, and it exerts a very powerful influence in many Islamic countries.

The author of these passages blurs together "secular humanism" as a philosophy of life with other meanings of the word "humanism," such as the educational term "the humanities," or humanism as any kind of engagement with humanity, which one can obviously have, whether one is a Christian, a Muslim, or a secular humanist. The true nature of secular humanism as a philosophy of life completely eludes the writer (and hence all of his readers).

Ring also attempts, in his book, to explain what science is. The topic is covered in roughly one half a page, and the section title is: "Can Science Tell Us What Is Right?" He describes science in the following manner:

> The word "science" means "knowledge." Someone who does scientific research arranges and organizes pieces of knowledge. This can be done by collecting pieces of knowledge through investigations, interviews, analyses of text, or observations of objects. A researcher can also gather knowledge by conducting experiments on people or animals. If you were to ask different researchers which of them is right, you would get many different answers. Some researchers wouldn't even wish to answer the question, while others would give vague, elusive answers. If even the researchers themselves don't know who is right, then how can anyone know?[5]

It's clear that anyone who is introduced to the notion of "science" through this bizarre description will not have the least interest in it. If "even the researchers themselves don't know," then why bother with it at all?

Of course Ring's halfhearted portrait of science is profoundly misleading. To be sure, there are controversies among scientists, but in the overwhelming majority of cases, scientists are in agreement on which theories work and which don't, and on how things in the world behave.

In his book *Religion och sånt* (Religion and Such Things, 2013), Ring had the chance to update his previous books, taking into account the new curriculum. But instead, in this new book, the blur around the notion of secular humanism becomes even blurrier. Take a look: "Islam is a religious philosophy of life, while humanism can be a religious or a nonreligious philosophy."[6]

This is true if by "humanism" you merely mean "the humanities" or "any sort of engagement with humanity," but if you mean "humanism" as a *philosophy of life*, then it is utterly false. Secular humanism seen as a philosophy of life, which was the intent under the new curriculum, is totally nonreligious, as most people realize.

Next, the author tackles the concept of atheism, which he also seems to wish to portray as a philosophy of life, even though this is certainly not the case. He makes a blatantly false statement, couched as a question posed to the students who will read his book: "Do you agree with atheism that nothing exists except the world that we see and can measure?"[7]

As we have seen, an atheist can believe in ghosts, homeopathy, or astrology, since atheists don't need to share the rich philosophy of life that secular humanism offers them. Moreover, atheists can obviously believe in all sorts of things that one cannot see or measure: love, equality, human rights, life on other planets . . .

It is clear, though, that Ring's question was formulated in such a way as to discourage students from calling themselves "atheists"; after all, who would want to be so deadly dull as to believe only in what one can see and measure?

In his 220-page book, Ring devotes just four pages to secular humanism (with the blurry definition he gives it, identifying it with atheism), while he devotes twenty pages to Buddhism. Also worth noticing in this connection is the fact that, according to an international survey of the world's philosophies of life,[8] there are roughly 2.2 billion Christians, 1.6 billion Muslims, and 900 million Hindus. In fourth place come nonreligious philosophies, with roughly 750 million people. Secular humanism of course falls under this rubric. Buddhism comes in only in seventh place, with roughly 350 million followers.

In an earlier (and still used) edition of the book, the New Age movement also comes in for criticism, but not for the reasons one might hope in a textbook. The movement is presented as "A protest against Christianity and other major religions": "New Age believers are opposed to most established religions, and especially Christianity."[9]

Now there are many grounds on which to criticize the New Age movement, but this is hardly one of them. It is not Christianity that is in fact the New Age movement's primary target but, instead, rational thinking and a scientific attitude in general. (In certain free churches in Sweden, the New Age movement is looked upon as an occult movement that is all about evil spirits, demons, and other unholy beings. Perhaps this is what troubles Börge Ring.)

In the updated version of Ring's book (2013), the just-quoted statement has been taken out. In its place, one reads about present-day communication with a spirit who supposedly lived thousands of years ago:

In Värmland [a Swedish county west of Stockholm] there lives a former carpenter named Sture Johansson, who can channel facts and experiences from a man named Ambres, who lived in ancient Egypt. When Sture Johansson receives messages from this man, he calls himself Ambres. He gives courses in which the participants become acquainted with a long-gone mode of life, but one that is now accessible through Sture Johansson.[10]

In Börge Ring's updated book *Religion och sammanhang*,[11] the presentation is not much better, although the book was published two years after the new curriculum had come into force. The publisher had been informed about all the errors in earlier editions but had chosen to ignore many of them before printing the revised edition.

In this book, Ring continues to be obsessed with Satanism, which is given an unreasonably large role, given its minimal relevance in the world. And secular humanism (once again blurred together with other meanings of "humanism"), along with atheism, takes up just four of the book's 350 pages, as compared to Buddhism's thirty-nine pages.

In the latest edition, the author considers whether atheism is a philosophy of life or not, and he winds up making the fuzzy statement that it is "above all a life orientation." This is progress, in the sense that he no longer describes atheism as a philosophy of life, but what does he mean by "a life orientation"? Presumably this would be something that would pervade a person's life. For instance, a researcher who has dedicated their life to seeking a cure for cancer would surely have research as a life orientation. And very likely marathon running is a life orientation for Eliud Kipchoge. But it wouldn't make sense to call *not* running marathons or *not* seeking a cure for cancer a "life orientation."

Börge Ring's irrational notions about atheism are most likely a consequence of his training as a priest, but they are hardly appropriate material to appear in textbooks in the year 2015.

The most absurd aspect of Ring's book *Religion och sammanhang*, however, has to be the following. In it (on page 326), Charles Darwin is held up as an idol and an inspirational thinker for Satanists! Even if the book's author were actually able to dig up some addle-headed Satanist who gleefully declared "Darwin is my hero!," it's not as if this is why Darwin's ideas have come to have such importance in science and in human thinking more generally.

There are many examples of errors, misleading formulations, and heavy-handed descriptions in Börge Ring's books. In 2014, I took part in a seminar in which the author's ideas were discussed with the author himself. It was filmed and can be seen in its entirety on the web.[12]

A Fresh New Philosophy of Life for Schools

Why have I gone through all this in such detail? Well, such matters are very important if we are concerned about what we are going to teach our children. It's not just which *facts* we hope to see taught but also which *values* we hope to see transmitted.

We need to move away from the common misconception that ethics is irrevocably linked with religion. Moral thinking can and should be carried out without reference to religious preconceptions. If in school we teach our children that the only route leading to morality is based on religion, then we are heading down a very dangerous pathway. Young people must be given the opportunity to encounter nonreligious philosophies of life out of which flow sensible attitudes concerning ethical and existential questions.

We also need to move away from the notion that atheism is a philosophy of life, as well as the notion that it is a dull and boring attitude toward the world.

Our children must be offered a broad, neutral, and objective view of the many diverse philosophies of life that exist and flourish all around our world. And last but not least, publishers of educational material have the important duty of being sure that what they publish reveals and celebrates the enormous variety of philosophies that can be found throughout the world.

~

INTERLUDE: ON THE NATURAL AND THE SUPERNATURAL

Can we really be sure that there isn't anything supernatural?

When I was in my early teens, I used to go visit an elderly couple who claimed that, using a big reel-to-reel tape recorder, they had recorded the voices of dead people on tape. On my visits to their home, I listened to many of their tapes, which crackled and hissed. Every once in a while you could make out the sound of voices.

We used their tape recorder for recording seances. At one point, they actually came out with a little book called "De döda talar på band" ("The Dead Speak on Tape"). I was deeply fascinated by their claims, and later I got a hold of their tape recorder and tried to conduct my own investigations of spirit voices. Unfortunately, though, I never managed to record any.

The idea of the supernatural is based on a misconception of what the word "natural" means. If the phenomena that are called "supernatural" in common parlance really existed, then they wouldn't be supernatural. They would simply be unexplained as of today. That would hold for ESP, for recordings of dead people talking, for telekinesis, and so forth. If any of them can really be shown to exist, then it will be a natural, not a supernatural, phenomenon.

When we investigate various phenomena, no matter of what type, we should first try to find evidence that the phenomena exist; after that, we seek explanations for them. Take mind reading, for example. If we wish to figure out for ourselves whether mind reading exists or not, we start with a laboratory situation in which we can test if it works. If we find evidence supporting it, then what we've found evidence for is not something supernatural but a genuinely existing though as-of-yet unexplained phenomenon.

The next step is to try to figure out how it works. Does this involve a never-before-seen force of nature? Or is it an unknown biological phenomenon? Only when we have worked out some plausible theories about the phenomenon can we begin to hope to explain it. But it would be meaningless to formulate a theory unless we have evidence that the phenomenon actually exists.

What truly exists is part of nature, and thus is natural, not supernatural. What doesn't exist is also not supernatural; it is simply nonexistent. If the souls of the deceased wound up on "the other side" and communicated with us, that would be a natural but as-of-yet unexplained phenomenon.

The "natural" is often contrasted with what is said to be "unnatural." Thus we have "natural" foods and herbal medicines, and some people believe them to be better than "unnatural" foods and medicines. But what does this really mean? Some medicines that are not particularly natural are nonetheless very effective. "Natural" food isn't always more nutritious than other types of food.

Imagine taking a city dweller into a Swedish forest with the sole mission of surviving on the basis of what the forest offers. The city dweller probably won't live very long. Natural products, such as mushrooms, can be extremely dangerous to eat.

I have often debated with the leaders of Swedish free churches (not the official state brand of Lutheranism) all around Sweden. Quite often they claim that homosexuality is not only abnormal but also unnatural.

The truth is quite the opposite. A certain percentage of people (and also of animals) are homosexual, so that homosexuality is as natural as you can get. "But homosexuals cannot make babies!" is a frequent rejoinder to this argument. But the earth is menaced by overpopulation, and it would not be harmed if there were fewer inhabitants on its surface. From that perspective, it would be distinctly better if more people were homosexual.

Moreover, it isn't so "natural" to drive cars, live in apartment buildings, or to take antibiotics like penicillin. So be careful when you advocate living "naturally."

AFTERWORD AND WORDS OF THANKS

"We live in the best of all possible worlds!" exclaims Master Pangloss in Voltaire's *Candide*. Is that really true? In the tale, Candide is forced out into the world, where he experiences one horror after another, each worse than the previous one: natural disasters, torture, war, murder, massacres, rapes, and betrayals. Voltaire's novel is of course a satire on optimism, as well as a confrontation with the problem of theodicy: if there is a good, omniscient, and almighty God, how can there be so much suffering in the world?

The flame of reason flickers faintly in the dark, and does so much too often in the world today. Bizarre conspiracy theories about a secret world order controlling humans with microchips injected with vaccine are flourishing. Religious fundamentalism is strengthening its position in many parts of the world.

But I am hopeful. Reason and science have helped most of us survive a pandemic that would have killed many more if it had taken place fifty years ago (but it probably wouldn't have spread so fast back in those days). And although information technology and communication technology are spreading bad and crazy ideas like wildfire, luckily they are also helping spread good ideas and scientific knowledge. Technology is a double-edged sword.

Those of us who have a secular outlook on life do not need to grapple with the dilemma of theodicy. Instead, we ourselves take responsibility for our world and our existence in it with the help of the tools we have at hand: reason, creativity, curiosity, compassion, and the ability to reconsider. By so doing, we may even grow into more authentic versions of ourselves.

> Fly me to the moon, let me play among the stars
> Let me see what spring is like, on Jupiter and Mars

The classic song *Fly Me to the Moon*, written by Bart Howard, conveys the feeling of existential vertigo and the happiness of existence, if only for a short time, in a world that becomes a better place for all of us with each passing century.

We should remember humanity's desire to discover worlds beyond our own; that is the ultimate example of the fantastic joy of discovery we all carry within us.

Although 2020 and 2021 were years of a global pandemic, they were also years in which we looked outward and explored our solar system. The United States, China, and the United Arab Emirates all executed (nonhuman) expeditions to Mars in 2021. It will be a while longer before humans travel to the Red Planet, but robots that study the planet's surface will in the meantime provide much new knowledge.

Perhaps David Bowie's iconic song "Life on Mars" will finally come true. Renewed human visits to the moon are also in the planning. Imagine what a wonderful experience it will be to observe such voyages from earth with today's audiovisual technology.

It's worth pointing out that global scientific cooperation in the effort to discover and produce effective vaccines against the pandemic is a magnificent testimony to the power of science. The discovery and development of various vaccines in the nineteenth century, and of penicillin in the early twentieth century, are estimated to have saved hundreds of millions of lives.

But what of war, populism, nationalism, and racism? Aren't things just getting worse? Fresh in our memories are two devastating world wars, as well as so many more recent wars, and today's religious fundamentalism and terrorism hardly seem to be "the best of all possible worlds." On the other hand, the risk of dying from violence today is lower than ever before in history. We live longer, we are healthier, more children are allowed to go to school, and fewer people are starving.

It is sometimes said that the Greek philosophers were the first to enjoy enough peace of mind to have the chance to ruminate on abstract ideas. Maybe that's true; in any case, we today are heading ever more in that direction.

If we take a bird's-eye point of view, it's fair to say that today we actually do live in the best of all possible worlds, or at least in the best of all eras of human history. In one's own personal life, such optimism can seem dubious, especially in the midst of a global pandemic. Music, however, can serve as a reminder—a benevolent and beautiful reminder urging us to look up with hope, even when life is at its lowest. As in the classic song composed by Charlie Chaplin:

> Smile, though your heart is aching
> Smile, even though it's breaking
> When there are clouds in the sky, you'll get by.
> If you smile through your fear and sorrow,
> smile—and maybe tomorrow,
> you'll see the sun come shining through for you.[1]

A song is a miniature world that can guide us in our own larger world. The character Oscar in Ingmar Bergman's movie *Fanny and Alexander* (1982) describes this well: "Outside is the big world, and sometimes the small world manages for a moment to reflect the big world so that we understand it better."

So I am hopeful. We need to light the flame of reason—and work hard to do so as well as in politics as in our personal lives and beliefs. If we do things right, the flame of reason will keep on burning—sometimes perhaps only faintly—but it will never die. That is what I believe.

This book project has been the writing project of my life. It summarizes everything I believe in and want to stand up for. I want to thank Douglas Hofstadter for a fantastic collaboration on this project. You were my intellectual hero when I was twenty years old. Now, a half life later, you have become a very dear friend. Not even Gödel would have guessed that! Thanks for the collaboration and all the extremely intellectually stimulating conversations we have had and will have in the future.

I also want to thank my love and life partner Victoria Larm for all her intellectual and emotional support throughout this project. Your strange and beautiful intellect, as well as your deep investigations into music, art, and the history of ideas have helped me so deeply in my thinking process. And all our games of blitz chess, almost ten a day, keep me on my feet (even though I'm quite worried that my percentage of wins is slowly but surely diminishing). Thank you, my love, for all this and everything else.

And finally, I want to thank my son Leonardo (named after the medieval Italian mathematician Leonardo Fibonacci) for giving my life the deepest of existential meaning. You are too young to read this book at the moment, but

THE FLAME OF REASON

I am sure that you will do so in the near future. I remember when you came home from school in your first year and said "Daddy, most of my classmates says they believe in ghosts! It is so irritating! What can I do?" The flame of reason had already been lit. It was so beautiful.

Stockholm, March 2021
Christer Sturmark

NOTES

Foreword by *Christer Sturmark*

1. And in one gentle e-mail during the process, he corrected not only some factual errors but also some grammatical errors I had made in Swedish—my mother tongue!

2. The term "secular" comes originally from the Latin word *saecularis*, meaning "worldly," as opposed to "ecclesiastic," meaning "churchly."

Prelude

1. Stefan Zweig, *The World of Yesterday* (Lincoln: University of Nebraska Press, 1964).

2. Zweig, *The World of Yesterday*.

3. The inventor of *in vitro* fertilization, Robert G. Edwards, was awarded the Nobel Prize in Medicine in 2010, which led to vehement protests from the Catholic Church, among others.

Chapter One. To Meet the World with an Open Mind

1. Pierre-Marc-Gaston, duc de Lévis, *Maximes et réflexions sur différents sujets de morale et de politique* (Paris, 1808).

2. Carroll, whose real name was Charles Lutwidge Dodgson, was an English writer, mathematician, logician, and photographer. His fantasy-filled books *Alice in Wonderland*, written in 1865, and *Alice through the Looking-Glass*, written seven years later, made him immortal.

3. "The Nobel Prize in Literature 1950," Nobel Prize, accessed August 16, 2021, https://www.nobelprize.org/prizes/literature/1950/summary/.

4. Raymond Smullyan (1919–2017) was an American mathematician, magician, concert pianist, logician, Taoist, and philosopher.

5. What I mean by "going overboard in nerdity" can be measured by your reaction to the following statement: "There are ten types of people in the world: those who can count in binary, and those who can't." If you laughed, you're a definite nerd.

6. See, for example, Michael Shermer's 1997 book *Why People Believe in Weird Things* (New York: Freeman) for a discussion of cold reading, pattern recognition, statistical trickery, and so forth.

7. William of Ockham (1285–1349), educated at Oxford University, was an English theologian and philosopher, as well as a Franciscan monk. His "razor" represents the idea that overly complex explanations of things should be "shaved away."

8. Plato, *Apology, Crito and Phaedo of Socrates* (ca. 399 BCE).

9. Francis Bacon, "Quotes: Quotable Quotes," Goodreads, accessed August 16, 2021, https://www.goodreads.com/quotes/1376291-for-the-mind-of-man-is-far-from-the-nature.

10. Translator's note: In the Swedish original, the author chose never to capitalize the word for "God." This is quite acceptable in contemporary Swedish. However, in English, it wouldn't work. For instance, the phrase "In god we trust" looks just as wrong as does the sentence "A book by christer sturmark was anglicized by douglas hofstadter." If we were to do in English what was done in Swedish, the author would come across as being aggressively scornful of all forms of religion, which he is decidedly not. We therefore opted for a different policy—namely, the word "God" is capitalized whenever it is acting as a singular definite proper noun. However, when the noun is used in an indefinite way, or is pluralized, then using a lowercase *g* looks fine, so that became our policy. In some cases, it required a judgment call, and we used our best judgment.

11. "Livsåskådning," Uppslagsverket, accessed August 16, 2021, https://www.ne.se/upp slagsverk/encyklopedi/l%C3%A5ng/livs%C3%A5sk%C3%A5dning.

12. The Sanskrit word "theravada" means "the oldest school." This original form of Buddhism is widely found today in Sri Lanka, Burma, Thailand, Laos, Cambodia, and Vietnam.

13. For more information on secular humanism, see, for example, Philip Kitcher's book *Life after Faith: The Case for Secular Humanism* (New Haven, CT: Yale University Press, 2014) or Stephen Law's book *Humanism: A Very Short Introduction* (Oxford, UK: Oxford University Press, 2011).

Chapter Two. I Believe That I Know

1. See Eugene Wigner, "The Unreasonable Effectiveness of Mathematics in Describing the Natural Sciences," *Communications in Pure and Applied Mathematics* 13, no. 1 (February 1960); Max Tegmark, *Our Mathematical Universe* (New York: Penguin, 2015).

2. Bertrand Russell, *The Analysis of Matter* (New York: Harcourt Brace, 1927), 201.

3. Aristotle, *Metaphysics*, quoted in Marian David, "The Correspondence of Truth," in *Stanford Encyclopedia of Philosophy*, last updated May 28, 2015, https://plato.stanford.edu/entries/truth-correspondence/#1.

4. For a good introduction to the mysteries of quantum mechanics, see Nicolas Gisin, *Quantum Chance: Nonlocality, Teleportation, and Other Quantum Marvels* (New York: Springer, 2014).

5. A good introduction to Rorty's ideas is given in Robert Brandom, *Rorty and His Critics* (Malden, MA: Blackwell, 2000).

6. See Paul Boghossian, *Fear of Knowledge: Against Relativism and Constructivism* (Oxford, UK: Oxford University Press, 2006).

7. See Bruno Latour, *Pandora's Hope: An Essay on the Reality of Science Studies* (Cambridge, MA: Harvard University Press, 1999).

8. See Ophelia Benson and Jeremy Stangroom, *Why Truth Matters* (New York: Continuum, 2006).

9. Today, we realize that some of Lysenko's ideas were not wrong. The study of what is called *epigenetics* shows that the environment can in fact exert a greater influence on what is inherited than had been thought.

10. For a critical examination of social constructivism, see the book by philosopher Ian Hacking titled *The Social Construction of What?* (Cambridge, MA: Harvard University Press, 1999).

11. Postmodernism is a very broad notion, and there are many interpretations of its meaning. A typical one is that there exist only subjective and personal truths, and that any belief in humanity's liberation through progress in science, technology, and rationality should be abandoned.

12. An excellent discussion of this famous affair and its aftermath is given by Alan Sokal, *Beyond the Hoax* (Oxford, UK: Oxford University Press, 2009).

13. See Moira von Wright, *Genus och text: När kan man tala om jämställdhet i fysikläromedel?* [Gender and text: When can one speak of equality in physics teaching materials?] (Stockholm: Skolverket, 1999), 24.

14. von Wright, *Genus och text*, 64.

15. von Wright, *Genus och text*, 65.

16. von Wright, *Genus och text*, 65.

Chapter Three. Beliefs Based on Good Reasons

1. A proof in mathematics is an argument—a sequence of purely logical steps—that starts from first principles (*axioms*), that follows strict formal rules (*rules of inference*), and that winds up in a final conclusion (*a theorem*).

2. As far as I know, however, no theologian has actually managed to explain how this alleged dependence actually comes about.

3. Pierre-Simon Laplace, *Théorie analytique des probabilités* (Paris, 1812); David Hume, "Of Miracles," in *An Enquiry concerning Human Understanding* (1748).

4. This example is adapted from the book by Sören Holst, *Tankar som ändrar allt* [Thoughts that change everything] (Stockholm: Fri Tanke, 2012).

5. As we saw earlier, Galileo wound up in a sharp clash with the Catholic Church when he claimed that certain statements about nature in the Bible were not true. One of these, he said, was the claim that the sun periodically circles the earth, rather than the other way around, which is what he himself believed.

Chapter Four. What Is Science?

1. In 2014, the Nobel Prize in Medicine was awarded for discoveries that reveal how our brains keep track of where we are in a room. Researchers John O'Keefe, in England, and May-Britt Moser and Edvard Moser, in Norway, discovered new types of cells in the brain

that are responsible for a person's sense of orientation in the world—namely, so-called "place cells" in the hippocampus and "grid cells" in the entorhinal cortex.

2. See, for instance, Nils J. Nilsson, *Understanding Beliefs* (Cambridge, MA: MIT Press, 2014).

3. Albert Einstein, *The Born-Einstein Letters, Correspondence between Albert Einstein and Max and Hedwig Born from 1916 to 1955*, trans. Irene Born (London: Macmillan, 1971).

4. String theory, briefly alluded to earlier, is a very abstract mathematical attempt to describe the tiniest building blocks of matter; as of yet, it has no empirical confirmation whatsoever. Nonetheless, there are many researchers who consider it to be highly promising. Only time will tell if it actually holds up.

5. Karl Popper, *Unended Quest: An Intellectual Autobiography* (New York: Routledge, 1976).

6. See Nilsson, *Understanding Beliefs*.

7. Popper, *Unended Quest*. (Italics in the original.)

Chapter Five. Ghosts in the Head

1. Daniel Kahneman, born in 1934 and now a professor emeritus at Princeton University, received the 2002 Nobel Memorial Prize in Economic Sciences "for having integrated insights from psychological research into economic science, especially concerning human judgment and decision-making under uncertainty." "Daniel Kahneman Facts," Nobel Prize, accessed August 16, 2021, https://www.nobelprize.org/prizes/economic-sciences/2002/kahneman/facts/.

2. See Daniel Kahneman and Amos Tversky, "Judgment under Uncertainty," *Science* 185, no. 4157 (1974): 1124–31.

3. See Nicholas Epley and Erin Whitchurch, "Mirror, Mirror on the Wall: Enhancement in Self-Recognition," *Personality and Social Psychology Bulletin* 34, no. 9 (2008): 1159–70.

4. See the video at my website: "The Monkey Business Illusion," Sturmark, accessed August 1, 2021, https://www.sturmark.se/bollspel.

5. The answer can be found here: "Zombieviruset," Sturmark, accessed August 1, 2021, https://www.sturmark.se/zombievirus.

6. If you feel that this is nonsense, as most readers do, then I invite you to check out the explanation here: "Mysteriet med de tre dörrarna," Sturmark, accessed August 1, 2021, https://www.sturmark.se/tredorrar.

7. Raymond Smullyan, *The Chess Mysteries of Sherlock Holmes* (New York: Knopf, 1979).

8. Smullyan, *The Chess Mysteries*.

9. Smullyan, *The Chess Mysteries*.

10. Smullyan, *The Chess Mysteries*.

11. For those readers who simply want to see the puzzle's answer, it can be found here: "Kortkontroll," Sturmark, accessed August 1, 2021, https://www.sturmark.se/kortspel.

12. The answer to the second riddle can be found here: https://www.sturmark.se/krogen.

13. The answer can be found here: "Eva och Daniel," Sturmark, accessed August 1, 2021, https://www.sturmark.se/eva.

14. The answer can be found here: "Vem föreställer porträttet?" Sturmark, accessed August 1, 2021, https://www.sturmark.se/vem.

15. See Daniel Dennett, *Breaking the Spell: Religion as a Natural Phenomenon* (New York: Viking, 2006).

16. See, for example, Todd Tremlin, *Minds and Gods: The Cognitive Foundations of Religion* (Oxford, UK: Oxford University Press, 2010).

17. Itzhak Fried, Charles L. Wilson, Katherine A. MacDonald, and Eric J. Behnke, "Electric Current Stimulates Laughter," *Nature* 391, no. 6668 (1998): 650, www.researchgate.net/publication/13742090_Electric_current_stimulates_laughter.

18. Justin Barrett, *Born Believers: The Science of Children's Religious Belief* (New York: Atria Books, 2012).

19. For our purposes, an agent can be thought of as something that has goals or intentions, in other words, a conscious entity. We tend naturally to distinguish between mere objects, such as stones, sticks, and stoves, and agents, such as people, animals, and possibly alien beings and/or gods.

20. Barrett, *Born Believers*. See Stephen Law, *Believing Bullshit: How Not to Get Sucked into an Intellectual Black Hole* (Lanham, MD: Prometheus, 2011), in which agent detection is explained in considerable detail.

21. Paul Bloom, "Is God an Accident?" *Atlantic Monthly*, December 2005.

Chapter Six. A Natural World

1. A prime number is a whole number divisible only by itself and 1; a few examples are 2, 3, 7, 37, and 2011. By contrast, 111, being the product of 3 and 37, is not prime.

2. Dwight D. Eisenhower, "Statement by the President upon Signing the Bill to Include the Words 'under God' in the Pledge to the Flag," American Presidency Project, June 14, 1954, https://www.presidency.ucsb.edu/documents/statement-the-president-upon-signing-bill-include-the-words-under-god-the-pledge-the-flag.

3. Jeffrey M. Jones, "Atheists, Muslims See Most Bias as Presidential Candidates," Gallup, June 12, 2012, https://news.gallup.com/poll/155285/atheists-muslims-bias-presidential-candidates-aspx.

4. Miles Godfrey, "Atheists Are Believers Who Hate God, Says Anglican Archbishop Peter Jensen," News.com.au, April 2, 2010, https://www.news.com.au/breaking-news/atheists-are-believers-who-hate-god-says-anglican-archbishop-peter-jensen/news-story/eae5d2eb21cc5c5d46cacf1cd01a1eb4.

5. Adolf Hitler, *Mein Kampf* (Berlin: Franz Eher Nachfolger GmbH, 1925). This standard English translation is unfortunately pretty atrocious, but hopefully you get the idea.

6. Hitler, *Mein Kampf*. Sorry once again for the atrocious translation of this "classic." But maybe atrocious books deserve atrocious translations.

7. Martin Gardner, *Fads and Fallacies in the Name of Science* (New York: Dover, 1952).

Chapter Seven. Being Good without Needing God

1. Derek Parfit, who was born in 1942 and died in 2017, was a professor of philosophy at Oxford University; he was widely respected as a great philosopher of mind. His most

celebrated book, *Reasons and Persons* (Oxford, UK: Oxford University Press, 1984), deals with questions of morality and human identity.

2. Plato, *Euthyphro* (389 BCE), Internet Classics Archive, accessed August 16, 2021, http://classics.mit.edu/Plato/euthyfro.html.

3. Plato, *Euthyphro*.

4. Simon Blackburn, *Ethics: A Very Short Introduction* (Oxford, UK: Oxford University Press, 2001).

5. For Hindus, Shiva is one of the most important incarnations of God, while Zeus, the son of Kronos and Rhea, was the highest of all the ancient Greek gods.

6. It's a bit like the American sound bite about guns: "Guns don't kill people; people kill people." Just substitute "religions" for "guns."

7. See, for example, Dutch ethologist Frans de Waal's books *Primates and Philosophers: How Morality Evolved* (Princeton, NJ: Princeton University Press, 2006) and *The Bonobo and the Atheist: In Search of Humanism among the Primates* (New York: Norton, 2014).

8. Douglas Hofstadter, "How We Live in Each Other," in *I Am a Strange Loop* (New York: Basic Books, 2007).

Chapter Eight. New Age Beliefs and the Crisis of Reason

1. See the journal *Skeptical Inquirer* at https://csicop.org/si.

2. See Scott O. Lilienfeld, Steven Jay Lynn, John Ruscio, and Barry L. Beyerstein, *Fifty Great Myths of Popular Psychology* (New York: Wiley-Blackwell, 2009).

3. Pyramidal sales schemes, also known as "multilevel marketing," are a notorious fraudulent method of selling and marketing, whereby sellers are recruited and are instructed to sell the company's products at the same time as they are also recruiting further sellers.

4. "An Indian Test of Indian Astrology," *Skeptical Inquirer* 37, no. 2 (March/April 2013), https://skepticalinquirer.org/2013/03/an-indian-test-of-indian-astrology/.

5. See Roger B. Culver and Philip A. Ianna, *Astrology: True or False?* (Lanham, MD: Prometheus, 1988).

6. See *Culture and Cosmos: A Journal of the History of Astrology and Astronomy* at https://www.cultureandcosmos.org.

7. See, for example, David J. Pittenger, "Measuring the MBTI . . . and Coming up Short," accessed July 29, 2021, https://jobtalk.indiana.edu/Articles/develop/mbti.pdf.

8. See, for example, Lennart Sjöberg, "En kritisk diskussion av Myers-Briggs testet" [A critical discussion of the Myers-Briggs test], DocPlayer, accessed August 16, 2021, https://docplayer.se/13823035-En-kritisk-diskussion-av-myers-briggs-testet.html.

9. As part of my research on this topic, I had an MBTI consultant test my personality. While we were discussing the test, she detected that I was skeptical, and she tried to counter my stance by saying that I simply had a skeptical personality type. It never seemed to cross her mind that my skepticism might be rooted in the fact that there was something to be skeptical about.

Chapter Nine. When Religion Runs off the Rails

1. Center for Reproductive Rights, "Center Appeals to Inter-American Commission on Human Rights for Release of Unjustly Imprisoned Salvadoran Women," August 8, 2020, https://reproductiverights.org/center-appeals-to-inter-american-commission-on-human-rights-for-release-of-unjustly-imprisoned-salvadoran-women/.

2. "Over Their Dead Bodies: Denial of Access to Emergency Obstetric Care and Therapeutic Abortion in Nicaragua," Human Rights Watch, October 1, 2007, https://www.hrw.org/report/2007/10/01/over-their-dead-bodies/denial-access-emergency-obstetric-care-and-therapeutic.

3. Ophelia Benson and Jeremy Stangroom, *Does God Hate Women?* (New York: Continuum, 2009).

4. *WWN* Editors Team, "Ultra-Orthodox Jewish Women Continue Protests against Segregation in Jerusalem," *Women News Network*, https://womennewsnetwork.net/2012/01/06/jewish-women-jerusalem-segregation.

5. *Sans Magazine* 4 (2012).

6. Farooq Hassan, "The Sources of Islamic Law," *Proceedings of the Annual Meeting of the American Society of International Law* 76 (1982).

7. "Cairo Declaration on Human Rights in Islam," Organization of the Islamic Conference, August 5, 1990, https://www.refworld.org/docid/3ae6b3822c.html.

8. "Islamic Law, the Nation State, and the Case of Pakistan," Wilson Center, October 26, 2018, https://www.wilsoncenter.org/event/islamic-law-the-nation-state-and-the-case-pakistan.

9. Organization of the Islamic Conference, "Cairo Declaration on Human Rights in Islam."

10. Organization of the Islamic Conference, "Cairo Declaration on Human Rights in Islam."

11. "Summary Record of the 20th Meeting, Held at the Palais des Nations, Geneva, on Monday, 10 February 1992: Commission on Human Rights, 48th Session," E.CN.4/1992/SR.20, paragraphs 17–20, United Nations Digital Library, https://digitallibrary.un.org/record/141520.

12. Lucy Carroll, "A Note on the Muslim Wife's Right to Divorce in Pakistan and Bangladesh," *New Community* 13, no. 1 (1986): 94–98, https://doi.org/10.1080/1369183X.1986.9975949.

13. "Report of the United Nations Working Group on Arbitrary Detention," E/CN.4/2004/3/Addition 2, United Nations, June 27, 2003, https://undocs.org/E/CN.4/2004/3.

14. See "Diya (Islam)," *Wikipedia*, last updated February 17, 2021, https://en.wikipedia.org/wiki/Diya_(Islam).

15. See Elyse Semerdjian, *Off the Straight Path: Illicit Sex, Law, and Community in Ottoman Aleppo* (Syracuse, NY: Syracuse University Press, 2008). See also "Iran: End Executions by Stoning," Amnesty International, January 2008, https://www.amnesty.org/download/Documents/56000/mde130012008en.pdf.

16. "Win-Gallup International Global Index of Religiosity and Atheism—2012," Scribd, https://www.scribd.com/document/136318147/Win-gallup-International-Global-Index-of-Religiosity-and-Atheism-2012.

17. Declan Walsh, "A Divided Pakistan Buries Salmaan [*sic*] Taseer and a Liberal Dream," *Guardian*, January 6, 2011, https://www.theguardian.com/world/2011/jan/05/pakistan-salman-taseer-liberal.

18. R. Upadhyay, "Barelvis and Deobandhis: 'Birds of the Same Feather,'" *Eurasia Review*, January 28, 2011, https://www.eurasiareview.com/28012011-barelvis-and-deobandhis-%E2%80%9Cbirds-of-the-same-feather%E2%80%9D/.

19. "Slain Salman Taseer's Son Kidnapped," *Dawn*, August 26, 2011, https://www.dawn.com/news/654867/slain-salman-taseers-son-kidnapped.

20. Hamid Shalizi and Jessica Donati, "Afghan Cleric and Others Defend Lynching of Woman in Kabul," Reuters, March 20, 2015, https://uk.reuters.com/article/uk-afghanistan-woman/afghan-cleric-and-others-defend-lynching-of-woman-in-kabul-idUKKBN0MG1ZA20150320.

21. Hassan Farhan and Ali Akbar, "Mardan University Student Lynched by Mob over Alleged Blasphemy: Police," *Dawn*, April 13, 2017, https://www.dawn.com/news/1326729/mardan-university-student-lynched-by-mob-over-alleged-blasphemy-police.

22. "A Devil's Design: Fanatics Used Fake Facebook Page to Run Rampage in Ramu," *Daily Star*, October 14, 2012, https://www.thedailystar.net/news-detail-253751.

23. "The Freedom of Thought Report 2018: Key Countries Edition," International Humanist and Ethical Union, 2018, https://demens.nu/wp-content/uploads/2019/02/FOT18-Key-Countries-edition-20181026-web.pdf.

24. Suhas Yellapantula, "FIR Filed against Babu Gogineni for Hurting Religious Sentiments," *Times of India*, June 27, 2018, https://timesofindia.indiatimes.com/city/hyderabad/fir-filed-against-babu-gogineni-for-hurting-religious-sentiments/articleshow/64751449.cms.

25. The term "caliphate" alludes to the realm that the prophet Muhammad created in the seventh century CE—supposedly a perfect society ruled by divine mandate.

26. For a more in-depth discussion of ISIS, see, among others, Loretta Napoleoni's book *ISIS: The Terror Nation* (New York: Seven Stories Press, 2014).

27. David Motadel, "The Ancestors of ISIS," *New York Times*, September 23, 2014, https://www.nytimes.com/2014/09/24/opinion/the-ancestors-of-isis.html.

28. Muhammed ibn Abd al-Wahhab, *Kitab al-Tawhid: The Book of Monotheism* (English translation) (Kuwait City: Kuwait al-Faisal Printing, 1986).

29. Jeffrey R. Macris, "Investigating the Ties between Muhammed ibn Abd al-Wahhab, Early Wahhabism, and ISIS," *Journal of the Middle East and Africa* 7, no. 3 (2016): 239–55.

30. "Islamic State Militants 'Destroy Palmyra Statues,'" BBC News, July 2, 2015, https://www.bbc.com/news/world-middle-east-33369701.

31. "Islamic State Destroys Ancient Mosul Mosque, the Third in a Week," *Guardian*, July 28, 2014, https://www.theguardian.com/world/2014/jul/28/islamic-state-destroys-ancient-mosul-mosque.

32. "Extrem tolkning av islam bakom IS destruktiva välde," *Dagens Nyheter*, August 27, 2014, https://www.dn.se/debatt/extrem-tolkning-av-islam-bakom-is-destruktiva-valde/.

33. "Anders Behring Breivik's Complete Manifesto '2083—a European Declaration of Independence,'" 371, Public Intelligence, July 28, 2011, https://publicintelligence.net/anders-behring-breiviks-complete-manifesto-2083-a-european-declaration-of-independence/.

34. Public Intelligence, "Anders Behring Breivik's Complete Manifesto," 707.

35. "2014 Religious Landscape Study (RLS-II): Topline," Pew Research Center, June 4–September 30, 2014, https://www.pewresearch.org/wp-content/uploads/sites/7/2015/11/201.11.03_RLS_II_topline.pdf.

36. Melanie Brewster, *Atheists in America* (New York: Columbia University Press, 2014).

37. Laurie Goodstein, "In Seven States, Atheists Push to End Largely Forgotten Ban," *New York Times*, December 6, 2014, https://www.nytimes.com/2014/12/07/us/in-seven-states-atheists-push-to-end-largely-forgotten-ban-.html.

38. It is instructive but quite frightening to watch the following documentary film about Jesus camps: https://vimeo.com/38531263.

39. Joshua Generation Ministries (website), accessed July 29, 2021, https://www.joshua generation.org.

40. Michelle Goldberg, *Kingdom Coming* (New York: Norton, 2007).

41. "Statement of Faith," Patrick Henry College, accessed July 29, 2021, https://www.phc.edu/statement-of-faith.

42. Patrick Henry College, "Statement of Faith."

43. The word "taliban" means "students" in the Pashto language, but today it refers to the Sunni Islamic fundamentalist political movement in Afghanistan.

44. Reformed Theology (website), accessed July 29, 2021, https://reformed-theology.org.

45. "Statement of Faith," American Vision, accessed July 29, 2021, https://americanvision.org/about/statement-of-faith/.

46. William O. Einwechter, "Stoning Disobedient Children," *Chalcedon Report*, January 1, 1999, https://chalcedon.edu/magazine/stoning-disobedient-children.

47. Jerry Falwell, "The Coming War with Russia," in *Nuclear War and the Second Coming of Jesus Christ*, Jerry Falwell Library, accessed August 16, 2021, https://liberty.contentdm.oclc.org/digital/collection/p17184coll4/id/1763/.

48. "Jerry Falwell Quotes on Life: 7 Memorable Statements from Evangelical Christian," Newsmax, accessed August 16, 2021, https://www.newsmax.com/fastfeatures/jerry-falwell-quotes-life-evangelical-christian/2015/05/01/id/642003/.

49. Christopher Reed, "The Rev. Jerry Falwell: Rabid Evangelical Leader of America's 'Moral Majority,'" *Guardian*, May 17, 2007, https://www.theguardian.com/media/2007/may/17/broadcasting.guardianobituaries.

50. Reed, "The Rev. Jerry Falwell."

51. "Robertson Letter Attacks Feminists," *New York Times*, August 26, 1992, https://www.nytimes.com/1992/08/26/us/robertson-letter-attacks-feminists.html.

Chapter Ten. The Battle over Our Origins

1. Stephen Jay Gould, *I Have Landed: The End of a Beginning in Natural History* (Cambridge, MA: Belknap Press, 2011).

2. Niall Shanks, *God, the Devil, and Darwin: A Critique of Intelligent Design Theory* (Oxford, UK: Oxford University Press, 2004).

3. Incidentally, in 2005, Thomas DeLay resigned from the House of Representatives after accusations of money laundering and conspiracy.

4. "Obama's Secular Humanist Upbringing Praised," American Humanist Association, January 19, 2009, https://americanhumanist.org/press-releases/2009-03-obamas-secular-humanist-upbringing-praised/.

5. "From George Washington to Edward Newenham, 22 June 1792," Founders Online, accessed April 16, 2021, https://founders.archives.gov/documents/Washington/05-10-02-0324.

6. "From John Adams to Thomas Jefferson, 19 April 1817," Founders Online, accessed April 16, 2021, https://founders.archives.gov/documents/Adams/99-02-02-6744.

7. "From Thomas Jefferson to John Adams, 11 April 1823," Founders Online, accessed April 16, 2021, https://founders.archives.gov/documents/Jefferson/98-01-02-3446.

8. Bryan Austin and Christian Cotz, "The Most Sacred Property," First Amendment Museum, accessed August 16, 2021, https://firstamendmentmuseum.org/the-most-sacred-property/.

9. "Tennessee Evolution Statutes, Public Acts of the State of Tennessee," University of Missouri–Kansas City, accessed September 3, 2021, http://law2.umkc.edu/faculty/projects/ftrials/scopes/tennstat.htm.

10. *Epperson v. Arkansas*, 393 U.S. 97 (1968).

11. "Programs," Discovery Institute, accessed July 30, 2021, https://www.discovery.org/about/programs.

12. If you can read Swedish, you might wish to check out this newspaper article: "Bläckfiskens ögen en gåta," *Dagens Nyheter*, April 8, 2007, www.dn.se/nyheter/vetenskap/blackfiskens-ogon-en-gata.

13. *Kitzmiller et al. v. Dover Area School District*, 400 F. Supp. 2d 707 (M.D. Pa. 2005).

14. A helpful website on this topic is National Center for Science Education (website), accessed August 16, 2021, https://ncse.ngo/association-science-education.

15. "Wedge Strategy," *Wikipedia*, last updated August 12, 2021, https://en.wikipedia.org/wiki/Wedge_strategy.

16. Pingstkyrkan, for example, is an evangelical Christian church distinct from the brand of Lutheranism known as the Church of Sweden, which for four centuries was Sweden's state church, although as of January 1, 2000, that was no longer the case.

17. "Allmän info om museet," Den Förhistoriska Världen, accessed August 16, 2021, https://www.dinosaurier.nu/om-museet/allman-info-om-museet/.

18. See Svante Pääbo's book, *Neanderthal Man: In Search of Lost Genes* (New York: Basic Books, 2014).

Chapter Eleven. The History of Ideas

1. Frans de Waal, *Primates and Philosophers: How Morality Evolved* (Princeton, NJ: Princeton University Press, 2009).

2. See Stephen Law, *Humanism: A Very Short Introduction* (Oxford, UK: Oxford University Press, 2011).

3. Taken from *The Analects of Confucius* (15, 23). These collections of Confucius's ideas were written down by his disciples around 400 BCE.

4. Birgitta Forsman, *Godless Morality*, trans. Douglas Hofstadter (Stockholm: Fri Tanke, 2011).

5. Averroës was actually named "Abū l-Walīd Muḥammad Ibn 'Aḥmad Ibn Rushd," and his first book dealt with medicine; however, his greatest contributions were made in philosophy.

6. Law, *Humanism*.

7. Some fifteen years after making this provocative statement, Cardinal Ratzinger was elected pope, and as such he adopted the name "Benedict XVI."

8. See Jacques Attali's biography *Diderot, ou Le bonheur de penser* (Paris: Pluriel, 2013).

9. Voltaire, *Traité sur la tolérance* [A treatise on tolerance] (1763).

10. Immanuel Kant, "Beantwortung der Frage: Was ist die Aufklärung?" [An answer to the question: What is enlightenment?] (1784).

11. Jeremy Bentham, *An Introduction to the Principles of Morals and Legislation* (1789).

12. Thomas Paine, *The Age of Reason* (1793).

13. See Susan Jacoby, *Freethinkers: A History of American Secularism* (New York: Metropolitan/Owl, 2005).

14. Vivian Green, *A New History of Christianity* (London: A. & C. Black, 2000).

15. Viktor Lennstrand, *Fritänkaren*, June 1, 1889, 1.

16. Viktor Lennstrand, "Six Theses of Örebro," *Orebro tidning*, 1888.

17. Douglas Hofstadter, *I Am a Strange Loop* (New York: Basic Books, 2007), chapter 23.

Chapter Twelve. Secular Voices in Our Day

1. Christopher Hitchens, *Slate Magazine*, October 20, 2003.

2. "Russell Tribunal," *Wikipedia*, last updated August 1, 2021, https://en.wikipedia.org/wiki/Russell_Tribunal.

3. *Wikipedia*, "Russell Tribunal."

4. Karl Popper, *Unended Quest: An Intellectual Autobiography* (London: Fontana Books, 1976).

5. Popper, *Unended Quest*.

6. "Humanist Manifesto II," American Humanist Association, accessed August 16, 2021, https://americanhumanist.org/what-is-humanism/manifesto2.

7. "Physics and Beyond: 'God Does Not Play Dice,' What Did Einstein Mean?" St. Mary's University, accessed August 16, 2021, https://www.stmarys.ac.uk/news/2014/09/physics-beyond-god-play-dice-einstein-mean/.

8. Albert Einstein, "The World as I See It," Einstein: Science and Religion, accessed August 16, 2021, http://www.einsteinandreligion.com/worldsee.html.

9. Einstein, "The World as I See It."

10. See Walter Isaacson's biography *Einstein: His Life and Universe* (New York: Simon & Schuster, 2007).

11. Albert Einstein, *Out of My Later Years* (New York: Philosophical Library, 1950).

12. See Helen Dukas and Banesh Hoffmann's loving memoir of their dear friend, *Albert Einstein: The Human Side* (Princeton, NJ: Princeton University Press, 1979).

13. Jeanna Bryner, "Auction for Einstein 'God Letter' Opens with Anonymous $3 Million Bid," October 8, 2012, NBC News, https://www.nbcnews.com/id/wbna49337533.

14. "Religious and Philosophical Views of Albert Einstein," *Wikipedia*, last updated August 4, 2021, https://en.wikipedia.org/wiki/Religious_and_philosophical_views_of_Albert_Einstein.

15. Daniel Dennett, *Breaking the Spell: Religion as a Natural Phenomenon* (New York: Viking, 2006).

16. Steven Pinker, "Science Is Not Your Enemy," *New Republic*, August 6, 2013.

17. Pinker, "Science Is Not Your Enemy."

18. Richard Dawkins, interview, *Sans Magazine* 1 (2013).

19. Christopher Hitchens, *God Is Not Great: How Religion Poisons Everything* (New York: Twelve, 2009).

20. Hitchens, *God Is Not Great.*

21. Hitchens, *God Is Not Great.*

22. Christopher Hitchens, *Mortality* (New York: Twelve, 2012).

23. Ricky Gervais (@rickygervais), Twitter, May 26, 2013, https://twitter.com/rickygervais/status/338596318144495617.

24. "Tim Minchin's *Storm the Animated Movie*," Sturmark, accessed July 30, 2021, https://www.sturmark.se/storm.

25. "Sekulära röster i Sverige," Sturmark, accessed July 30, 2021, www.sturmark.se/sekulararoster.

26. Birgitta Forsman, *Godless Morality*, trans. Douglas Hofstadter (Stockholm: Fri Tanke, 2011).

27. This passage is taken from Christer Sturmark's book *Personligt: Samtal med Fritänkare* [Personally speaking: Conversations with freethinkers], trans. Douglas Hofstadter (Stockholm: Fri Tanke, 2008).

Chapter Thirteen. Enlightenment

1. See Christian Welzel, *Freedom Rising: Human Empowerment and the Quest for Emancipation* (Cambridge: Cambridge University Press, 2013).

2. John Rawls, *A Theory of Justice* (Cambridge, MA: Belknap Press, 1971).

3. "Diskrimineringslagen," Diskriminerings Ombudsmannen, accessed August 16, 2021, https://www.do.se/lag-och-ratt/diskrimineringslagen/.

4. "Universal Declaration of Human Rights," United Nations, accessed August 16, 2021, https://www.un.org/en/about-us/universal-declaration-of-human-rights.

5. See Ronald Dworkin's book *Religion without God* (New Haven, CT: Harvard University Press, 2013).

6. See the case *Multani v. Commission scolaire Marguerite-Bourgeoys*, 1 SCR 246 (2006), https://scc-csc.lexum.com/scc-csc/scc-csc/en/item/15/index.do.

7. See Brian Leitner, *Why Tolerate Religion?* (Princeton, NJ: Princeton University Press, 2013).

8. Dworkin, *Religion without God.*

9. Melanie Brewster, *Atheists in America* (New York: Columbia University Press, 2014).

10. See "Win-Gallup International Global Index of Religiosity and Atheism, 2012," Scribd, accessed July 31, 2021, https://www.scribd.com/document/136318147/Win-gallup-International-Global-Index-of-Religiosity-and-Atheism-2012.

11. See "Win-Gallup International Global Index of Religiosity and Atheism, 2012."

12. See "Human Development Reports," United Nations Development Programme, accessed July 31, 2021, www.hdr.undp.org/en/.

13. See Philip Zuckerman, "Atheism, Secularity, and Well-Being: How the Findings of Social Science Counter Negative Stereotypes and Assumptions," *Sociology Compass* 3, no. 6 (2009): 949–71.

14. See "The Urban Disadvantage: State of the World's Mothers 2015," Save the Children, accessed July 31, 2021, https://www.savethechildren.org/content/dam/usa/reports/advocacy/sowm/sowm-2015.pdf.

15. See Philip Zuckerman's book *Living the Secular Life* (New York: Penguin, 2014).

16. See Philip Zuckerman's book *Society without God* (New York: New York University Press, 2010).

17. Zuckerman, *Living the Secular Life*, 9.

18. Taken from Scribd, "Win-Gallup International Global Index of Religiosity and Atheism—2012."

Chapter Fourteen. What Should We Teach Our Children?

1. "Convention on the Rights of the Child," United Nations, Human Rights, Office of the High Commissioner, accessed August 16, 2021, https://www.ohchr.org/en/professional interest/pages/crc.aspx.

2. "Kommentarmaterial till kursplanen i religionskunskap," Skolverket, accessed August 16, 2021, https://skolverket.se/publikationsserier/kommentarmaterial/2021/kommentarmaterial -till-kursplanen-i-religionskunskap.

3. Börge Ring, *Religion—helt enkelt* [Religion, simply presented] (Stockholm: Liber, 2012), 231.

4. Ring, *Religion—helt enkelt*.

5. Ring, *Religion—helt enkelt*, 10.

6. Börge Ring, *Religion och sånt* [Religion and such things] (Stockholm: Liber, 2013).

7. Ring, *Religion och sånt*.

8. *New Scientist*, 2012.

9. Börge Ring, *Religion och sånt* [Religion and such things] (Stockholm: Liber, n.d.), 196.

10. Ring, *Religion och sånt* (2013), 201.

11. Börge Ring, *Religion och sammanhang* [Religion and related matters] (Stockholm: Liber, 2013).

12. "Enfald eller mångfald i läroboken? Om hur tros- och livsåskådningar beskrivs i skolan" [Simplicity or diversity in the textbook? About how beliefs and views of life are described in school], Katharina Foundation, March 3, 2014, https://www.sturmark.se/ring.

Afterword and Words of Thanks

1. A perfect album of songs helping us reach this state of mind has been recorded by my good friend, jazz singer Isabella Lundgren. See https://www.isabellalundgren.com/.

INDEX

Note: The page numbers in *italics* refer to tables and figures.